Conservation Tillage in Temperate Agroecosystems

Edited by

Martin R. Carter

LEWIS PUBLISHERS
Boca Raton Ann Arbor London Tokyo

Library of Congress Cataloging-in-Publication Data

Conservation tillage in temperate agroecosystems: development and adaptation to soil, climatic, and biological constraints/edited by M.R. Carter.

p. cm.

Includes bibliographical references and index.

ISBN 0-87371-571-3

1. Conservation tillage. 2. Agricultural ecology. I. Carter, M. R. (Martin R.) II. Title: Temperate agroecosystems.

S604.C65 1994

631.4′51—dc20 94-26512

CIP

Direct all inquiries to CRC Press, Inc., 2000 Corporate Blvd., N.W., Boca Raton, Florida 33431.

Lewis Publishers is an imprint of CRC Press

No claim to original U.S. Government works

International Standard Book Number 0-87371-571-3

Library of Congress Card Number 94-26512

Printed in the United States of America 1 2 3 4 5 6 7 8 9 0

Printed on acid-free paper

Preface

Modern agriculture has seen the advent and increasing application of conservation tillage systems in temperate regions. The introduction of conservation tillage can be viewed as one of the major recent developments to improve soil and water conservation and to prevent soil degradation in crop production systems.There is, however, no universal prescription for the adoption of conservation tillage to any one location or region. Invariably, specific constraints impede the implementation of tillage practices which have the potential to conserve soil and energy and to sustain the agricultural ecosystem. Some soil types have a high cultivation requirement to maintain optimum soil structure. Climatic factors, such as level of precipitation, can influence both plant response to soil compaction and timing of crop establishment. Furthermore, biological constraints such as plant diseases or specific weed species can become controlling factors governing the successful adoption of new agricultural tillage practices.

Climate plays a direct role in many facets of agriculture, especially the choice of tillage system. This book covers the development of conservation tillage in temperate or middle latitudes (above 30°N and below 30°S). Three broad climatic regions are found within the temperate zone according to the Trewartha/Köppen classification: humid micro-thermal, humid meso-thermal, and semi-arid climates. The first two are mainly differentiated by the temperature range of the coldest month, while the latter represents mainly sub-humid to semi-arid continental interiors where evaporation greatly exceeds precipitation in the summer months.

Over the last four decades extensive literature has been produced about conservation tillage. Thus, there is a need to evaluate the existing information and to characterize present and future research needs. Moreover, there is a need to address the role of conservation tillage in sustainable agricultural systems. In this book, leading agronomic researchers outline present conservation tillage practices in temperate regions and how these systems have been developed and adapted to fit specific environments. Current and past research is summarized to show how soil, climatic, and biological constraints have been overcome or circumvented. The book provides a synthesis of existing work on conservation tillage in temperate agroecosystems and points towards future directions.

M. R. Carter
Editor

The Editor

Martin R. Carter holds degrees in Agriculture and Soil Science from the University of Alberta and obtained a Ph.D. degree in Soil Science from the University of Saskatchewan in 1983. Since 1977, he has held agricultural research positions in Alberta and Saskatchewan, and is currently a Research Scientist with Research Branch, Agriculture Canada at the Charlottetown Research Station in Prince Edward Island. Dr. Carter's present research activities involve soil tillage, specifically soil biological and physical properties in conservation farming systems. He has authored over 60 scientific papers in the area of soil management and tillage and recently edited *Soil Sampling and Methods of Analysis* (Lewis Publishers, 1993) for the Canadian Society of Soil Science. Dr. Carter holds appointment as Adjunct Professor in the Department of Chemistry and Soil Science, Nova Scotia Agricultural College, and currently serves on the Editorial Advisory Board of *Soil & Tillage Research.*

Contributors

Raymond R. Allmaras, Ph.D.
Soil Scientist
Soil Science Department
U.S. Department of Agriculture
Agriculture Research Service
University of Minnesota
St. Paul, Minnesota

C. J. Baker, Ph.D.
Chartered Engineer
Department of Agricultural Engineering
Massey University
Palmerston North
New Zealand

Bruce C. Ball, Ph.D.
Scottish Agricultural College
Edinburgh, Scotland

Jean Marie Bodet, Ph.D.
Agronomist
Department of Crop Management
Institut Technique des Cereales et des Fourrages
Station Experimentale de la Jailliere
La Chapelle St. Sauveur, France

Denis Boisgontier, Ph.D.
Head
Department of Machinery and Technology
Institut Technique des Cereales et des Fourrage
Station Experimentale
Boigneville, France

Trond Børresen, Ph.D.
Associate Professor
Department of Soil and Water Sciences
Agricultural University of Norway
Ås, Norway

Andelko Butorac, Ph.D.
Professor
Faculty of Agriculture
University of Zagreb
Zagreb, Croatia

David L. Carter, Ph.D.
Supervisory Soil Scientist
Soil and Water Management Research
Agricultural Research Service
U.S. Department of Agriculture
Kimberly, Idaho

Martin R. Carter, Ph.D.
Research Scientist
Agriculture Canada Research Station
Charlottetown, Prince Edward Island
Canada

M. A. Choudhary, Ph.D.
Department of Agricultural Engineering
Massey University
Palmerston North
New Zealand

Dudley G. Christian, B.Sc.
Crop Management Department
Institute of Arable Crops Research
Rothamsted Experimental Station
Harpenden, Herts
England

Wilhelm Claupein, Ph.D.
Institute of Agronomy and Plant Breeding
Georg August University
Goettingen, Germany

S. M. Copeland, M.Sc.
Agricultural Engineer
Agricultural Research Service
U.S. Department of Agriculture
Soil Science Department
University of Minnesota
St. Paul, Minnesota

Warren A. Dick, Ph.D.
Professor
Department of Agronomy
The Ohio State University
Wooster, Ohio

Donald J. Eckert, Ph.D.
Professor
Department of Agronomy
The Ohio State University
Columbus, Ohio

Wilfried Ehlers, Ph.D.
Professor
Institute of Agronomy and Plant Breeding
Georg August University
Goettingen, Germany

Egil Ekeberg, Ph.D.
Research Officer
Norwegian State Agricultural Research Stations
Apelsvoll Division Kise
Nes på Hedmark, Norway

Anthony Ellington, M.Sc.
Research Scientist
Department of Agriculture
Rutherglen Research Institute
Rutherglen, Victoria
Australia

Jean Paul Gillet, Ph.D.
Department of Crop Management
Institut Technique des Cereales et des Fourrages
Station Experimentale de la Jailliere
La Chapelle St. Sauveur, France

William L. Hargrove, Ph.D.
Department of Crop Sciences
University of Georgia
Griffin, Georgia

R. César Izaurralde, Ph.D.
Associate Professor
Department of Soil Science
University of Alberta
Edmonton, Alberta
Canada

K. Janovicek, Ph.D.
Research Associate
Department of Crop Science
University of Guelph
Guelph, Ontario
Canada

Rattan Lal, Ph.D.
Professor
Department of Agronomy
The Ohio State University
Columbus, Ohio

Francis J. Larney, Ph.D.
Soil Conservationist
Land Resource Sciences Section
Agriculture Canada Research Station
Lethbridge, Alberta
Canada

C. Wayne Lindwall, Ph.D.
Tillage Engineer and Head
Soil Science Section
Agriculture Canada Research Station
Lethbridge, Alberta
Canada

Terry J. Logan, Ph.D.
Professor
Department of Agronomy
The Ohio State University
Columbus, Ohio

Jack Massé, Ph.D.
Head
Department of Crop Management
Institut Technique des Cereales et des Fourrages
Station Experimentale
Boigneville, France

Daniel V. McCracken, Ph.D.
Assistant Research Scientist
Department of Crop and Soil Sciences
University of Georiga
Griffin, Georgia

A. P. Moulin, Ph.D.
Research Scientist
Soils Section
Agriculture Canada Research Station
Melfort, Saskatchewan
Canada

J. F. Power, Ph.D.
Soil Scientist
Agricultural Research Service
U.S. Department of Agriculture
Department of Agronomy
University of Nebraska
Lincoln, Nebraska

James E. Pratley, Ph.D.
Associate Professor
Faculty of Science and Agriculture
Charles Sturt University
Wagga Wagga, New South Wales
Australia

Hugh Riley, Ph.D.
Research Officer
Norwegian State Agricultural Research Stations
Apelsvoll Division Kise
Nes på Hedmark, Norway

Tomas Rydberg, Ph.D.
Senior Research Leader
Department of Soil Sciences
Swedish University of Agricultural Sciences
Uppsala, Sweden

M. J. Shipitalo, Ph.D.
Research Soil Scientist
Agricultural Research Service
U.S. Department of Agriculture
North Appalachian Experimental Water Sheds
Coshocton, Ohio

E. L. Skidmore, Ph.D.
Soil Scientist
Agricultural Research Service
U.S. Department of Agriculture
Department of Wind Erosion Research
Kansas State University
Manhattan, Kansas

Robert E. Sojka, Ph.D.
Soil Scientist
Soil and Water Management Research
Agricultural Research Service
U.S. Department of Agriculture
Kimberly, Idaho

Graham R. Steed, Ph.D.
Senior Research Officer
Department of Agriculture
Rutherglen Research Institute
Rutherglen, Victoria
Australia

Donald L. Tanaka, Ph.D.
Soil Scientist
Agricultural Research Service
U.S. Department of Agriculture
Northern Great Plains Research Laboratory
Mandan, North Dakota

Donald D. Tyler, Ph.D.
West Tennessee Experimental Station
Department of Plant and Soil Science
University of Tennessee
Jackson, Tennessee

Paul W. Unger, Ph.D.
Supervisory Soil Scientist
Soil and Crop Management Research Unit
Agricultural Research Service
U.S. Department of Agriculture
Bushland, Texas

Tony J. Vyn, Ph.D.
Associate Professor
Department of Crop Science
University of Guelph
Guelph, Ontario
Canada

Michael G. Wagger, Ph.D.
Associate Professor
Soil Science Department
North Carolina State University
Raleigh, North Carolina

Table of Contents

Humid Meso-Thermal Climates

Mainly Sub-Humid to Semi-Arid Continental Climates

Introduction

CHAPTER 1

Strategies to Overcome Impediments to Adoption of Conservation Tillage

Martin R. Carter
Agriculture Canada Research Station; Charlottetown, Prince Edward Island, Canada

TABLE OF CONTENTS

0-87371-571-3/94/$0.00+$.50

1.1. INTRODUCTION

The location of a specific agroecosystem is dependent on many factors that can be characterized using physical, biological, and socio-economic principles. At present, a unified approach to describe agricultural systems, based on the above principles, remains elusive. However, in most inventories of agricultural systems, climate and soil type are major factors that exert a powerful influence on the location and productivity of any specific farming practice.[1]

In the development of sustainable agricultural systems, the balance between conservation and degradative practices is modified by climate and soil type. For example, hot and arid climates can accelerate degradation by reducing plant biomass production and consequently crop residues, while humid climates can accelerate leaching, soil acidification, waterlogging, and erosion. In like manner, soil type can also influence plant productivity through differences in water-holding capacity, rooting depth, and chemical, biological, and physical properties. In general, a sustainable system can be created when benefits from conservation practices outweigh the negative effects of any soil degradative processes.

Tillage practices can play an important role in sustainable farming systems. As a soil management component, tillage functions as a subsystem directly influencing system performance or productivity mainly through crop establishment, modification of soil structure, incorporation of crop residues or amendments, and weed control. More importantly, tillage can be used directly to modify or alleviate both climatic and soil constraints. Lal[2] illustrated the ability of changing tillage technology to improve both water use and nutrient use efficiency within farming systems, and the important role tillage can play in sustainable agriculture.

Conservation tillage is required for many areas and soil types to prevent soil degradation. Mechanical manipulation of the soil can result in soil structure decline, subsoil compaction, decrease in soil stability, and increased soil erosion. Generally, excessive tillage leading to degradation of the soil physical condition reduces the range of plant-available water, due to increases in soil bulk density and mechanical resistance.[3] Conservation tillage in combination with improved stubble management techniques is essential in many semiarid regions for water conservation and reduction in soil fallowing. Conservation tillage is also beneficial for double-cropping technology.[4] Overall, successful adoption of conservation tillage usually requires a time factor, as many benefits (e.g., improved soil structure) associated with changes in tillage are a result of many factors (e.g., climate, soil type, agronomy) and are thus time dependent.

In intensive agricultural systems, commonly found in temperate regions, sustainability is now emphasized because of food surpluses, limiting nonrenewable resources, high soil degradation, and pollution of groundwater. Adaptation of tillage practices to the above notion of sustainability invariably entails a reduction in both intensity and frequency of cultivation. This is because tillage in temperate climes is mainly a resource consumer (e.g., expensive fossil fuels) and also, when used in excess, a cause of soil degradation. The reduction in tillage, however, must be geared to soil tillage requirement and crop establishment needs. This development ensures that the resultant conservation or sustainable tillage practice will be both soil and crop specific in nature.[2] In some cases constraints imposed by climate and soil type impede the direct application of conservation tillage techniques. Such constraints invoke the need for strategies to overcome or circumvent obstacles to conservation tillage adoption.

The objective of this introductory chapter is to characterize strategies that have been used to overcome constraints to the adoption of conservation tillage practices, especially in the context of soil and climate constraints in temperate regions. It is recognized, however, that public policy, machinery requirements, and both sociological and economical factors can also influence adoption of conservation tillage.[5]

1.2. DEFINING CONSERVATION TILLAGE

Attempts to devise an acceptable definition for conservation tillage are confounded by the fact that tillage requirement is dependent on both soil and crop conditions. Climatic variations also play an important role in dictating the tillage requirement of any one farming system. For example, climate can influence crop residue production while soil moisture regimes can be a factor in soil response to compaction forces. Thus, precise definitions of conservation tillage are only possible within the context of known crop species and varieties, soil types and conditions, and climates. These considerations lead to the concepts of prescription tillage[6] and farming-by-soil,[7] where tillage is matched to local agroecosystems in both space and time.

Conservation tillage, therefore, becomes an umbrella term, covering a range of tillage practices from direct drilling to full soil profile or deep tillage. Thus, conservation tillage does not of necessity mean less tillage. However, as suggested by Mannering and Fenster[8] a common characteristic of any conservation tillage is its potential to reduce soil and water loss relative to conventional tillage. This comparative description could also be expanded to address conservation of input costs, including both energy and time.

Classification of farming systems indicate that most arable agriculture occurs where potential evapotranspiration is relatively high and precipitation is moderate.[1] Successful conservation tillage practices within this broad climatic zone would be characterized by their ability to conserve soil moisture and reduce soil erosion (mainly wind erosion). Thus, a practical field measurement, or indicator, presently used to gauge conservation tillage is the degree of soil cover (e.g., a minimum of 30%) provided by retention of plant residues at the soil surface from the previous crop at the time of planting[4,9] or during the noncrop period.[10] Movement of arable farming into climatic regions where potential evapotranspiration declines and precipitation increases would tend to shift the practical indicator of conservation tillage from surface residues to a continuum of live cover (e.g., cover crops), degree of residue incorporation (where high residue yields are a problem), and speed and efficiency of crop establishment. The latter would allow the potential for conservation of input costs and reduce time duration of bare soil.

1.3. CHARACTERIZING CONSTRAINTS

1.3.1. Soil Tillage Requirement

In some locations (southern Australia,[11] for instance) cultivation damages soil structure and leads to excessive soil compaction and erosion. Such a situation has encouraged the rapid acceptance of reduced tillage techniques. Many soils, however, require regular tillage to prevent or ameliorate excessive soil compaction and poor structure which can result from both natural consolidation and vehicular traffic and cause restricted permeability for both water and air.[12] For example, relatively high contents of silt and fine sand can lead to soil instability and poor structure, while a predominance of nonexpanding clay minerals can limit the ability of the soil to restore structure by shrinkage and swelling cycles.[13] These conditions can present a soil tillage requirement even under relatively undisturbed soil management systems (for example, grassland). The need for tillage has been assessed in a range of soil types by grouping soils on the basis of aggregate stability and shrinkage and on soil compactability indices.[14] In some cases the soil tillage requirement may also be related to specific soil biological factors. For example, sandy soils may require relatively deep and regular tillage to fully incorporate and bury crop residues or organic amendments, so that the potential for residue decomposition can be reduced.[15]

In other cases the soil tillage requirement is not a constant based on soil properties, but a variable parameter related to soil management — for example, the result of deleterious conventional tillage practices. In Germany, for example, excessive compaction (partly subsoil compaction) and poor structure in many medium-textured soils are associated with traditional moldboard plowing and concomitant use of heavy machinery.[15] Here, intensive mechanization of the tillage event coupled with the propensity and increased capability for deeper tillage results in overloosening of the topsoil and a reduction in its bearing capacity. Subsequent employment of heavy machinery loads rapidly creates a new compaction state and thus completes the "circle" of tillage-induced soil compaction. Under these circumstances, amelioration of the soil compaction followed by controlled vehicular traffic (e.g., zone tillage; see Section 1.4.2) may be required before reduced tillage techniques can be employed.

Various land classifications have been devised to characterize the soil tillage requirement.[4,16-21] Some are based on soil suitability for direct drilling. In addition to soil properties, land classifications take into consideration soil profile characteristics such as texture, subsoil permeability, stoniness, drainage class, and soil depth, as well as site factors such as slope. These efforts underline the need to establish soil science databases for tillage. In this regard, links between soil survey and agricultural engineering have been recommended to better characterize rate trafficability, compactibility, and tillage requirement in specific soil types.[22]

1.3.2. Interactions of Soil and Climate

General climatic constraints such as short growing season and an excess or deficiency of temperature and/or precipitation will have an important impact on the feasibility of any agricultural system. In some cases conservation tillage techniques, such as direct drilling, can potentially overcome the constraints of a short growing season by providing timely and rapid crop establishment.[23]

For considerations of conservation tillage adoption, however, the interaction between soil type and climate is of prime importance. Cultivation investigations have demonstrated that plant performance is not related to the tillage implement used, but to the soil environment created.[24,25] This environment is dynamic, being subject to both temporal and spatial changes in soil properties and soil climate. Furthermore, the soil condition does not directly control plant productivity; it only influences plant behavior.[26]

Various procedures have been developed to predict soil workability and trafficability based on both soil properties and climatic components.[27,28] Excess precipitation, as modified by soil internal drainage and presence of a shallow water table, is the main climatic constraint to soil workability. This is reflected in the period of field capacity moisture over the potential growing season, especially the autumnal return to field capacity. The latter is an important criterion in several soil suitability classifications for tillage.[16,19] Soil factors such as clay content, mineralogy, and organic matter also influence soil workability mainly through their control on water retention, plasticity, and strength properties. The lower plastic limit, for example, is controlled by both organic matter and clay content, while only the latter controls the upper plastic limit.[29] For some soil types, the relationship between field capacity moisture and plastic limit is a critical measure of soil workability.[27] For example, a field capacity value below the plastic limit in the topsoil would permit a degree of tillage to occur. However, trafficability could still be impeded if the inverse of the above condition occurred in the subsoil.[13]

Interactions between soil moisture regime and soil compactability can also pose a constraint on the adoption of conservation tillage, if the level of cultivation is reduced. Soils can respond differently to a specific applied stress at a given moisture content due to differences in the amount and type of clay and the organic matter content.[30] In addition, soil organic matter and presence of living roots can increase the resistance of a soil to deformation by influencing soil elasticity and aggregate stability, respectively,[31] while cropping practices can modify soil compaction characteristics at a given moisture content by influencing soil aggregate tensile strength and interaggregate porosity.[32] All these interactions may affect soil response to conservation tillage practices, especially if wet soil conditions coincide with the use of heavy field machinery.

Some soils in temperate climates are subject to freeze-thaw effects which can be modified by soil aggregation characteristics and soil water content.[33] Both parameters are significantly influenced by tillage practices. Conservation tillage systems that increase organic matter near the soil surface can improve aggregate stability in a short time frame,[34,35] although multiple freeze-thaw cycles over the winter period will cause aggregate instability.[36] Temporal variations in aggregate stability under conservation tillage, however, showed that aggregates developed at the surface of direct drilled soils maintained an increased resistance to slaking and freeze-thaw events, in comparison to aggregates under conventional moldboard plowing.[35] Freeze-thaw cycles along with the formation of ice lenses in the soil are usually seen as a positive force in the reduction of soil compaction, however, the latter phenomenon is dependent on the water stability of the soil aggregates and the orientation of the ice lenses.[37]

1.3.3. Biological Constraints

Biological constraints are common to all agronomic systems but especially to forms of conservation tillage where crop residues are concentrated at the soil surface. The juxtaposition of plant roots and crop residues, as mainly demonstrated under certain forms of direct drilling, can produce decreased plant growth and vigor due to both water-soluble toxins in the residue and toxins derived from microbial decomposition of the residue.[38,39] In some cases tillage is required to either physically separate residues from the seed row or dilute the residue concentration in the topsoil. Burning crop residues is also practiced; however, this strategy can have a detrimental effect on both soil carbon and soil physical properties.[40,41] In some humid climates, crop residues maintained near the soil surface may have a positive effect on soil structural stability[34,42] and soil properties in general, especially via earthworm activity.[43,44]

In humid regions under intensive farming systems, the production of relatively high yields of crop residues can present problems for optimum crop establishment. This is mainly a mechanical constraint, wherein excessive residues interfere with seeding operations and thus necessitate improved residue management or residue removal. Methods of residue management can involve partial width tillage in row crops to remove residues from the seed row; adoption of technologies to grain harvesters to allow uniform residue distribution;[45] burning where permitted or desirable for small grains; and, more recently, the use of incorporation methods.[46] With regard to the latter, depth distribution of incorporated or buried residue is strongly related to the type of tillage tool.[47] In some cases, excess crop residue or macroorganic matter in and above the seedbed have necessitated the return to use of spring crops to allow an adequate time span for residue decomposition.[48]

Conservation tillage can also influence both plant disease and populations of insects and other pests. Overall, insect response to reduced cultivation is varied, and most increases are associated with direct drilling.[49] Plant diseases are mainly influenced through residue management, the concentration of microbes at the soil surface layer, and changes in soil chemical and physical properties, especially the latter.[50,51] The disease phenomenon can be highly variable under conservation tillage due to the various soil conditions and factors that can be involved.[51] Extra tillage is sometimes needed to ensure the success of conservation tillage systems. For example, in parts of Australia the effect of rhizoctonia root rot under direct drilling requires a tillage input to disrupt hyphae networks and reduce propagule formation.[50]

Differences in tillage practices can influence the number of viable weed seeds in the soil as well as their vertical distribution.[52-54] In addition, tillage differences can affect weed emergence patterns,[55] incidence of volunteer crops, and type of weed species.[49] All of these factors can dictate the tillage requirement or role of a specific cultivation within a rotation. In many cases, changes in the cropping system are required to control unwanted vegetation, either volunteer crops or weeds, under conservation tillage practices.[48]

Concentration of crop residues and organic carbon accumulation near the soil surface can influence the efficacy of pesticides,[49] fate of nitrogen fertilizer,[56] and immobilization of nitrogen.[48] Improved management can overcome these potential constraints, however. Placement of nitrogen below the residue layer can increase nitrogen efficiency as compared to broadcast applications.[56] In dry soils this is related to improved soil moisture use efficiency.[57] Knowledge of the relationship between tillage tool and resulting spatial distribution of crop residue in the soil generally will improve the efficiency of placement technology.[47]

1.4. STRATEGIES TO OVERCOME CONSTRAINTS

Various approaches can be utilized to remove soil and climatic impediments to conservation tillage adoption. Nearly all involve two strategies: removing or circumventing the constraint and utilizing new or alternative technologies. This section attempts to characterize how some of the above responses have been integrated into temperate farming systems.

1.4.1. Minimum Tillage

Minimum tillage involves the strategy of matching tillage inputs with a specific soils tillage requirement and/or the tillage necessary for crop production. The idea is based on the fact that many soils have an ongoing cultivation requirement to maintain adequate soil conditions for plant growth. Minimizing the tillage input can conserve energy, time, and labor.

Minimum tillage is very adaptable to the objectives of most agricultural systems with regard to tillage requirement. Tillage can be applied to various soil depths to alleviate adverse conditions. For example, specific seed drill openers can loosen crusted or compacted soil at the soil surface, while chisel plows can accommodate full soil profile loosening but with reduced soil inversion, as compared to moldboard and disc plows. Shallow tillage has wide application to soils that require regular loosening within the seedbed only. In these cultivation systems, tillage is generally confined to the 0 to 10 cm soil depth.[58,59] Minimum tillage is also applicable to moldboard plowed systems when the plowing depth is minimized and when both the amount and degree of secondary tillage are reduced, including the number of vehicular passes over the field. Use of one-pass moldboard plowed techniques, for example, can reduce operating and energy costs for certain soil types and increase the time the soil surface remains covered with crop residues or live cover.[60]

The use of slant-legged soil looseners allows full or partial-width tillage of the soil profile at depth, but retains crop residues in a relatively undisturbed state at the soil surface.[13,61-63] This technique opens the possibility of using direct drilling practices on soils prone to compaction. The frequency of soil loosening (e.g., on an annual or biannual basis) can be geared to the soil tillage requirement. Recent studies have shown that deep loosening a compacted loess soil in Germany with a slant-legged subsoiler followed by a minimum cultivation system (using a rotary cultivator) prevented recompaction of the subsoil, in comparison to deep loosening followed by moldboard plowing.[15]

Tillage inputs may also be required for mechanical weed control and to facilitate crop establishment requirements by removal of mechanical and biological constraints. High amounts of crop residues, for example, can adversely affect seeding operations and directly or indirectly generate phytotoxins. In these situations shallow incorporation of the residues, using a wide range of tillage tools, can improve soil conservation and workability over time.[46]

1.4.2. Zone Tillage

Cropping systems that include row crops allow tillage practices to accommodate both soil and crop requirements.[64] A different tillage method can be employed in the interrow and row zone. Thus, in these partial-width systems a uniform method of tillage is not utilized across the interrow and row areas.[4] The two main variants of zone tillage are strip tillage and ridge tillage. The latter involves the formation of slightly elevated ridges with residues removed over the seed row zone. Lal[65] outlined the benefits of ridge tillage as a form of conservation tillage.

Various studies have utilized strip tillage as a conservation tillage technique for row crop production. Most of these involve the use of a form of deep tillage in either the interrow or row zone to remove subsoil or lower soil horizon compaction.[63,66]

Forms of strip tillage have also been used for root crops such as sugar beet (*Beta vulgaris* L.) and potato (*Solanum tuberosum* L.), where untilled strips provide temporary or permanent wheel tracks and thus allow cropped areas to be traffic free.[67,68] Chamen et al.[69] discussed the benefits of zero traffic zones and reduced ground pressure systems in a conservation tillage context.

1.4.3. Rotational Tillage

Many of the early studies on conservation tillage indicated that soil conditions may change gradually, either adversely or beneficially, over time. Climatic vagaries can also present different soil conditions on an annual basis. Furthermore, crop tillage requirements (e.g., small grain vs root crops) can vary within a crop rotation. Thus, the soil tillage requirement is not constant over time. Under these conditions, tillage inputs can be opportunistic and geared to soil or crop requirements at any one time.[70]

The practical outcome of the above scenario is that specific forms of conservation tillage, such as direct drilling, can be accommodated on a rotational basis within crop sequences. In a similar manner, deep tillage to remove soil barriers to root penetration in the subsoil can be used intermittently. Pierce et al.[63] showed that rotational tillage with a subsoiler on a loam soil had the potential to sustain the conservation benefits of direct planted corn (*Zea mays* L.). Sommer and Zach[71] demonstrated the potential of using rotational tillage, in a conservation tillage context, to control both soil erosion and traffic-induced soil compaction on a loamy sand in Germany.

1.4.4. Tillage Timing

Soil suitability classifications for conservation tillage have noted that timing of tillage and crop establishment can be an important strategy for the removal of constraints to crop productivity.[16] Climatic conditions that increase the propensity for soil compaction can be circumvented by shifting the time of the tillage event. For example, in the Atlantic Provinces of Canada, direct drilling for winter cereals can result in less excessive soil compaction, in comparison to spring cereals, because of the increased probability of field work days with optimum soil moisture conditions in the early autumn as opposed to late spring.[72] Use of cover crops also can be used to shift the tillage event to a season or time when the soil erosion hazard is relatively low.[4]

1.4.5. Crop Rotations

Many of the initial studies with conservation tillage concentrated on a form of sequential monoculture which over time presented various constraints to sustainable crop production. A conscious choice of crop sequences can allow a positive "rotation effect" as demonstrated by improved soil conditions, plant nutrition, root growth, and weed, pest, and disease control.[73,74] Similar to conventional tillage systems, use of crop rotations within conservation tillage can prevent the decreasing crop productivity often associated with monoculture.[75] Rotations of grain legumes with cereals tended to remove the differences in accumulation of nitrogen in the cereal grain previously found in direct drilling and moldboard plowing comparisons.[76] A combination of crop rotation with conservation tillage, especially when the latter maintains crop residues within the surface soil, can have a positive effect on soil biological activity, suppress the buildup of deleterious rhizobacteria that are indirectly pathogenic,[77] and provide the needed biodiversity to maintain optimum soil and crop productivity in different soil and climatic zones.[78]

1.4.6. Adoption of New Technology

In conservation tillage new technologies have been developed primarily to improve plant establishment. Initial attempts at reducing tillage and maintaining crop residues at the soil surface were problematic due to fertilizer placement or distribution, reduction in efficiency of agrochemicals, and inadequate seed placement.[79] Research has demonstrated the importance of seed microenvironment and new seed openers have been developed for direct drilling,[80] which improved seedling establishment, even if wet soil conditions followed seeding.[81] The technology that allows uniform distribution of chaff and straw behind grain harvesters has also been developed, with obvious benefits for residue-related problems in conservation tillage.[82]

1.5. CONSERVATION TILLAGE AND SUSTAINABLE AGRICULTURE

Although tillage may function as a small component of a much larger agricultural system, its dynamic nature can strongly influence system performance and productivity. Soil structure tends to function as a major organizer for terrestrial ecosystems.[83] In agricultural systems tillage is the principal agent resulting in soil perturbation and subsequent modification in soil structure. From an ecological viewpoint, such perturbations strongly influence the distribution of energy-rich organic substances within the soil and thus impact on energy flow and the dynamics of soil geochemical cycles. However, tillage can also adversely affect soil structure and cause excessive breakdown of aggregates leading to a potential for soil movement via erosion. Thus, tillage is one system component that is easily subject to manipulation leading to either positive or negative consequences for agricultural sustainability.

It is generally agreed that the basic principles underlying sustainable agriculture await development and demonstration;[84,85] however, tillage will play a pivotal role in sustainability due to its ability to manipulate soil properties and processes[2] and to change soil quality.[86,87] With regard to the latter, tillage can influence the capacity of the soil to both store and cycle water, nutrients, and energy. Therefore, looking at conservation tillage in the context of sustainable agriculture emphasizes the need to expand the definition of the former to include conservation or sustainability of soil functions related to the improvement of soil quality.

The concern over agricultural sustainability accentuates the need to characterize energy and material flows (input:output ratios) and ecological processes for individual conservation tillage practices. Of necessity this will involve evaluating conservation tillage within the broader context of farming systems because, in contrast to natural ecosystems, agroecosystems operate under nonequilibrium conditions. For example, agroecosystems are subject to variable additions (e.g. manure, fertilizers) and continuous outputs (e.g., harvesting of plant biomass) of nutrients and energy. This can cause a wide dispersal and availability of nutrients and agricultural chemicals within the system resulting in the potential for "leakage" by leaching, gaseous losses, and erosion. In view of the above considerations recent investigations in conservation tillage, in addition to monitoring conservation of organic matter and water, have attempted to evaluate tillage effects on groundwater quality,[88] and integrating tillage with livestock and pasture management systems.[11]

1.6. GENERAL SUMMARY AND CONCLUSIONS

Soil tillage is involved in a wide range of agricultural aspects, such as crop management, regulation of soil conditions, tillage implements, and system economics.[24] In modern agriculture, systems of tillage are required which will prevent degradation and erosion of the soil resource. Development of such systems over time has demonstrated that conservation tillage has the potential to conserve soil, time, energy, and labor. Moreover, conservation tillage is flexible in different soil and crop environments, although its degree of success is dependent on the concomitant utilization of sound farming practices and new technologies. Improvement in soil and water conservation through the adoption of new tillage practices, relative to a specific form of conventional tillage, can be viewed as a general definition of conservation tillage. Within particular climatic regions a more detailed definition is possible relative to the dominant forces (e.g., wind erosion) that have an impact on soil conservation. Present concerns over sustainable agriculture, however, are tending to enlarge the focus of conservation tillage to address the wider concerns of conservation farming systems.

The various constraints to conservation tillage adoption can be characterized by: soil tillage requirement, soil and climatic interactions, and biological constraints. These constraints can be overcome in many circumstances by judicious use of tillage to meet the soil tillage requirement, manipulation of crop residues, and application of new technology.

1.7. ACKNOWLEDGMENTS

The author is grateful to the following individuals for their constructive criticisms and comments on an early draft of this chapter: Dr. R. R. Allmaras, USDA-ARS, St. Paul, MN; Drs. W. Ehlers, W. Claupein, and K. Baeumer, Georg-August University, Goettingen, Germany; and Prof. J. E. Pratley, Charles Sturt University, NSW, Australia.

1.8. REFERENCES

1. Spedding, C. R. W. *An Introduction to Agricultural Systems* (London: Elsevier, 1988), pp. 110–114.
2. Lal, R. "Tillage and Agricultural Sustainability," *Soil Tillage Res.* 20:133–146 (1991).
3. Letey, J. "Relationship between Soil Physical Properties and Crop Production," *Adv. Soil Sci.* 1:277–294 (1985).
4. Allmaras, R. R., G. W. Langdale, P. W. Unger, R. H. Dowdy and D. M. Van Doren. "Adoption of Conservation Tillage and Associated Planting Systems," in *Soil Management for Sustainability*, R. Lal, and F. J. Pierce (Eds.) (Ankeny, IA: Soil and Water Conservation Society, 1991), pp. 53–83.
5. D'Itri, F. M., Ed. *A Systems Approach to Conservation Tillage* (Chelsea, MI: Lewis Publishers, 1985), p. 384.
6. Schafer, R. L., C. E. Johnston, C. B. Elkins and J. G. Hendrick. "Prescription Tillage: The Concept and Example," *J. Agric. Eng. Res.* 32:123–129 (1985).
7. Larson, W. E., and P. C. Robert. "Farming by Soil," in *Soil Management for Sustainability,* R. Lal, and F. J. Pierce (Eds.) (Ankeny, IA: Soil and Water Conservation Society, 1991), pp. 103–112.
8. Mannering, J. V., and C. R. Fenster. "What is Conservation Tillage?", *J. Soil Water Conserv.* 38:140–143 (1983).
9. Mannering, J. V., D. L. Schertz and B. A. Julian. "Overview of Conservation Tillage," in *Effects of Conservation Tillage on Groundwater Quality*, T. J. Logan, J. M. Davidson, J. L. Baker and M. R. Overcash (Eds.) (Chelsea, MI: Lewis Publishers, 1987), pp. 3–17.
10. Felton, W. L., D. M. Freebairn, N. A. Fettell and J. B. Thomas. "Crop Residue Management," in *Tillage — New Directions in Australian Agriculture*, P. S. Cornish, and J. E. Pratley (Eds.) (Melbourne, Australia: Inkata Press, 1987), pp. 171–193.
11. Cornish, P. S., and J. E. Pratley. "Tillage Practices in Sustainable Farming Systems," in *Dryland Farming A System Approach: An Analysis of Dryland Agriculture in Australia*, V. Squires, and P. Tow (Eds.) (Sydney, Australia: Sydney University Press, 1991), pp. 54–71.
12. Soane, B. D., and J. D. Pidgeon. "Tillage Requirement in Relation to Soil Physical Properties," *Soil Sci.* 119:376–384 (1975).
13. Carter, M. R. "Physical Properties of Some Prince Edward Island Soils in Relation to Their Tillage Requirement and Suitability for Direct Drilling," *Can. J. Soil Sci.* 67:473–487 (1987).

14. Stengel, P., J. T. Douglas, J. Guérif, M. J. Goss, G. Monnier and R. Q. Cannell. "Factors Influencing the Variation of Some Properties of Soils in Relation to Their Suitability for Direct Drilling," *Soil Tillage Res.* 4:35–53 (1984).
15. Ehlers, W. Personal communication (1993).
16. Cannell, R. Q., D. B. Davies, D. MacKney and J. D. Pidgeon. "The Suitability of Soils for Sequential Direct Drilling of Combine-Harvested Crops in Britain: A Provisional Classification," *Outlook Agric.* 9:306–316 (1978).
17. Ross, C. A., and A. D. Wilson. "Conservation Tillage — Soils and Conservation," *N. Z. J. Agric. Sci.* 17:283–287 (1982).
18. Canarache, A. "Romanian Experience with Land Classification Related to Tillage," *Soil Tillage Res.* 10:39–54 (1987).
19. Ball, B. C., and M. F. O'Sullivan. "A Land Grouping System for Cultivation and Sowing Requirements for Winter Barley: Evaluation and Modification of Recommendations," *J. Sci. Food Agric.* 39:25–34 (1987).
20. Mannering, J. V., D. R. Griffith, S. D. Parsons and C. R. Meyer. "The Use of Expert Systems to Provide Conservation Tillage Recommendations," in *Proc. 11th Conf. International Soil Tillage Research Organization* (Edinburgh: 1988), pp. 763–768.
21. Heinonen, R. "Plowing and Non-Plowing Under Nordic Conditions," *Soil Tillage Res.* 21:185–189 (1991).
22. Wilson, G. "Pedotechnology and Tillage Research," *Soil Tillage Res.* 10:55–65 (1987).
23. Carter, M. R., H. T. Kunelius, R. P. White and A. J. Campbell. "Development of Direct Drilling Systems for Sandy Loam Soils in the Cool Humid Climate of Atlantic Canada," *Soil Tillage Res.* 16:371–387 (1990).
24. Kuipers, H. "The Objectives of Soil Tillage," *Neth. J. Agric. Sci.* 11:91–96 (1963).
25. Spoor, G. "Fundamental Aspects of Cultivations," in *Soil Physical Conditions and Crop Production* (London: Her Majesty's Stationery Office, 1975), pp. 128–144.
26. Gill, W. R., and G. E. Vanden Berg. *Soil Dynamics in Tillage and Traction* (Washington, DC; U.S. Government Printing Office), pp. 300–304.
27. Smedema, L. K. "Drainage Criteria for Soil Workability," *Neth. J. Agric. Sci.* 27:27–35 (1979).
28. Thomasson, A. J. "Soil and Climatic Aspects of Workability and Trafficability," in *Proc. 9th Conf. International Soil Tillage Research Organization,* Osijek, Croatia, (1982), pp. 551–557.
29. McBride, R. A. "A Re-Examination of Alternative Test Procedures for Soil Consistency Limit Determination. II. A Stimulation Desorption Procedure," *Soil Sci. Soc. Am. J.* 53:184–191 (1989).
30. Voorhees, W. B. "Assessment of Soil Susceptibility to Compaction Using Soil and Climatic Data Bases," *Soil Tillage Res.* 10:29–38 (1987).
31. Soane, B. D. "The Role of Organic Matter in Soil Compactibility: A Review of Some Practical Aspects," *Soil Tillage Res.* 16:179–201 (1990).
32. Angers, D. A., B. D. Kay and P. H. Groenevelt. "Compaction Characteristics of a Soil Cropped to Corn and Bromegrass," *Soil Sci. Soc. Am. J.* 51:779–783 (1987).
33. Benoit, G. R., and J. Bornstein. "Freezing and Thawing Effects on Drainage," *Soil Sci. Soc. Am. Proc.* 34:551–557 (1970).

34. Carter, M. R. "Influence of Reduced Tillage Systems on Organic Matter, Microbial Biomass, Macro-Aggregate Distribution and Structural Stability of the Surface Soil in a Humid Climate," *Soil Tillage Res.* 23:361–372 (1992).
35. Angers, D. A., N. Samson and A. Légére. "Early Changes in Water-Stable Aggregation Induced by Rotation and Tillage in a Soil Under Barley Production," *Can. J. Soil Sci.* 73:51–59 (1993).
36. Edwards, L. M. "The Effect of Alternate Freezing and Thawing on Aggregate Stability and Aggregate Size Distribution of Some Prince Edward Island Soils," *J. Soil Sci.* 42:193–204 (1991).
37. Kay, B. D., C. D. Grant and P. H. Groenevelt, "Significance of Ground Freezing on Soil Bulk Density Under Zero Tillage," *Soil Sci. Soc. Am. J.* 49:973–978 (1985).
38. Kimber, R. W. L. "Phytotoxicity from Plant Residues. I. The Influence of Rotted Straw on Seedling Growth," *Aust. J. Agric. Res.* 18:361–374 (1967).
39. Elliott, L. F., and R. I. Papendick. "Crop Residue Management for Improved Productivity," *Biol. Agric. Horticult.* 3:131–142 (1986).
40. Biederbeck, V. O., C. A. Campbell, K. E. Bowren, M. Schnitzer and R. N. McIver. "Effect of Burning Cereal Straw on Soil Properties and Grain Yields in Saskatchewan," *Soil Sci. Soc. Am. J.* 44:103–111 (1980).
41. Pikul, J. L., Jr., and R. R. Allmaras. "Physical and Chemical Properties of a Haploxeroll After 50 Years of Residue Management," *Soil Sci. Soc. Am. J.* 50:214–219 (1986).
42. Mele, P. M., and M. R. Carter. "Effect of Climatic Factors on the Use of Microbial Biomass as an Indicator of Changes in Soil Organic Matter," in *Soil Organic Matter Dynamics and Sustainability of Tropical Agriculture,* K. Mulongoy and R. Merkx (Ed.) (London: John Wiley & Sons, 1993), pp. 57–63.
43. Douglas, L. A. "Effects of Cultivation and Pesticide Use on Soil Biology," in *Tillage — New Directions in Australian Agriculture*, P. S. Cornish, and J. E. Pratley (Eds.) (Melbourne, Australia: Inkata Press, 1987), pp. 308–317.
44. Logsdon, S. D., and D. R. Linden. "Interactions of Earthworms with Soil Physical Conditions Influencing Plant Growth," *Soil Sci.* 154:330–337 (1992).
45. Douglas, C. L., Jr., P. E. Rasmussen and R. R. Allmaras. "Cutting Height, Yield Level, and Equipment Modification Effects on Residue Distribution by Combines," *Trans. Am. Soc. Agric. Eng.* 32:1258–1262 (1989).
46. Ball, B. C., and E. A. G. Robertson. "Straw Incorporation and Tillage Methods: Straw Decomposition, Denitrification and Growth and Yield of Winter Barley," *J. Agric. Eng. Res.* 46:223–243 (1990).
47. Staricka, J. A., R. R. Allmaras and W. W. Nelson. "Spatial Variation of Crop Residue Incorporated by Tillage," *Soil Sci. Soc. Am. J.* 55:1668–1674 (1991).
48. Baeumer, K., and W. Ehlers, (Eds.) *Agriculture: Energy Saving by Reduced Soil Tillage,* Report EUR 11258 (Luxembourg: Commission of European Communities, 1989).
49. Fawcett, R. S. "Overview of Pest Management for Conservation Tillage," in *Effects of Conservation Tillage on Groundwater Quality,* T. J. Logan, J. M. Davidson, J. L. Baker and M. R. Overcash (Eds.) (Chelsea, MI: Lewis Publishers, 1987), pp. 19–37.
50. Rovira, A. D., L. F. Elliott, and R. J. Cook. "The Impact of Cropping Systems on Rhizosphere Organisms Affecting Plant Health," in *The Rhizosphere*, J. M. Lynch (Ed.) (London: John Wiley & Sons, 1990), pp. 389–436.

51. Rothrock, C. S. "Tillage Systems and Plant Disease," *Soil Sci.* 154:308–315 (1992).
52. Staricka, J. A., P. M. Burford, R. R. Allmaras and W. W. Nelson. "Tracing the Vertical Distribution of Simulated Shattered Seeds as Related to Tillage," *Agron. J.* 82:1131–1134 (1990).
53. Cardina, J., E. Regnier and K. Harrison. "Long-Term Tillage Effects on Seed Banks in Three Ohio Soils," *Weed Sci.* 39:186–194 (1991).
54. Yenish, J. P., J. D. Doll and D. D. Buhler. "Effects of Tillage on Vertical Distribution and Viability of Weed Seed in Soil," *Weed Sci.* 40:429–433 (1992).
55. Roberts, H. A. "Crop and Weed Emergence Patterns in Relation to Time of Cultivation and Rainfall," *Ann. Appl. Biol.* 105:263–275 (1984).
56. Randall, G. W., and V. A. Bandel. "Overview of Nitrogen Management for Conservation Tillage Systems," in *Effects of Conservation Tillage on Groundwater Quality*, T. J. Logan, J. M. Davidson, J. L. Baker and M. R. Overcash (Eds.) (Chelsea, MI: Lewis Publishers, 1987), pp. 39–63.
57. Carter, M. R., and D. A. Rennie. "Crop Utilization of Placed and Broadcast ^{15}N-Urea Fertilizer Under Zero and Conventional Tillage," *Can. J. Soil Sci.* 64:563–570 (1984).
58. Carter, M. R. "Evaluation of Shallow Tillage for Spring Cereals on a Fine Sandy Loam. I. Growth and Yield Components, N Accumulation and Tillage Economics," *Soil Tillage Res.* 21:23–35 (1991).
59. Rydberg, T. "Ploughless Tillage in Sweden. Results and Experiences from 15 Years of Field Trials," *Soil Tillage Res.* 22:253–264 (1992).
60. Carter, M. R., R. P. White and R. G. Andrew. "Reduction of Secondary Tillage in Mouldboard Ploughed Systems for Silage Corn and Spring Cereals in Medium Textured Soils," *Can. J. Soil Sci.* 70:1–9 (1990).
61. Braim, M. A., K. Chaney and D. R. Hodgson. "Preliminary Investigation on the Response of Spring Barley (*Hordeum sativum*) to Soil Cultivation with the 'Paraplow,'" *Soil Tillage Res.* 4:277–293 (1984).
62. Ehlers, W., and K. Baeumer. "Effect of the Paraplow on Soil Properties and Plant Performance," in *Proc. 11th Conf. International Soil Tillage Research Organization* (Edinburgh: 1988), pp. 637–642.
63. Pierce, F. J., M. J. Stanton and M. C. Fortin. "Immediate and Residual Effects of Zone-Tillage in Rotation with No-Tillage on Soil Physical Properties and Corn Performance," *Soil Tillage Res.* 24:149–165 (1992).
64. Larson, W. E. "Soil Parameters for Evaluating Tillage Needs and Operations," *Soil Sci. Soc. Am. Proc.* 28:119–122 (1964).
65. Lal, R. "Ridge Tillage," *Soil Tillage Res.* 18:107–111 (1990).
66. Reeves, D. W., and J. T. Touchton. "Effects of In-Row and Interrow Subsoiling and Time of Nitrogen Application on Growth, Stomatal Conductance and Yield of Strip-Tilled Corn," *Soil Tillage Res.* 7:327–340 (1986).
67. Cannell, R. Q. "Reduced Tillage in North-West Europe — A Review," *Soil Tillage Res.* 5:129–177 (1985).
68. Dickson, J. W., D. J. Campbell and R. M. Ritchie. "Zero and Conventional Traffic Systems for Potatoes in Scotland, 1987–1989," *Soil Tillage Res.* 24:397–419 (1992).
69. Chamen, W. C. T., G. D. Vermeulen, D. J. Campbell and C. Sommer. "Reduction of Traffic-Induced Soil Compaction: A Synthesis," *Soil Tillage Res.* 24:303–318 (1992).

70. van Ouwerkerk, C. "Rational Tillage," in *Semaine d'Etude, Agriculture et Environment,* Bull. Rech. Agron. (Gembloux, Belgium, Hors sér., 1974), pp. 695–709.
71. Sommer, C., and M. Zach. "Managing Traffic-Induced Soil Compaction by Using Conservation Tillage," *Soil Tillage Res.* 24:319–336 (1992).
72. Carter, M. R. "Characterizing the Soil Physical Condition in Reduced Tillage Systems for Winter Wheat on A Fine Sandy Loam Using Small Cores," *Can. J. Soil Sci.* 72:395–402 (1992).
73. Francis, C. A., and M. D. Clegg. "Crop Rotations in Sustainable Agricultural Systems," in *Sustainable Agriculture Systems,* C. A. Edwards, R. Lal, P. Madden, R. H. Miller and G. House (Eds.) (Ankeny, IA: Soil and Water Conservation Society, 1990), pp. 107–122.
74. Crookston, R. K., J. E. Kurle, P. J. Copeland, J. H. Ford and W. E. Leuschen. "Rotational Cropping Sequence Affects Yield of Corn and Soybean," *Agron. J.* 83:108–113 (1991).
75. Griffith, D. R., E. J. Kladivko, J. V. Mannering, T. D. West and S. D. Parsons. "Long-Term Tillage and Rotation Effects on Corn Growth and Yield on High and Low Organic Matter, Poorly Drained Soil," *Agron. J.* 80:599–605 (1988).
76. Carter, M. R. "Influence of Tillage Practices on Nitrogen Yield in Barley and Soybean Rotations in Prince Edward Island," unpublished results (1993).
77. Dick, R. P. "A Review: Long-Term Effects of Agricultural Systems on Soil Biochemical and Microbial Parameters," *Agric. Ecosyst. Environ.* 40:25–36 (1992).
78. Cornish, P. S., and J. E. Pratley, (Eds.) *Tillage — New Directions in Australian Agriculture* (Melbourne, Australia: Inkata Press, 1987).
79. Boone, F. R. "Weather and Other Environmental Factors Influencing Crop Responses to Tillage and Traffic," *Soil Tillage Res.* 11:283–324 (1988).
80. Choudhary, M. A., and C. J. Baker. "Effects of Drill Coulter Design and Soil Moisture Status on Emergence of Wheat Seedlings," *Soil Tillage Res.* 2:131–142 (1982).
81. Chaudhry, A. D., and C. J. Baker. "Barley Seedling Establishment by Direct Drilling in a Wet Soil. I. Effects of Openers Under Simulated Rainfall and High Water-Table Conditions," *Soil Tillage Res.* 11:43–61 (1988).
82. Douglas, C. L., Jr., P. E. Rasmussen and R. R. Allmaras. "Nutrient Distribution Following Wheat-Residue Disposal by Combines," *Soil Sci. Soc. Am. J.* 56:1171–1177 (1992).
83. Coleman, D. C., E. P. Odum and D. A. Crossley, Jr. "Soil Biology, Soil Ecology, and Global Change," *Biol. Fertil. Soils* 14:104–111 (1992).
84. Lockeretz, W. "Open Questions in Sustainable Agriculture," *Am. J. Alternative Agric.* 3:174–181 (1988).
85. Keeney, D. "Sustainable Agriculture: Definition and Concepts," *J. Prod. Agric.* 3:281–285 (1990).
86. Anderson, D. W. and E. G. Gregorich. "Effect of Soil Erosion on Soil Quality and Productivity," in *Soil Erosion and Land Degradation. Proc. 2nd Annu. Western Provincial Conf. Rationalization of Water and Soil Research and Management* (Saskatoon, Saskatchewan: 1983), pp. 105–113.

87. Larson, W. E., and F. J. Pierce. "Conservation and Enhancement of Soil Quality," in *International Workshop on Evaluation for Sustainable Land Management in the Developing World, Vol. 2, Technical Papers,* C. R. Elliot, M. Latham and J. Dumanski (Eds.) (Chiang Mai, Thailand: 1991) International Board for Soil Research and Management.
88. Logan, T. J., J. M. Davidson, J. L. Baker and M. R. Overcash (Eds.) *Effects of Conservation Tillage on Groundwater Quality* (Chelsea, MI: Lewis Publishers, 1987), p. 292.

Humid Micro-Thermal Climates

CHAPTER 2

Trends in Reduced Tillage Research and Practice in Scandinavia

Hugh Riley
Norwegian State Agricultural Research Stations; Nes på Hedmark, Norway

Trond Børresen
Agricultural University of Norway; Ås, Norway

Egil Ekeberg
Norwegian State Agricultural Research Stations; Nes på Hedmark, Norway

Tomas Rydberg
Swedish University of Agricultural Sciences; Uppsala, Sweden

TABLE OF CONTENTS

0-87371-571-3/94/$0.00+$.50

2.1. INTRODUCTION

As in many other temperate regions, the conventional tillage system in Scandinavia comprises autumn plowing to combat weeds and bury plant residues, followed by secondary cultivations in the spring in order to produce a seedbed. Inherent in such systems is the contradiction that the soil is first loosened and thereafter recompacted before the crop is sown. Besides being wasteful of energy and labor input, this leads to a long period during which the soil is left in a condition highly susceptible to erosion and nutrient loss.

The trend in recent decades toward heavier and more powerful tractors and harvesting equipment has resulted in a general increase in the depth of plowing, and also in the depth to which soil compaction may be expected to take place. The avoidance of compaction has therefore been a major research priority for many years, as has been the investigation of how to achieve optimum seedbed conditions.

The need to investigate the possibility of plowless tillage systems emerged in the mid-1970s, following the advent of effective herbicides and spurred on by the "energy crisis." Growing environmental awareness in recent years has resulted in renewed public interest in this research, especially as a result of the North Sea Treaty of 1987, which is aimed at halving the nutrient losses from agriculture by 1995.

2.2. CLIMATIC AND EDAPHIC FEATURES OF THE AGROECOSYSTEM

From an agricultural viewpoint, the dominant climatic features of most of Scandinavia are the shortness of the growing season (May to September) and the long period during which the ground is either snow covered or frozen (November to April) and therefore inaccessible for tillage operations. Rainfall deficits are common in early summer in the major arable areas, whereas in late summer and autumn both harvesting and tillage are often hampered by excess rainfall. The provision of suitable conditions for rapid plant establishment is therefore of importance, both to ensure the use of available moisture in spring and to avoid harvesting losses in autumn. Often, insufficient time is available for the establishment of autumn-sown crops, and spring-sown cereals therefore dominate, except in the southernmost areas. Cereals are grown either in monoculture or in rotation with potatoes (*Solanum tuberosum* L.), oilseed rape (*Brassica napus* L.), or annual root or forage crops. The proportion of grassland in arable areas is fairly low, as dairy and animal production is mostly concentrated in more marginal areas.

Soil conditions vary widely, according to the nature and mode of deposition of the parent material. A common feature is that the soils are relatively young (postglacial), being derived from glacial tills, alluvial silts, and sands or marine clays. The tills normally give rise to soils of loamy texture with few management problems, apart from their high stone and boulder content. The marine clays present a greater challenge, being dominated by nonswelling, illitic clay minerals. High silt contents give these and many other alluvial soils a poor structural stability.

A special problem is encountered in parts of Norway, where low organic matter contents have arisen as a result of land leveling by bulldozer. This practice has become common in order to allow mechanization in steeply undulating landscapes. Soil, topography, and climatic conditions combine to make such land especially prone to erosion, particularly during the snow-melt period.

2.3. DEFINING CONSERVATION TILLAGE FOR SCANDINAVIA

Conservation tillage may be interpreted as any system that promotes good crop yields while at the same time maintaining soil fertility and minimizing soil and nutrient losses. A more comprehensive definition should take account of the energy inputs in machinery, fuel, and herbicide production. The greatest benefits in terms of both soil loss reductions and machinery/fuel savings are undoubtedly associated with the abandonment of annual plowing. Unfortunately, this often leads to a greater use of pesticides, which reduces the savings in energy and which may be undesirable from an environmental viewpoint.

A switch from autumn to spring plowing would alleviate the risk of erosion, but it is often considered undesirable for other reasons. The main disadvantage of spring plowing is the inevitable delay in sowing that it causes. An exception may be made in the case of excessively moist alluvial silts, where spring plowing often enhances soil drying in the spring, thus allowing earlier sowing. Experiments on loam soils have shown no difference in yields between spring and autumn plowing for crops sown on the same date.[1] On heavier soils spring plowing usually gives rise to the formation of clods. This impairs the quality of the resulting seedbed and can seriously reduce yield.[2] Research in both Sweden and Norway[3,4] has clearly demonstrated the importance of aggregate size distribution on moisture conservation and resulting germination.

In general, conservation tillage in Scandinavia normally involves some form of reduced tillage. This expression covers a wide range of alternatives, ranging from systems that include thorough stubble cultivations in autumn followed by harrowing in spring, to direct drilling systems with no cultivations at all prior to sowing. Research in Scandinavia has included work on several alternatives, but most emphasis has been placed on systems that allow the use of conventional seed drills, as this is deemed most acceptable from an economic viewpoint. The following sections of this chapter review the problems and potentialities of the various systems.

Numerous questions arise in the transition from conventional to reduced tillage systems:

- How can perennial weeds be controlled and plant residues disposed of?
- Which pests and diseases can be expected to increase and what rotation should be followed to avoid them?
- How are soil physical conditions affected and what are the implications of such changes under different soil and climatic conditions?
- Will fertilizer requirements remain the same?
- What changes in machinery should be considered?
- Can small-plot trial results be reproduced on a field scale?
- What long-term trends may be expected?

Field trials have provided answers to many of these questions, and in some cases have highlighted the pitfalls and problems that are likely to be encountered. In other cases, unexpectedly positive results have been obtained, as in the case of direct planting of potatoes in Norway.[5] A summary of selected research findings is given here.

2.4. CROP PERFORMANCE UNDER REDUCED TILLAGE

The most striking feature of many trials has been how yields have been affected relatively little by the omission of tillage practices which until recently were considered essential. Average yields of well over 90% of those achieved with conventional tillage have normally been attained (Table 2.1). Taking into account the potential savings in machinery, labor and fuel, this in itself is remarkable.

It would be misleading, however, to give the impression that such "simplified" systems present no problems to the grower; indeed, they probably place greater demands upon the level of husbandry and general awareness of soil and crop conditions. The most obvious problems are concerned with weed control and crop residues, but the avoidance of compaction is also of importance.

2.4.1. Weed Problems

Experience with weed problems includes the spread of perennials such as couch grass (*Elymus repens* L. Gould) and sow thistle (*Sonchus arvensis* L.),

Table 2.1. Average Grain Yields (t/ha) in Scandinavian Tillage Trials

Country	No. of Harvests	Plow Tillage	Shallow Tillage	Direct Drilling	Proportion Autumn-Sown (%)
DENMARK[a]	163	4.45 (100)	4.20 (94)	—	24
	80	4.91 (100)	—	4.54 (92)	49
SWEDEN[b]	464	4.47 (100)	4.40 (99)	—	36
	22	4.89 (100)	—	4.45 (91)	27
NORWAY[c]	216	4.76 (100)	4.57 (96)	—	0
	27	5.17 (100)	—	5.11 (99)	19

Note: Conventional plow tillage has been compared with shallow tillage (harrowing) and direct drilling. Relative yields (%) are shown in parentheses.

a Rasmussen;[6-9] Rasmussen and Olsen.[10]
b Henriksson;[11] Rydberg.[12]
c Børresen;[13] Ekeberg[14,15] and unpublished data; Ekeberg, Riley, and Njøs;[16,17] Marti;[18] Riley[19,20] and unpublished data.

biennials such as marsh foxtail (*Alopecurus geniculatus* L.), and winter annuals such as bluegrass (*Poa annua* L.) and mayweed (*Matricaria inodora* L.). Of these, couch is undoubtedly the most difficult to control without plowing. While the others may be effectively controlled by the use of selective herbicides during the growing season, couch must be controlled in autumn or early spring. Low temperatures and frequent rainfall often limit the effectiveness of herbicide (glyphosate) treatment at such times. The use of glyphosate before harvesting is restricted by law throughout Scandinavia.

Yield levels generally decline in the presence of couch, and losses of 20 to 25% are common on unplowed plots with high couch infestation.[16,18] It is not easy to distinguish between cause and effect in this relationship because increased weediness may result from poor crop growth caused by other factors. This often happens when germination is restricted due to inadequate seedbed conditions. Nevertheless, it is clear that satisfactory yields cannot be maintained in couch-infested soil.

Stubble cultivation provides some couch control, but is unlikely to be sufficient in cases of severe infestation. In such cases, glyphosate treatment in spring is an alternative that has been investigated to some extent in Norway, where the development of couch plants is rapid on unplowed soil at this time. Satisfactory results have been obtained, at least in the year of treatment, provided that tillage and sowing are delayed for a period of 3 to 4 days after spraying.[21] Reduced tillage appears to present few problems with annual weed species. Indeed, a lower species diversity is sometimes encountered.[22]

2.4.2. Straw Residues

Straw residues left on the surface may present problems both at sowing and later in the growing season. Straw in the proximity of the seed may present a physical barrier to moisture uptake under dry conditions, or it may aggravate problems of anaerobism (such as the accumulation of phytotoxins) under wet conditions. Straw residues cause a marked lowering of early season soil

Table 2.2. Effects of Straw Disposal Method on Yields of Direct Drilled Crops, Relative to Conventional Tillage with Straw Removed

	No. of Harvests	Straw		
		Burned	Baled	Chopped
Winter rape[a]				
After cereals	47	113	102	96
Winter wheat[b]				
Good rotation	11	98	96	96
Poor rotation	11	94	80	77
Spring cereals[c]				
Loam soil	4	104	100	100
Silty clay	3	102	90	76
Silt	4	91	84	76

[a] Cedell[23]
[b] Henriksson[11]
[c] Riley[19] and unpublished data

temperatures[13] and there is also the possibility of an effect on the incidence of certain fungal diseases.

In many of the tillage trials cited above, straw has been removed or incorporated into the soil by stubble cultivation, and thus seldom presents a problem.[12,21] When autumn crops are direct drilled, the best results have been achieved when straw is burned (Table 2.2). Little evidence has been found in Sweden for the presence of phytotoxins in autumn-sown crops.[23,24] In spring cereals, some adverse effects of straw residues have been found on poorly drained soils in Norway[19] which were associated with very low air porosity and permeability. Less adverse effects of plant residues have been found when wheat (*Triticum aestivum* L.) and barley (*Hordeum vulgare* L.) are grown in "favorable" rotations (with oilseeds, beans, and oats) than when they are grown in monoculture.[11]

2.4.3. Crop Rotation

Crop rotation indeed may be of greater importance with reduced tillage than in conventional systems, both in order to minimize problems associated with fungal diseases, and as a means of reducing weed problems without resorting to herbicides. However, opinions differ as to the severity of the "rotational" problems that are likely to be encountered.

Severe yield reductions were found in a Swedish trial series,[25] when winter wheat was direct drilled either in monoculture or in rotation with barley. This was attributed primarily to attacks of eyespot (*Pseudocercosporella herpotrichoides*) and glume blotch (*Septoria nodorum*). Other factors were probably also of importance, as even rotation with oats (*Avena sativa* L.) gave about 20% yield loss. In Danish trials,[9,26] an increased incidence of leaf blotch (*Rhyncosporium secalis*) has been reported in direct drilled barley, but eyespot and take-all (*Gaeumannomyces graminis*) appear to be unaffected by tillage.

Moreover, the grouping of a large number of Swedish trials according to the preceding crop revealed only minor yield differences that could be associated with rotation.[12]

2.4.4. Fertilizer Requirements

Fertilizer requirements of crops grown with reduced tillage have received relatively little attention. One might expect that more nitrogen would be applied in order to compensate for any suboptimal physical or biological conditions resulting from such systems. On the other hand, over the long term, requirements may even be expected to decline as a result of organic matter accumulation. Broadly similar response curves for nitrogen fertilization have been found on both plowed and unplowed soil.[9,20] In one long-term trial in Norway, yields have been higher without plowing in 10 of 13 years at a low nitrogen level, whereas little difference between tillage methods has been found at a higher fertilizer level.

A clear increase in crop lodging in the absence of plowing has also been found in several Norwegian trials[15,18,20] (Figure 2.1). The reason for this is not clear, but it is considered unlikely that the increase is caused by an increase in fungal diseases. One possible explanation may be that plant uptake of nitrogen continues for a longer period in unplowed soil, due to nitrogen mineralization later in the season. Higher plant nitrogen concentrations at harvest on unplowed soil support this theory, but it is too early to say whether lower fertilizer rates can be recommended.

Fertilizer placement in the seedbed has been found to be equally important in plowless systems as with conventional tillage.[12,21]

2.4.5. Plowless Tillage of Nongrain Crops

Plowless tillage of crops such as vegetables and potatoes is of less interest to growers, perhaps because the costs involved in tillage are relatively small as compared to the value of the crops concerned. Little advantage was found, however, for plowing in trials with cabbage (*Brassica oleracea* L.) and turnip (*Brassica rapa* L.) on loam soil.[27,28] In the case of onions (*Allium cepa* L.) and carrots (*Daucus carota* L.), on the other hand, plowless tillage has been found to reduce salable yields considerably.

As mentioned earlier, unexpectedly positive results have been found in the case of potatoes, which are traditionally thought to require a deep, loose seedbed.[5] Despite a somewhat slower rate of development early in the growing season, it appears that potatoes planted on unplowed soil continue to grow longer in the autumn, and often give higher final yields than those grown using conventional tillage (Figure 2.2). Such a system is unsuitable for the production of early potatoes but otherwise appears perfectly feasible, provided the grower has suitable equipment for planting and ridging.

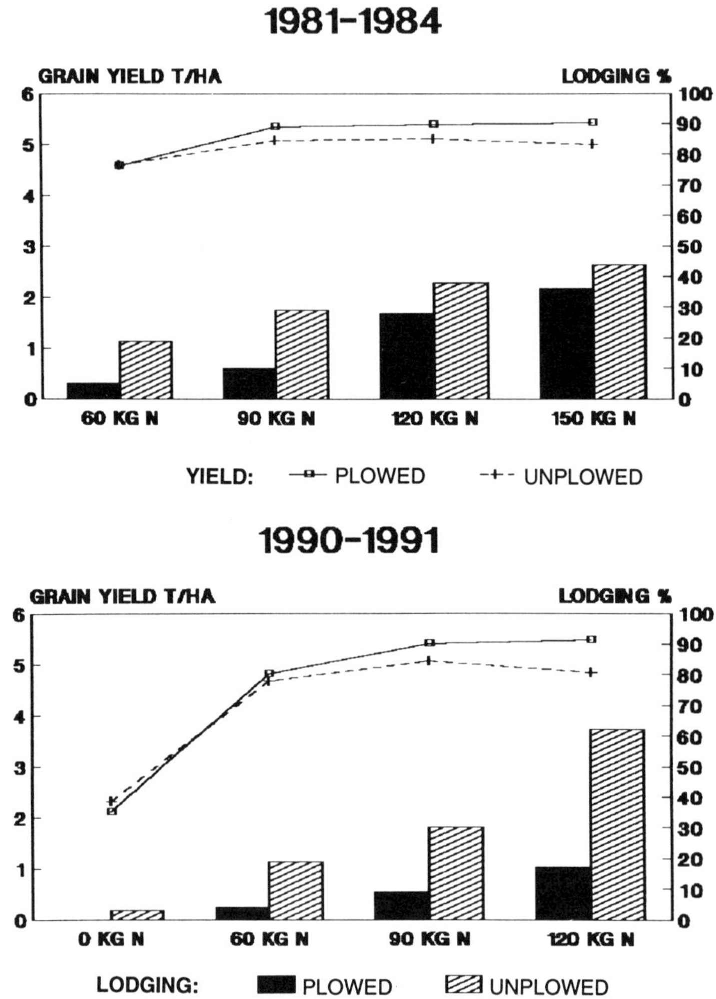

Figure 2.1. Grain yields (lines) and percentage lodging (bars) with increasing levels of nitrogen-fertilization on plowed and unplowed soil during two periods of a long-term tillage trial. (Compiled from Riley[20] and unpublished material.)

2.4.6. Influence of Soil Type and Weather

The influence of soil type on crop performance with reduced tillage usually has been of minor importance, in comparison to factors such as weediness. Greater difficulties nevertheless may be expected on poorly drained, heavy soils than on well-drained, lighter soils.

Poor results have been found after early sowing on silt soil in Norway,[20] probably due to waterlogging at germination. Alternately, silty clay soils in Sweden have been among those for which plowless tillage has been most

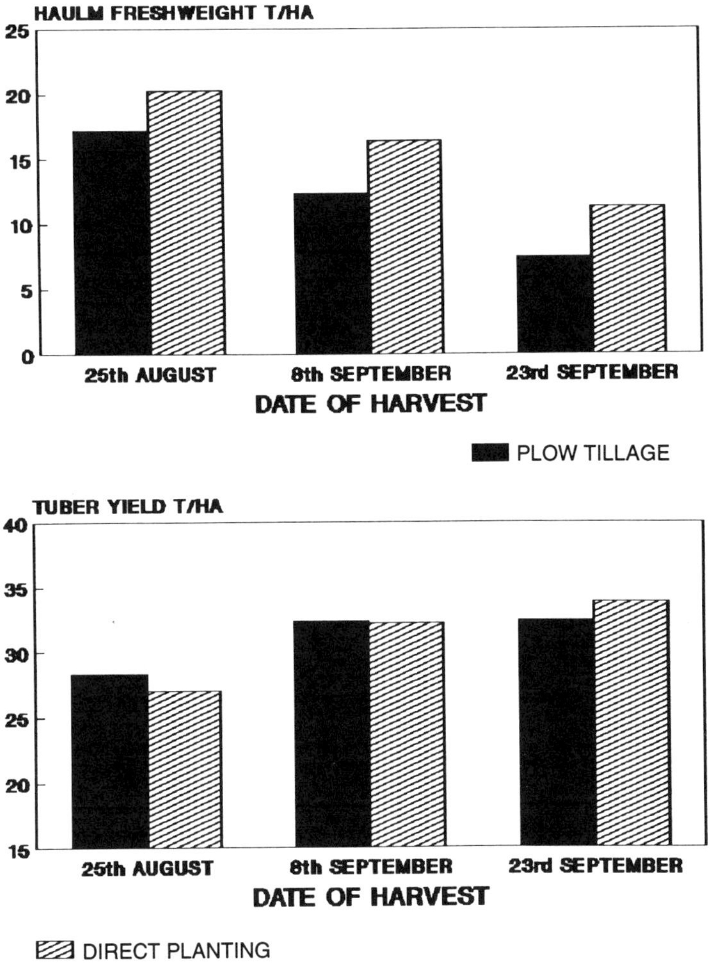

Figure 2.2. The effect of direct-planting potatoes on haulm freshweight and tuber yields at different dates of harvest, compared with conventional planting. Mean data for 5 years. (Unpublished material of Ekeberg).

successful.[29] The latter finding is thought to be due to improved surface moisture relations and possibly also to better rooting in unplowed soil.[30,31]

Weather conditions in the growing season also appear to play a part in the success of plowless tillage. In Norway, better results often have been observed in dry years than in wet years[5,15,18,19] (Figure 2.3). This is in accordance with the changes in soil physical conditions that are discussed below. Danish research,[32] however, revealed no effect of tillage on crop water use efficiency during the growing season. In any case, taking into account the variations in weather conditions that are encountered in Scandinavia, it is unlikely that any net gains or losses would be the result of weather effects over a longer period.

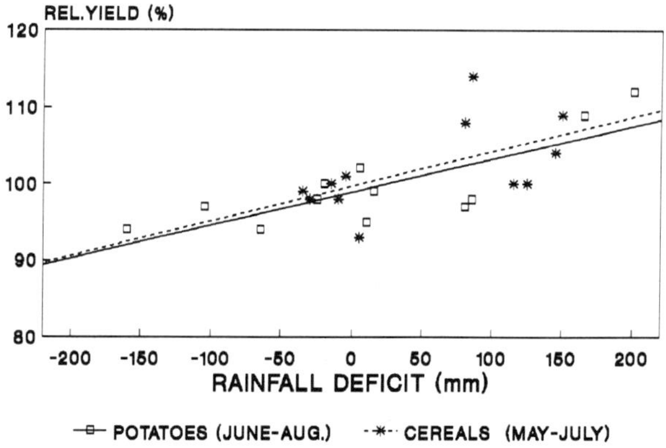

Figure 2.3. The effect of weather conditions during the growing season on relative yields of direct-planted potatoes and cereals grown with reduced tillage, relative to plow tillage. (Compiled from Riley,[19] Ekeberg,[5,15] and unpublished results).

2.4.7. Large-Scale Trials

Large-scale trials with plowless tillage performed by practicing farmers have been successful generally in both Sweden[33] and Norway.[17] Long-term trials in Norway have shown no negative trend in yields over a period of 15 years, provided that basic management requirements such as weed control have been mastered.

2.5. ENVIRONMENTAL ISSUES IN REDUCED TILLAGE ADOPTION

2.5.1. Changes in Soil Properties and Processes

The effects of plowless tillage on soil properties has received wide attention wherever such systems have been studied. Generally the same trends have been found in Scandinavia as in many other temperate regions. Many of the changes are interrelated, and their consequences may be of greater or lesser importance, depending on the soil type and on the external constraints of the climate. They may be summarized as follows:

- Accumulation of available nutrients (phosphorus and potassium) and organic matter near the soil surface;
- Increased bulk density and penetration resistance in upper and central topsoil layers;
- Lower air-filled porosity and gaseous exchange and, sometimes, higher water-holding capacity;

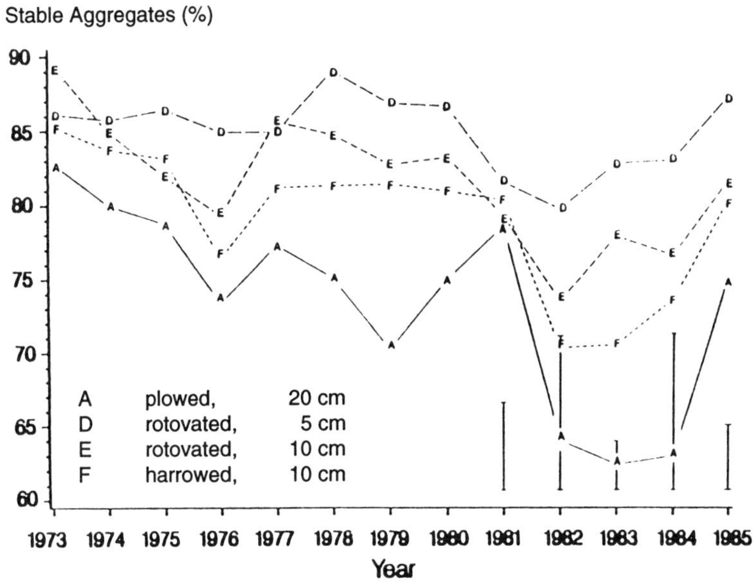

Figure 2.4. The influence of different tillage practices on aggregate stability over time in a Danish experiment. (Adapted from Schønning and Rasmussen, *Soil Tillage Res.,* 15: 84, 1989. With permission.)

- Lower surface infiltration rates, but in some cases, increased hydraulic conductivity between topsoil and subsoil; and
- Greater aggregate stability, greater earthworm activity, and more favorable conditions for promoting pore continuity.

Changes in the distribution of nutrients within the topsoil have not been found to affect their uptake by plants. Increased organic matter contents are undoubtedly responsible for the higher aggregate stability that has been found after some years without plowing (Figure 2.4). In Norway, such increases have been particularly noticeable on soils having low organic matter reserves[18] such as silty clays and silt, which are often regarded as "problem" soils in terms of soil structure.

Increases in organic matter content may also account for somewhat higher water-holding capacity in the seedbed. This may, however, also be a result of the redistribution of pore sizes caused by compaction (Table 2.3). Increases in bulk density are particularly noticeable at about 10 to 15 cm, just beneath the depth of tillage. The general increase in the compaction state of the upper topsoil layers has several implications.

Initial root development is sometimes retarded, but this may be compensated for in cases in which plow pans are a problem with conventional tillage.[31] The downward movement of water through the soil may be similarly affected (Figure 2.5). Aeration may be reduced to critical levels on some soils (e.g., silts), while on others (loams and well-structured clays) this is considered

Table 2.3. Changes Caused after 3 Years of Shallow Tillage and Direct Drilling in Soil Organic Matter, Bulk Density, Air-Filled Pore Space, Air Permeability and Available Moisture-Holding Capacity in the Upper Topsoil

	Tillage			
	Depth (cm)	Plow	Shallow	Direct Drilling
Organic matter (%)	0–5	6.9	+0.4	+0.5
	10–15	6.8	–0.2	+0.4
Bulk density (t/m^3)	0–5	1.17	–0.01	+0.03
	10–15	1.27	+0.08	+0.02
Air-filled porosity (%)	0–5	19.2	–1.1	–3.7
	10–15	12.7	–3.5	–2.9
Air permeability (μm^2)	0–5	20.5	+0.2	–3.8
	10–15	12.4	–7.4	–6.4
Available water (%)	0–5	25.6	+2.2	+3.9
	10–15	27.0	+0.2	+2.3

Note: Mean values for trials on loam, silty clay and silt soils; Riley.[37]

unlikely[20,34,36,38,39] (Figure 2.6). Greater mechanical strength and improved aggregate stability combine to reduce erosion risk, and also give rise to higher bearing capacity.[35] Important processes such as heat transfer and surface evaporation may also be affected. Both processes have been the subject of detailed studies.

The effects on evaporation by replacing autumn plowing by stubble cultivation were investigated in Sweden using undisturbed soil-core lysimeters.[40] Evaporation was monitored with and without the incorporation of crop residues, under both dry surface conditions and following simulated precipitation. Cumulative evaporation was lower from unplowed soil than from plowed soil, and the effect was particularly noticeable when the soil surface was intermittently moistened (Figure 2.7). This result is of considerable significance on silty-clay soils, whose high capillarity often leads to excessive seedbed drying. Crop residues left on the soil also reduced moisture loss, as may be expected.

Reduced tillage and direct drilling appeared to have little direct influence on soil temperature in a Norwegian study, although both practices had some effect on soil thermal properties, increasing both heat capacity and thermal flux.[41] The presence of straw residues, however, caused a marked reduction in both mean temperature and temperature amplitude in the topsoil, sufficient to account for the slower germination that is sometimes observed with these systems. The low thermal conductivity of crop residues has also been found to restrict frost depth, whereas the effect of plowing itself on this property seems to be slight.[42]

2.5.2. Environmental Benefits and Conflicts

The most obvious advantage of reduced tillage from an environmental viewpoint is its role in minimizing the risk of erosion. Erosion caused by

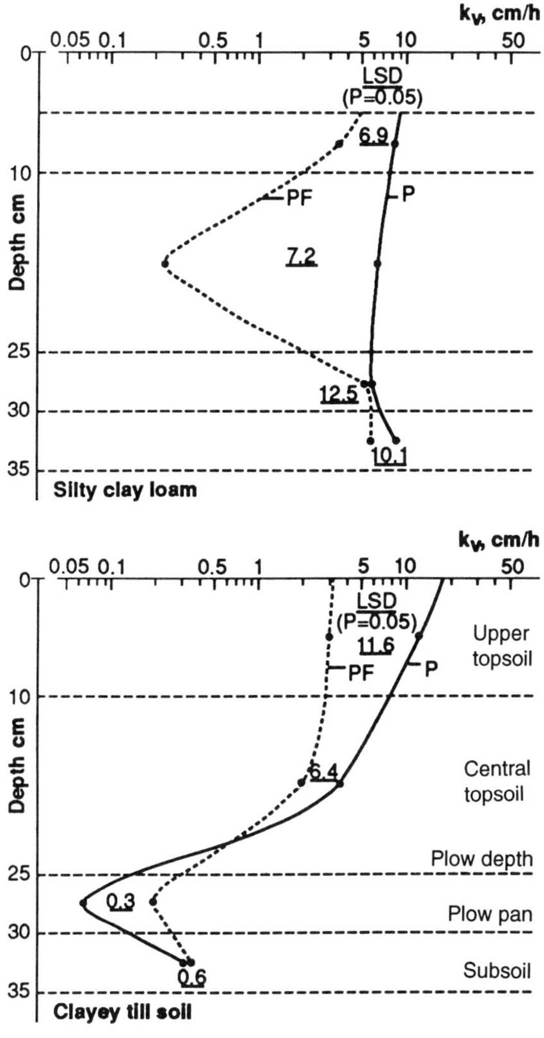

Figure 2.5. Modification of saturated hydraulic conductivity in clay soils caused by compaction beneath the harrow layer in unplowed soil (top) and by plow pan alleviation (bottom). (Solid line, plowed soil; dotted line, unplowed soil.) (Adapted from Rydberg, T., *Div. Soil Manage. Rep. No. 70,* Agricultural University of Sweden, 1986, p. 15. With permission.)

surface water runoff is a serious problem on sloping, silty-clay soils in some parts of Scandinavia, particularly during the snow-melt period and after the heavy autumn rains.[43] Of particular concern is the high amount of inorganic phosphorus that is deposited in freshwater reserves, thereby encouraging algal growth and eutrophication.

Shallow tillage has been shown to reduce erosion in such situations by about one half to two thirds by comparison with conventional tillage (Figure 2.8),

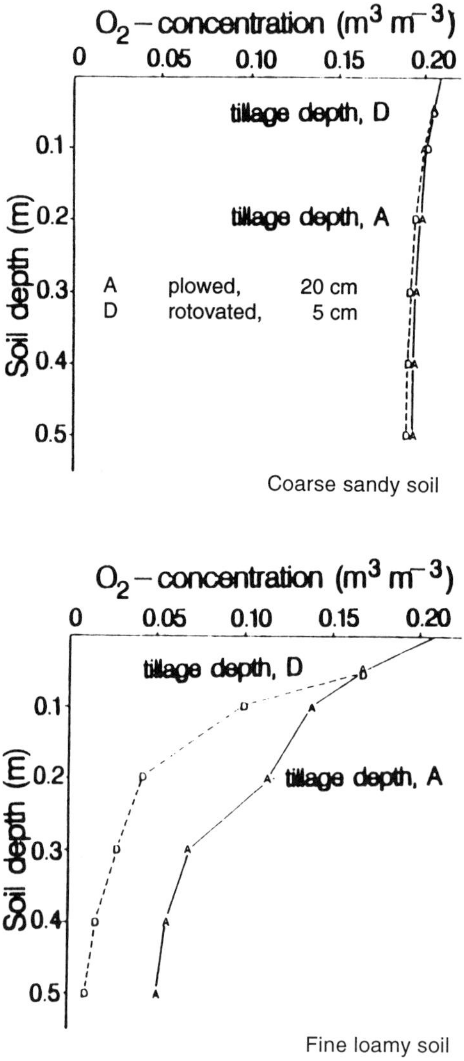

Figure 2.6. Effect of tillage on calculated profiles of oxygen concentration in contrasting Danish soils (solid line, plowed soil; dotted line, unplowed soil.) (Adapted from Schønning, P., *Soil Tillage Res.*, 15: 99, 1989. With permission.)

although this is still well above the levels found in grasslands.[44,45] Unfortunately, the soils on which erosion is most severe are also those that are most difficult to manage without plowing, owing to their poor internal drainage and high susceptibility to compaction.

A reduction in nitrate leaching might also be expected in the absence of plowing, because nitrification is probably encouraged whenever the soil is disturbed. Past evidence in Denmark[46,47] indicates that this effect is only slight, as compared with the benefits that may be obtained from autumn catch-cropping.[48,49]

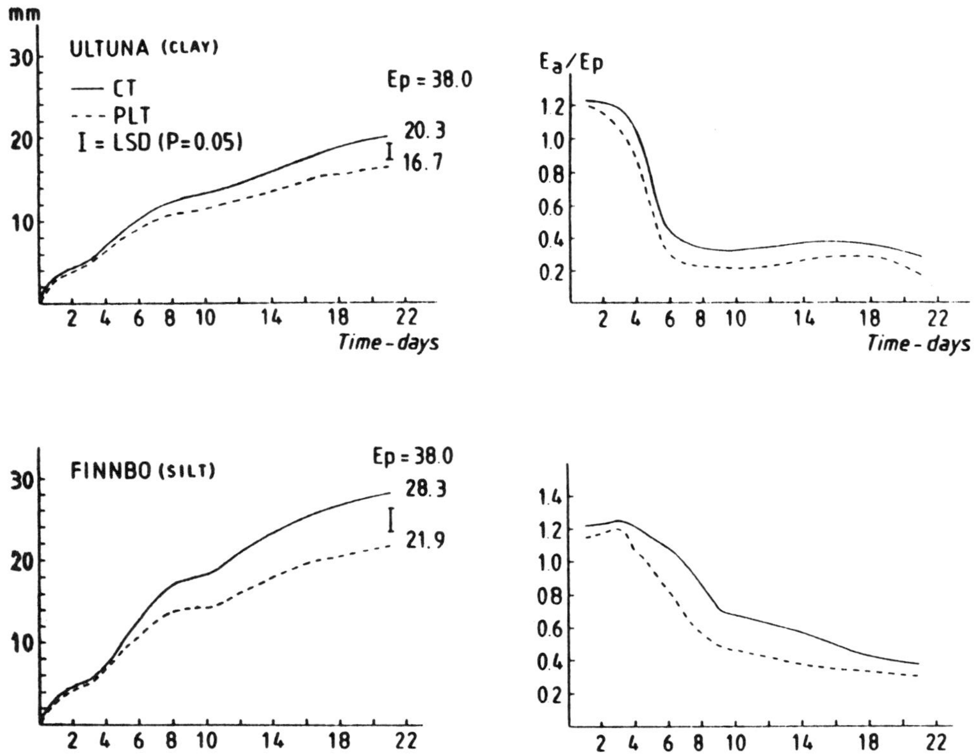

Figure 2.7. The effect of tillage treatments (solid line, plowed soil; dotted line, unplowed soil) on cumulative evaporation (left) and relative evaporation rate (right) following irrigation of bare clay and silt soils. (Adapted from Rydberg, T., *Soil Tillage Res.*, 17, 309, 1990. With permission.)

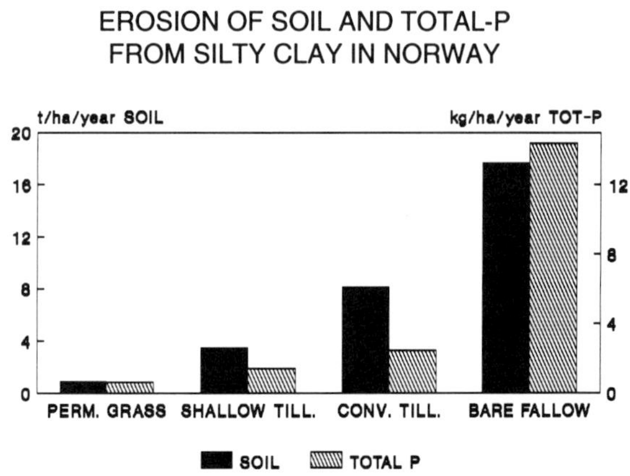

Figure 2.8. The effect of tillage and crop cover on losses of suspended solids (soil) and total phosphorus from sloping (1:8) silty clay soils in southern Norway. Data are means of 13 years in the case of soil losses and of 10 years for phosphorus losses. (Data compiled from Skøien, S., *Norw. Agric. Res.,* 2: 207–218, 1988. With permission.)

However, recent research indicates that both plowing and stubble cultivations may be responsible for increased nitrate losses.[50] Plowing has in some cases also been shown to result in an increase in soil particle and phosphorus content in drainage water from unstable soil.[51] Normally, however, drainage water from arable land contains only low concentrations of phosphorus.[52,53]

The potential environmental benefits of reduced tillage, in terms of avoiding nutrient loss, must be weighed against the possible undesirable side effects of such systems. A reduction in tillage intensity will almost certainly lead to a greater reliance upon the use of herbicides and other plant protection chemicals. Even if these are shown to be rapidly immobilized or degraded in the soil, their production involves considerable energy consumption.

Another undesirable consequence of reduced tillage is the probable increase in straw burning that may result in order to overcome difficulties with sowing or disease control or both. This may have a relatively minor influence on the maintenance of soil organic matter, because the increases associated with straw incorporation[54-56] are no larger than those that may be expected to result from the omission of plowing. Nevertheless, public opposition to straw burning for health and other reasons is widespread, and a ban has already been enforced in Denmark. Straw burning involves the loss of 25 to 35 kg/ha of nitrogen to the atmosphere, and its effects in terms of NO_x gas emission are still unknown.

2.6. ECONOMIC ASPECTS OF REDUCED TILLAGE SYSTEMS

A reduction in plowing depth and frequency clearly creates large savings in fuel consumption. The overall use of fuel in spring direct drilling systems in

Scandinavia is about 20 to 25% of that found with plow tillage.[57-59] Fuel consumption represents, however, only a minor part of the total tillage costs, while in terms of energy the savings are partly offset by the increase in herbicide use.

Of greater interest, therefore, are the savings to be made in time (labor) and machinery investment. Plowing alone represents about 40% of the total time required for traditional tillage and sowing. Direct drilling systems provide, under average conditions, overall savings of 60 to 75% in labor requirement, even when extra spraying is necessary.

A lower labor requirement for tillage means that the average sowing date will also be brought forward, and this in turn results in a higher yield potential. How much this means in practice depends upon the time available for tillage, which is governed by soil moisture conditions. Figure 2.9 shows the results of model calculations of such relationships in southeastern Norway. The highest overall spring work rates (3.5 to 5 ha per man per day, depending upon fertilizer placement method) may be achieved by direct drilling, while the lowest overall work rate (1.5 ha per man per day) applies to a system with spring plowing. Systems with autumn plowing permit work rates of 2 to 2.5 ha per man per day, while shallow tillage gives values of 2.5 to 3 ha per man per day.

Practical experience shows that the soil is seldom trafficable until the surface layer has dried to at least 90% of field capacity. This implies a reduction in yield potential of 5% for shallow tillage, 10% for autumn plowing and 15% for spring plowing, relative to the yield potential obtainable with direct drilling. This potential gain from direct drilling is seldom taken into account in tillage trials.

The economic consequences of different tillage systems are more difficult to assess, as they depend greatly on farm size, investment patterns, and fluctuating cost-price relations. An assessment of the costs involved in five alternative tillage systems is shown in Table 2.4 for a typical Norwegian farm of 30 ha. The profitability of reduced tillage depends to a large extent upon whether machinery investment is cut to the minimum essential for a particular system, or whether the full range of equipment is maintained.[60] In the latter case, the purchase of a direct drill will obviously involve increased total costs. Under Danish conditions it has been estimated that the break-even farm size to support the use of a direct drill is 60 ha.[61]

In terms of social economy (taking into account all consequences of changes in agricultural practices and land use), both shallow tillage and direct drilling compare favorably with other methods, according to a cost-benefit analysis performed in Norway to assess various alternatives for reducing phosphorus losses to waterways.[62] Only fertilizer planning that accounts for phosphorus recycled in plant and animal wastes gives lower costs per ton of phosphorus "removed" from runoff, because of the savings in fertilizer use. A tax on fertilizer would increase production costs significantly, as would the spreading of manure in spring instead of in autumn, because of the need to increase manure storage capacity. Spring plowing is judged to reduce yield potential because of delayed

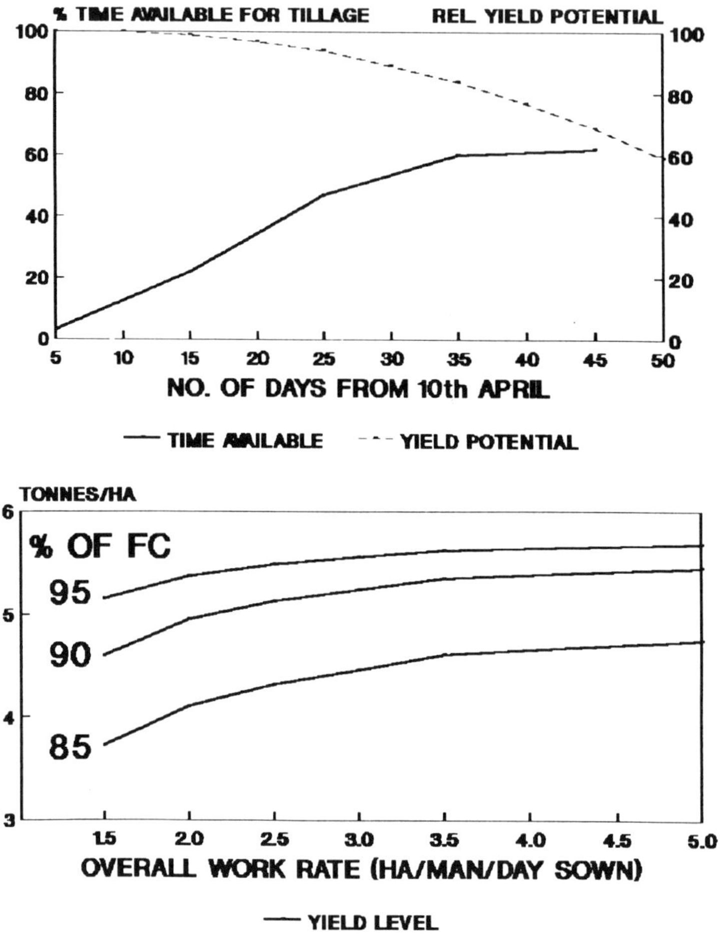

Figure 2.9. The decline in yield potential with delayed sowing and the proportion of time available for tillage (topsoil moisture <90% of field capacity) (top) and the effect of work rate (tillage plus sowing) on overall yields obtainable assuming a maximum potential yield of 6 t/ha, at three limiting levels of soil moisture (bottom). Based on 25 years of weather data in southeastern Norway. (Adapted from Riley, H., *Div. Soil Manage. Rep. No. 77,* Agricultural University of Sweden, 1988.)

sowing, while changes in land use such as the use of grassed waterways or afforestation would cause major losses of revenue to the agricultural community.

2.7. ADOPTION OF REDUCED TILLAGE AND FUTURE REQUIREMENTS

The adoption of reduced tillage by farmers has been slow throughout Scandinavia. This may be attributed partly to a natural reluctance to change

Table 2.4. Total Investment Requirement and Annual Fuel Consumption, Labor Requirement, and Total Costs for Five Alternative Systems of Tillage and Sowing in Norway (farm size 30 ha)[60]

	System[a]				
	1	2	3	4	5
Machinery investment (NOK[b] 1000)	370	318	261	221	343
Labor requirement (man-days)	14.6	11.6	9.0	9.0	5.3
Fuel consumption (1000 L)	1.68	0.92	0.30	0.24	0.19
Total annual costs (NOK 1000)[c]					
50%	92.3	86.6	82.0	82.0	99.1
100%	81.8	62.9	45.4	39.5	54.0

[a] 1, Conventional tillage with autumn plowing; 2, shallow tillage in both autumn and spring; 3, shallow tillage in spring with combine seed drill; 4, shallow tillage in spring with separate seed/fertilizer; and 5, direct drilling without tillage.

[b] NOK = Norwegian kroner.

[c] 50% = Conservation tillage on half of farm, and equipment for conventional tillage maintained on farm. 100% = Conservation tillage on entire farm, and machinery investment reduced to the minimum requirement.

time-proven tillage practices and partly to fears of income losses due to yield reductions. Despite the success of many trials, the uncertainties associated with weed control and straw residue management have often weighed in such systems' disfavor. Many farmers have pointed to a lack of suitable seedbed and sowing equipment, despite the fact that the majority of trials with shallow tillage have been performed using existing types of seedbed harrow and standard seed drills. Few farmers have taken advantage of the systems' full economic potential by reducing their overall investment in tillage machinery.

No statistics are available of the extent to which reduced tillage is practiced in Scandinavia. It is, however, probably least widespread in Denmark, where it is in conflict with the official policy of maintaining a high proportion of "green" crops over the winter, in order to avoid nitrate contamination of groundwater. This entails the need for such crops to be plowed down in spring. In Sweden, reduced tillage is accepted as the most suitable system on certain soils in which a lack of seedbed moisture is a problem. Direct drilling of autumn-sown oilseed crops has also been very successful on heavy soils. The recent changes in Swedish agricultural policy, which include drastic cuts in grain prices and the fallowing of up to 1 million ha each year, is likely to have a large impact on future tillage trends.

The most active official interest in reduced tillage has been seen in Norway, where erosion risks are greatest due to a combination of topographic, edaphic and climatic factors. Government subsidies have been introduced recently for farmers who refrain from autumn plowing on erosion-prone land. The definition of such areas is aided by soil mapping and, on the whole, the scheme has been met with approval by the farming community. Much needs to be learned, including:

- How to integrate reduced tillage in mixed farming systems.
- How to handle straw residues if burning is banned and no market exists for straw products.
- How to overcome the need for increased use of herbicides and plant protection chemicals.

The answers to these questions may well be decisive in whether reduced tillage becomes an important new direction in Scandinavian agriculture, or whether it will be forgotten as a well-intentioned but impracticable sidetrack to the mainstream of development. Research can play an important part in encouraging farmers to consider new alternatives, but it will be their own experiences that determine their choice.

2.8. GENERAL SUMMARY AND CONCLUSIONS

Conservation tillage in Scandinavia usually involves reduced tillage in order to maintain an undisturbed soil surface during winter and early spring, thereby reducing erosion risk. Reduced tillage has been shown in trials to be successful under a wide range of soil conditions. The main problems that arise are the control of perennial weeds, the disposal of straw residues, and the avoidance of compaction. These problems have prevented widespread adoption of the system in practice.

Savings in production costs and labor requirement may prove to be a more important incentive to farmers in the future. For this reason, systems that do not involve investment in specialized new equipment (such as direct drills) will be the most likely to be adopted in practice.

2.9. REFERENCES

1. Njøs, A., and E. Ekeberg. "Trials with Two Depths of Plowing in Autumn and Spring on a Morainic Loam Soil in Stange, S. Norway, during the Years 1969–1975," *Res. Norw. Agric.* 31:221–242 (1980).
2. Njøs, A., and T. Børresen. "Long-Term Experiment with Straw Management, Stubble Cultivation, Autumn and Spring Plowing on a Clay Soil in S.E. Norway," *Soil Tillage Res.* 53:53–66 (1991).
3. Håkansson I., and J. von Polgar. "Experiments on the Effects of Seedbed Characteristics on Seedling Emergence in a Dry Weather Situation," *Soil Tillage Res.* 4:115–135 (1984).
4. Njøs, A. "Aggregate Size Distribution in the Seedbed: Effects on Soil Temperature, Matric Suction and Emergence of Barley," *Proc. 8th Conf. of the International Soil Tillage Research Organization*, Vol. 1 (Stuttgart: Hohenheim, 1979), pp. 121–129.
5. Ekeberg, E. "Reduced Tillage on Loam Soil. II. Potato," *Norw. Agric. Res.* 1:7–14 (1987).

6. Rasmussen, K. J. "Reduced Cultivation in Barley Monoculture," *Dan. J. Plant Soil Sci.* 85:171–183 (1981).
7. Rasmussen, K. J. "Soil Tillage Systems for Winter Cereals," *Dan. J. Plant Soil Sci.* 86:531–541 (1982).
8. Rasmussen, K. J. "Methods of Tillage for Spring Barley on Coarse Sandy Soils," *Dan. J. Plant Soil Sci.* 88:443–452 (1984).
9. Rasmussen, K. J. "Ploughing, Direct Drilling and Reduced Cultivation for Cereals," *Dan. J. Plant Soil Sci.* 92:223–248 (1988).
10. Rasmussen, K. J., and C. C. Olsen. "Soil Tillage and Catch Crops by Growth of Barley," *Dan. J. Plant Soil Sci.* 87:193–215 (1983).
11. Henriksson L. "Direct Drilling of Cereals and Leys," Div. of Soil Manage. Rep. No. 77 (Uppsala, Agricultural University of Sweden, 1988), pp. 148–153.
12. Rydberg, T. "Studies in Ploughless Tillage in Sweden 1975–1986," Div. of Soil Manage. Rep. No. 74 (Uppsala, Agricultural University of Sweden, 1987), pp. 35.
13. Børresen, T. "Effects of Three Tillage Systems Combined with Different Compaction and Mulching Treatments, on Cereal Yields, Soil Temperature and Physical Properties on Clay Soil in South-Eastern Norway," Dr.Sci. thesis, (Ås: Agricultural University of Norway, 1986).
14. Ekeberg, E. "Autumn and Spring Tillage for Spring Cereals," *Res. Norw. Agric.* 36:133–139 (1985).
15. Ekeberg, E. "Reduced Tillage on Loam Soil. I. Cereals," *Norw. Agric. Res.* 1:1–6 (1987).
16. Ekeberg, E., H. Riley and A. Njøs. "Ploughless Cultivation of Spring Cereals. I. Yields and Couch Grass," *Res. Norw. Agric.* 36:45–51 (1985).
17. Ekeberg, E., and H. Riley. "Ploughless Tillage in Large-Scale Trials. I. Yields, Grain Quality and Couch Grass," *Norw. Agric. Res.* 3:97–105 (1989).
18. Marti, M. "Cereal Monoculture without Ploughing in South-East Norway — Effects on Yields and on Soil Physical and Chemical Properties," Dr.Sci. thesis, (Ås: Agricultural University of Norway, 1984).
19. Riley, H. "Reduced Cultivations and Straw Disposal Systems with Spring Cereals on Various Soil Types. I. Yields and Weed Incidence," *Res. Norw. Agric.* 34:209–219 (1983).
20. Riley, H. "Reduced Tillage for Spring Cereals. Different Seed Drills and Sowing Dates," *Res. Norw. Agric.* 36:61–70 (1985).
21. Ekeberg, E. "Yield Results and Aspects of Cultivation with Shallow Tillage in Norway," Div. Soil Manage. Rep. No. 77 (Uppsala, Agricultural University of Sweden, 1988), pp. 154–166.
22. Andersen, A. "Effects of Direct Drilling and Ploughing on Weed Populations," *Dan. J. Plant Soil Sci.* 91:243–254 (1987).
23. Cedell, T. "Direct Drilling of Oilseed Crops," Div. Soil Manage. Rep. No. 77 (Uppsala, Agricultural University of Sweden, 1988), pp. 138–147.
24. Olsson, S., and L. Ohlander. "Early Growth Disturbance in Direct Drilled Winter Rape and Winter Wheat," Dept. Plant Husbandry Rep. No. 164 Stockholm: University of Sweden, 1986).
25. Olofsson, S., and B. Wallgren. "Rotational Aspects Concerning Direct Drilling of Winter Wheat," Div. Soil Manage. Rep. No. 77 (Uppsala, Agricultural University of Sweden, 1988), pp. 111–116.

26. Jensen, A. "Plant Pathological Aspects of Ploughless Tillage," Div. Soil Manage. Rep. No. 77 (Uppsala, Agricultural University of Sweden, 1988), pp. 117–127.
27. Ekeberg, E. "Reduced Tillage on Loam Soil. III. Brassica Crops," *Norw. Agric. Res.* 1:15–21 (1987).
28. Dragland, S. "The Effects of Tillage and Soil Compaction on Cabbage, Onion and Swede," *Norw. Agric. Res.* 3:145-152 (1989).
29. Rydberg, T. "Field Experiments with Ploughless Tillage in Sweden, 1976–81," *Proc. 9th Conf. International Soil Tillage Research Organization* (Osijek, Croatia, 1982), pp. 125–130.
30. Rydberg, T. "Effects of Ploughless Tillage on Soil Physical and Soil Chemical Properties in Sweden," Div. Soil Manage. Rep. No. 70 (Uppsala, Agricultural University of Sweden, 1986).
31. Rydberg, T. and T. Öckerman. "The Effects of Ploughless Tillage on Root Development and Evaporation," Div. Soil Manage. Rep. No. 74 (Uppsala, Agricultural University of Sweden, 1987).
32. Djurhuus, J. "Actual Evapotranspiration in Spring Barley in Relation to Soil Tillage and Straw Incorporation," *Dan. J. Plant Soil Sci.* 89:47–59 (1985).
33. Rydberg, T. "Big-Plot Experiments with Ploughless Farming, 1976–78," Div. Soil Manage. Rep. No. 59 (Uppsala, Agricultural University of Sweden, 1980).
34. Schønning, P. "Soil Pore Characteristics. II. Effect of Incorporation of Straw and Soil Tillage," *Dan. J. Plant Soil Sci.* 89:425–433 (1985).
35. Schønning, P., and K. J. Rasmussen. "Long Term Reduced Cultivation. I. Soil Strength and Stability," *Soil Tillage Res.* 15:79–90 (1989).
36. Schønning, P. "Long Term Reduced Cultivation. II. Soil Pore Characteristics as Shown by Gas Diffusivities and Permeabilities and Air-Filled Porosities," *Soil Tillage Res.* 15:91–103 (1989).
37. Riley, H. "Reduced Cultivations and Straw Disposal Systems with Spring Cereals on Various Soil Types. II. Soil Physical Conditions," *Res. Norw. Agric.* 34:221–228 (1983).
38. Riley, H., A. Njøs and E. Ekeberg. "Ploughless Cultivation of Spring Cereals. II. Soil Investigations," *Res. Norw. Agric.* 36:53–59 (1985).
39. Riley, H., and E. Ekeberg. "Ploughless Tillage in Large-Scale Trials. II. Studies of Soil Chemical and Physical Properties," *Norw. Agric. Res.* 3:107–115 (1989).
40. Rydberg, T. "Effects of Ploughless Tillage and Straw Incorporation on Evaporation," *Soil Tillage Res.* 17:303–314 (1990).
41. Børresen, T., and A. Njøs. "The Effects of Three Tillage Systems Combined with Different Compaction and Mulching Treatments on Soil Temperature and Soil Thermal Properties," *Norw. J. Agric. Sci.* 4:363–371 (1990).
42. Thunholm, B., and I. Håkansson. "Influence of Tillage on Frost Depth in Heavy Clay Soil," *Swed. J. Agric. Res.* 18:61–65 (1988).
43. Skøien, S. "Extent and Geographical Distribution of Soil Erosion in Norway," *Norw. Agric. Res.* 2:199–205 (1988).
44. Skøien, S. "Soil Erosion and Runoff Losses of Phosphorus, Effect of Tillage and Plant Cover," *Norw. Agric. Res.* 2:207–218 (1988).
45. Njøs, A., and P. Hove. "Studies of Soil Erosion by Water," Final Rep. No. 655 (Oslo: Norrway Agricultural Research Council, 1986).
46. Djurhuus, J., and S. E. Simmelsgaard. "N-Leaching with Direct Drilling." Div. Soil Manage. Rep. No. 77 (Uppsala, Agricultural University of Sweden, 1988), pp. 101–110.

47. Rasmussen, K. J. "Effect of Tillage on Soil Nitrogen and Leaching Losses," *Dan. J. Plant and Soil Sci.* S 1669: 82–88 (1983).
48. Nielsen, N. E., and H. E. Jensen. "Soil Mineral Nitrogen as Affected by Undersown Catch Crops," in *Assessment of Nitrogen Fertilizer Requirement,* (Haren, Germany: Institute for Soil Fertility, 1985), pp. 101–110.
49. Lewan, L. "Catch Crops to Reduce Nitrate Leaching," NJF Sem. No. 162, "Hydrological Effects of Agricultural Practices and Impacts on Water Quality," held at Tune, Denmark (1989).
50. Møller Hansen, E. Personal communication (1991).
51. Lundekvam, H. "Arable Land and Erosion Problems," paper presented at Conf. on Agricultural Policy and Environmental Impacts, Drammen, Norway (1990).
52. Uhlen, G. "Nutrient Leaching and Surface Runoff in Field Lysimeters on a Cultivated Soil," *Norw. J. Agric. Sci.* 3:33–46 (1989).
53. Sibbesen, E. "Leaching of Phosphorus," *Ugeskrift for Jordbrug* No. 49:1543–1547 (1987).
54. Uhlen, G. "The Effect of Ploughed in Cereal Straw on Yields and Soil Properties," *Sci. Rep. Agric. Univ. of Nor.* 52:10 (1973).
55. Wølner, K., L. Sogn and N. H. Hauge. "Crop Rotation Trials at Bjørke, Hagan, Hellerud and Staur 1951–1975," *Res. Norw. Agric. Suppl.* No. 4: 313–361 (1978).
56. Uhlen, G. "Long-term Effects of Fertilizers, Manure, Straw and Crop Rotation on Total-N and Total-C in Soil," *Acta Agric. Scand.* 41(2):119–127 (1991).
57. Nielsen, V. "Energy Consumption and Labor Requirements of Direct Drilling and Conventional Tillage," Dan. Inst. Agric. Eng. Rep. No. 37 (1987).
58. Danfors, B. "Fuel Consumption and Capacity in Different Systems of Soil Tillage and Sowing," Swed. Inst. Agric. Eng. Rep. No. 420 (1988).
59. Riley, H. "Energy and Labour Use with Various Tillage Systems," Div. Soil Manage. Rep. No. 77 (Uppsala, Agricultural University of Sweden, (1988), pp. 196–206.
60. Ekeberg, E. "Economic Assessment of Reduced Tillage in Norway," Div. Soil Manage. Rep. No. 77 (Uppsala, Agricultural University of Sweden, (1988), pp. 228–235.
61. Iversen, K. K., and A. S. Møller. "Reduced Tillage — Economy and Energy Effectiveness," Dan. Inst. for Agric. Econ. Rep. No. 29 (1987).
62. Johnsen, F. H. "Economic Consequences of Measures for Controlling Erosion and Phosphorus Losses from Cultivated Land," in *Assessment of Erosion and Control Measures* (Ås: Norway State Agriculture Advisory Service, 1990), pp. 129–136.

CHAPTER 3

Tillage Requirements for Annual Crop Production in Eastern Canada

Tony J. Vyn and K. Janovicek
University of Guelph; Guelph, Ontario, Canada

Martin R. Carter
Agriculture Canada Research Station; Charlottetown, Prince Edward Island, Canada

TABLE OF CONTENTS

0-87371-571-3/94/$0.00+$.50

3.1. INTRODUCTION

In eastern Canada, soil traditionally has been prepared for seeding field crops by autumn moldboard plowing followed by spring secondary tillage. Prompted by concerns about soil erosion and high capital and labor investments associated with intensive tillage systems, grain and oilseed producers are increasingly adopting less intensive tillage systems. However, according to a survey conducted in 1991, 65% of the cropping area in southern Ontario was still moldboard plowed.[1] A chisel plow-based system was the predominant alternative to moldboard plowing and was utilized on 22% of the crop land. No-till tillage systems were utilized on only 4% of the field crop area. A similar situation exists for other areas of eastern Canada, such as the Atlantic Provinces of Canada.[2]

This chapter reviews research that has been conducted to assess tillage requirements for the predominant annual crops in eastern Canada. The focus is on tillage systems for corn (*Zea mays* L.), small grains, and soybean (*Glycine max* (L.) Merr.) in southern Ontario and Prince Edward Island. Tillage-induced changes on soil properties and their effects on soil erosion and crop performance are also discussed.

No-till refers to planting crops in total absence of tillage, except for a single coulter in front of the seed disc openers, and possibly another coulter in front of the fertilizer disc openers, if side dress fertilizer is applied at planting. Unless otherwise stated, all moldboard and chisel plow tillage systems referred to include secondary tillage just prior to crop planting with either a field cultivator or tandem disc.

3.2. DESCRIPTION OF AGROECOSYSTEM

3.2.1. Climate

The agricultural producing region of southern Ontario has a continental-type climate that is modified by the Great Lakes. The length of the frost-free period ranges from 120 to 180 days and is dependent primarily on elevation and proximity of the Great Lakes.[3] An Ontario system for classifying the length of the growing season based on a measure of air temperature accumulated during the frost-free period has rated the growing season at 2200 to 3500 Ontario corn heat units.[4] This range is similar to 75 to 115 days on the Minnesota Relative Maturity Scale. The average precipitation is 900 mm, a total that is generally evenly distributed throughout the year (i.e., monthly precipitation of approximately 75 mm).[3] During the growing season, the long-term average water deficiencies are about 25 to 50 mm, with some regions in the extreme southwest reporting water deficiencies as high as 125 mm. As a result, crop productivity occasionally can be limited by the availability of water, particularly on soils that have low water-holding capacity and/or during drier than normal years.

Prince Edward Island, located in the Maritime region of eastern Canada, has a modified continental-type climate owing to the prevailing westerly air movements from the interior of the North American continent.[5] The growing season in Prince Edward Island lasts about 190 days (end of April to end of October) and is characterized by a wet, cool spring and autumn and a warm, moist summer with occasional periods of drought. The long-term average frost-free period is 154 days.

The soil climate is characterized by mainly cool or moderately cool boreal soil temperature classes.[6] Soil moisture regimes are dominantly perhumid or humid. The long-term growing season precipitation (416 mm, May to September) is slightly below the long-term (1941 to 1970) growing season potential evapotranspiration (447 mm). This results in varying periods of soil water deficits during the growing season.

Relatively high precipitation in the spring is a major climatic constraint for agricultural systems because wet soil conditions limit the number of field work days for both primary and secondary tillage and seeding operations.[7] This can influence the yield and quality potential of small grains and soybean if seeding is delayed under conventional tillage systems due to a lack of field work days. A second major climatic constraint for some crops is relatively low soil temperatures in the spring. Although soil temperatures, at a depth of 5 cm, may be within the optimum range for cereals in the germinating phase they can significantly affect the planting date for corn. In addition, the relatively short time period between planting and the onset of freezing temperatures in the autumn can affect physiological maturity. Generally, planting of corn is not recommended until the average soil temperature, at a depth of 5 cm, has reached 10°C for at least 2 to 3 days, which usually occurs between May 20 and June 5.[7]

3.2.2. Soil Types

The parent material for the majority of soils in southern Ontario is the product of glacial deposition.[8] The texture of the parent material is diverse, ranging from coarse sands to clay. These soils were formed under temperate deciduous forests and belong to either the gray brown luvisolic, brunisolic,or gleysolic soil order.[6] The main limitation of soils belonging to the gleysolic and medium- to finer-textured luvisolic and brunisolic soil orders is imperfect or poor internal drainage. When this limitation is overcome by artificial drainage, these soils are considered highly productive. The majority of Ontario's fine-textured soils used in annual crop production are systematically tile drained.

Most of the soils in Prince Edward Island belong to the podzolic and luvisolic orders.[6] For agriculture their main limitation is low fertility followed by poor structure, which is mainly evident as a dense, compact subsoil.[9] The combination of poor subsoil structure with perhumid climatic conditions results in excessive soil moisture, especially in the spring and autumn. Many of the soils are susceptible to soil compaction. Soil compactibility is related to both soil particle size and organic carbon content.[10] With regard to particle size, the relatively high silt and fine sand content is especially conducive to soil structural instability. Another important consideration is the relatively low presence of expansible minerals in the clay fraction. Relationships between soil compaction and soil composition (e.g., particle size) can be significantly influenced by expanding-type clays, which can play an important role in natural regeneration of soil structure.[7]

3.3. TILLAGE REQUIREMENTS FOR MONOCULTURE CORN

The majority of tillage research conducted in eastern Canada, particularly in Ontario, has focused on assessing the tillage requirements for corn production. Experiments comparing tillage systems for corn production were initiated in the early 1960s. Much of the research in the first two decades focused on tillage for monoculture corn production because it was often more profitable than other crop sequences[11] and because experiments involving ten or more tillage treatments were easier to perform and analyze over periods of 5 or more years with a constant previous crop residue. Vyn et al.[12] reviewed and combined the results from numerous tillage experiments that were conducted in southern Ontario from 1964 to 1978. They reported that corn yields, averaged over all site/years, were 11% lower when planted without tillage compared to after autumn moldboard plowing. However, they observed that corn yield differences between these tillage systems were significant only when corn followed corn on finer-textured imperfectly drained soils, while yield differences were not significant when corn followed corn on well-drained gravely loam soils or following alfalfa (*Medicago sativa* L.)/timothy (*Phleum pratense* L.) sod mixtures. Reduced grain corn yields in no-till compared to the autumn moldboard plowing system has also been reported in later Ontario research wherever corn followed corn in a rotation.[13-18]

Vyn et al.[12] reported that when corn followed corn in a rotation, reduced grain yields in no-till compared to autumn plow tillage systems could not be attributed to differences in rates of emergence or corn plant populations. However, they observed that when no-till grain corn yields were reduced, often there were slower rates of growth and development of corn plants from the four- to eight-leaf stage. Furthermore, the reductions in final grain yields associated with no-tillage were not a result of inadequate fertility,[12,19] inadequate weed control, or improper planter adjustment.[12]

Early season soil temperature has been reported to be lower in no-till when compared to autumn plowed treatments and was implicated as a possible reason for inferior early season growth and final grain yields associated with no-till tillage systems.[20] However, in earlier research soil temperature differences among these tillage systems were not detected on a consistent basis and temperature was dismissed as the primary explanation for inferior no-till corn performance.[12] Although greater amounts of surface-placed residue may depress soil temperatures in no-till tillage systems, the relatively small differences in corn performance among tillage systems when planted after high residue crops such as alfalfa/timothy mixtures[12,15] and barley (*Hordeum vulgare* L.),[15] suggests that soil temperature does not totally explain the differences in early season growth and grain yields. This is further supported by a study in which a relatively small proportion of the variability in the mass of corn plants and number of emerged leaves 6 weeks after planting was explained by the amount of accumulated soil growing degree days (GDD), when planted no-till into various crop residues.[21] Also in this study, no correlation was found between the amount of soil GDD accumulated during the first six weeks after planting and final grain yield.

Soil moisture contents have been reported to be higher in no-till compared to autumn moldboard plow tillage systems.[18,20,22] Despite higher soil moisture contents in no-till tillage systems, corn performance was severely inhibited in an unusually dry growing season.[13,18] Relatively poor no-till corn performance in drier growing seasons, particularly when the drought stress occurs well before corn flowering, suggests that lack of tillage has resulted in changes in soil properties that reduced the availability of water and/or inhibited corn plant root growth.

3.3.1. Tillage Effects on Corn Growth

The yield reductions in monoculture corn production associated with no-tillage do not appear to be eliminated by the maintenance of no-tillage for long periods of time (Figure 3.1). Vyn and Raimbault[18] reported that after 15 years of continuous no-till corn production, higher soil resistance, greater bulk densities, and a lower proportion of fine aggregates were still observed as compared to moldboard plowing. They concluded that soil physical conditions with continuous corn production in no-till tillage systems on finer-textured soils in Ontario will not improve with time to match the physical structure following the traditional moldboard plowing system.

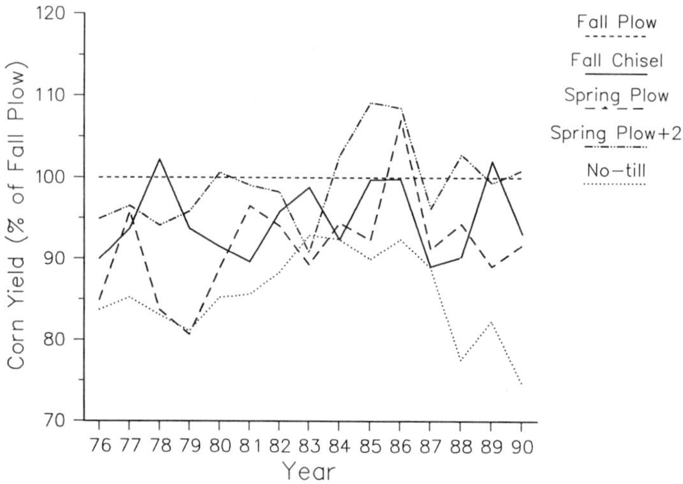

Figure 3.1. Grain corn yields in response to autumn moldboard, autumn chisel, spring moldboard, spring moldboard with no secondary tillage, and no-till tillage systems over 15 years of continuous corn production on a silt loam soil. Grain yields are expressed as a percentage of autumn moldboard yields.

Although corn yields after chisel plowing are often higher than those after no-tillage, the extent to which chisel plowing can result in yields similar to moldboard plowing is dependent on soil type and crop sequence. In a long-term study involving continuous corn production on a silt loam soil, corn yields in an autumn chisel plow tillage system averaged 5% lower than in an autumn moldboard plow tillage system (Figure 3.1). Other studies have also reported yields after chisel plowing that were intermediate to autumn moldboard plowing and no-till tillage systems when corn followed corn in a rotation.[13,15,23] Seedbed fineness has been reported to be similar between autumn chisel and autumn moldboard plow tillage systems,[18,23] but superior to that observed with no-till systems.[15,18] These studies suggested that the finer seedbeds in chisel plowing were partially responsible for superior corn performance when compared to the no-till tillage system.

Some research has examined the potential of delaying moldboard plowing until the spring on a silt-loam soil. Averaged over 15 years, little difference was found between a spring and an autumn moldboard tillage system.[18] The main advantage of the spring moldboard compared to the autumn moldboard tillage system is that the potential for late winter and early spring erosion is reduced because surface-placed residues remain undisturbed until just prior to corn planting. However, a spring moldboard tillage system may affect the timeliness of corn planting, especially in springs in which excessive amounts of rain fall. Excessive soil clodiness may also be a negative effect of delaying moldboard plowing until spring, especially on fine-textured soils and in dry springs.

Eliminating secondary tillage using a one-pass spring moldboard plow tillage system resulted in corn yields that averaged about 8% lower than spring moldboard with secondary tillage[18] (Figure 3.1). The absence of secondary tillage was associated with coarser seedbeds, which may have been partially responsible for inferior corn performance. However, the one-pass spring moldboard plow tillage system did result in finer seedbeds and higher corn yields than the no-till tillage system. On a fine sandy loam in Prince Edward Island, silage corn yields over 4 years in a one-pass moldboard plow tillage system were similar to an autumn moldboard plow system and superior to the no-till tillage system.[24]

3.3.2. Strip-Tillage Systems

In light of the relatively poor corn performance when corn follows corn in no-till tillage systems, some research has been conducted to investigate the potential for modifying the no-till tillage system to provide a more favorable soil environment. The modifications have involved various combinations of in-row soil loosening, in-row surface residue removal, and ridge planting systems.

Initial strip-tillage experiments investigated the width and depth of tillage necessary to eliminate the yield reductions associated with no-tillage systems when corn followed corn in a rotation. Vyn et al.[14] reported that planting corn in a roto-tilled zone that was only 8 cm deep × 13 cm wide resulted in corn yields that were similar to spring moldboard plowing with secondary tillage.

In a later study, strip-tillage experiments were initiated to evaluate the potential for various commercially available power take-off (PTO) powered tillage implements that were capable of tilling a strip that was 25 cm wide × 8 to 10 cm deep just prior to planting operations.[17] Corn was planted in the center of the tilled strips. A row cultivator was also modified to perform strip-tillage using high-clearance tines.

On sandy loam soils, some strip tillage treatments yielded similarly to autumn moldboard plowing and greater than no-till tillage systems (Table 3.1). On a silt loam soil, strip-tillage yields were intermediate to autumn moldboard plowing and no-till tillage systems. On a clay loam soil, strip-tillage used as a secondary tillage option after autumn offset discing resulted in corn yields that tended to be intermediate to autumn moldboard and no-till tillage systems. Often when there was a difference in corn yields among the various strip-tillage systems evaluated, the implement that prepared a seedbed with the greatest proportion of fine aggregates (<5 mm in diameter at planting time in the seedbed zone) resulted in the highest grain corn yield. Vyn and Raimbault[17] concluded that a potential existed for favorable corn performance in strip-tillage systems provided that the strip-tillage implement was capable of producing a seedbed with proportions of fine aggregates which were not different from autumn moldboard plowing.

Table 3.1. Grain Corn Yield in Response to Autumn Moldboard, No-Till, and Various Strip-Till Tillage Systems on a Sandy Loam Soil

Tillage Treatment[a]	Yields per Year (Mg/ha)		
	1982	1983	1984
Moldboard plow	9.23 a[b]	10.38 a	8.49 a
Howard Rotaspike	8.16 ab	10.19 a	8.70 a
Howard Rotavator	8.42 ab	10.53 a	7.62 ab
Lely Roterra	8.25 ab	10.25 a	7.02 b
Cultivator	8.05 b	10.13 a	7.79 ab
No-till	6.35 c	8.74 b	7.70 ab
Contrasts:			
MP vs ST	*	NS	NS
NT vs ST	**	**	NS
CH vs ST	NS	NS	NS

Note: **, *, NS — Significance at 1%, 5%, and nonsignificant at the 5% level of probability, respectively.

[a] MP = moldboard plow, ST = strip tillage treatments, NT = no-till, CH = chisel plow.

[b] Within each column, data followed by the same letter are not statistically different according to the protected LSD (least significant difference) test.

A reluctance to adopt PTO powered strip-tillage systems by corn producers due to a combination of high capital investment, intensive maintenance requirements, and relatively slow planting speeds prompted research into the potential for planter-mounted coulters to loosen in-row soil.[25] They reported that seed zone soil macroporosity was higher and bulk density lower when coulters were used to loosen in-row soil while no-till planting corn after corn. However, their study found no difference in the 1991 corn yields between spring moldboard, no-till with in-row soil loosened, and no-till with minimal soil loosened treatments due to ideal growing conditions. They concluded that using coulters to create a more favorable seed zone soil condition may not necessarily improve corn yields.

3.3.3. Reducing In-Row Corn Residue Cover

Vyn[26] reported that occasionally no-till corn yields following corn and barley were increased by as much as 10% when a 15-cm band of corn residue was cleared from the row area using planter-mounted disc furrowers. Similar results were obtained from another tillage experiment in which clearing corn residue out of the row area increased no-till grain yields by at least 9%, 2 out of 4 years.[58] Residue removal from the row area was not beneficial to corn yields following soybeans and red clover (*Trifolium pratense* L.).

Although using disc furrowers on occasion increased no-till corn yields, they were still often inferior to corn yields in autumn moldboard plow tillage systems. Janovicek et al.[21] reported that clearing corn residue out of the row area was associated with higher soil temperatures. However, in their study higher soil temperatures did not result in superior corn growth and final yields due to poor seedbed conditions caused by excessive soil movement by disc furrowers. Stewart and Vyn[25] reported that using disc furrowers to clear corn residues resulted in higher in-row seed zone bulk density and soil resistance when compared to either no-till with no residue removal or spring moldboard plow tillage systems. Using disc furrowers to clear residue from the row area may be beneficial only if excessive soil movement and seed zone compaction are minimized. Proper adjustment appears to be critical; superior corn establishment has been observed when surface residues are removed without displacing surface soil from the row zone.

3.3.4. Ridge Planting Systems

Research on ridge-till planting systems has been limited to the poorly drained clay loam soils of southern Ontario. Stone et al.[27] suggested that expansion of ridge-till planting systems will be restricted to the warmer regions of Ontario, where the predominant crops are corn and soybeans. A greater diversity of crops, such as cereals and forages, which are not well suited for ridge planting systems, and significant soybean yield reductions associated with row spacings >38 cm, will limit adoption in the cooler regions of eastern Canada.

Soil temperature and moisture conditions in the ridges of ridge-till planting systems more closely resemble the conditions in autumn moldboard plow tillage systems, rather than no-till tillage systems in which ridges are not formed. Stone et al.[20] reported that early season soil temperature, 5 cm below the ridge crown, was higher than no-tillage and similar to after autumn moldboard plowing, even though surface residue cover was significantly higher. They also reported that early season soil moisture content was significantly lower on the ridges than in the no-till tillage system.

Vyn et al.[16] suggested that faster early season growth when corn was planted after corn in the ridge-till system was partially due to higher soil temperatures. However, the best ridge-till planting option (top of ridge roto-tilled prior to planting) still resulted in significant yield reductions as compared to the autumn moldboard tillage system (Table 3.2). These authors suggested that a ridge planting system in which some type of strip-tillage to disturb the ridge top would result in the best corn performance.

In Prince Edward Island, preliminary studies with ridge planting suggested that soil temperature differences at a depth of 5 cm between ridges and unridged treatments were minor (<1°C) and would not support earlier seeding. Planting on ridges as compared with flat conventional soil surfaces did not significantly influence corn silage dry matter yields, but did slightly increase the growth rate to silking when planting occurred under optimum soil temperatures.[7]

Table 3.2. Grain Corn Yield in Response to Autumn Moldboard, No-Till and Various Ridge Planting Tillage Systems on a Clay Loam Soil

Tillage Treatment	Yields per Year (Mg/ha)	
	1984[a]	1985[a]
No-till	6.05 a[b]	9.10 a
Autumn moldboard	8.02 c	8.61 ab
Ridge; no-till	6.14 a	7.78 bc
Ridge; top scraped	6.67 ab	8.34 ab
Ridge; top roto-tilled	7.00 b	8.29 ab

[a] Preceding year's crop was corn and soybeans in 1984 and 1985, respectively.
[b] Within-year data followed by the same letter are not statistically different according to the protected LSD test ($p = 0.05$).

3.4. TILLAGE REQUIREMENTS FOR CORN IN ROTATION

Raimbault and Vyn[23] observed that there was little difference in corn performance between an autumn chisel and an autumn moldboard tillage system when the preceding year's crop was soybeans, wheat (*Triticum aestivum* L.), barley, or alfalfa. In shorter term rotation trials, similar results were reported with smaller corn yield differences among no-till, autumn chisel plow, and autumn moldboard plow tillage systems following soybeans, barley, and alfalfa/timothy sod mixtures as compared to after corn.[15] On a poorly drained clay loam soil, Vyn et al.[28] reported that grain corn yields following soybean were not different among no-till, autumn moldboard and ridge-till (with in-row soil loosening) tillage systems (Table 3.2).

The corn yield response to increasing intensity of tillage observed when corn follows corn in a rotation has been less evident when the crops of the preceding year was either soybean or alfalfa. However, reduced tillage systems have resulted in occasional severe yield reductions following either red clover or fall rye.[15,29]

3.4.1. Corn Following a Red Clover Cover Crop

Initial studies with no-till planting corn into spring-killed red clover resulted in yield reductions that were occasionally as high as 50% when compared to either autumn chisel plow or autumn moldboard plow tillage systems.[15] The severe yield reductions on the no-till tillage system was associated with reduced plant populations that were caused by either poor seed placement or a higher incidence of early season pests. In this study, red clover was desiccated in the no-till treatments by spraying with phenoxy herbicides 1 day prior to corn planting. However, in another study in which pest and planting problems did not occur, corn yields were similar for no-till and moldboard plow tillage systems in 2 out of 4 years (Table 3.3). No-till corn yields after red clover were lower in 1990 and 1992, years with cool, wet springs (Table 3.3).

After red clover is chemically killed, the soil surface is almost completely covered by a layer of fine residue which can dramatically slow the rate of soil warming and drying. Stewart and Vyn[25] expressed concern that relatively slower rates of soil drying following red clover, resulting in relatively high early season soil moisture contents, may affect timeliness of planting or planter performance or both in no-till tillage systems. The cool, moist soil environment provides ideal conditions for slugs (*Agriolimax reticulatus* Muller), which have been reported to be major pests affecting the early season performance of corn planted no-till into red clover residues.[26,30]

The use of planter-mounted disc furrowers to clear a band of red clover residue out of the row area resulted in superior early season growth and development and final grain yields whenever the use of disc furrowers was not associated with soil movement which caused seed placement problems and reduced plant stands.[21] Superior early season corn performance and final corn yields could be partially attributed to faster early season rates of GDD accumulation where residue was cleared. However, Stewart and Vyn[25] reported that using disc furrowers to clear red clover underseeded into winter wheat resulted in corn yields that were similar to treatments in which residue was not cleared. They reported that using two fluted coulters to loosen in-row soil increased corn yields as compared to the single coulter no-till tillage system in one out of three site/years.

The probability of obtaining an acceptable seedbed, especially where disc furrowers or coulters are utilized in an attempt to improve seed zone soil conditions, would be higher when desiccation occurs at least 2 weeks before planting. Desiccation at least 2 weeks prior to planting rather than just prior to planting has often been associated with faster rates of early season growth and development, earlier flowering dates, and lower harvest grain moistures.[28]

Spring moldboard plowing or autumn chisel plowing are two tillage alternatives to autumn moldboard plowing which can help reduce the potential for soil erosion by maintaining higher amounts of winter residue cover. Janovicek[30] reported that corn yields were similar after spring and autumn moldboard plowing. However, autumn chisel plowing red clover has been associated with corn yield reductions of 5% as compared to autumn moldboard plowing.[23]

3.4.2. Corn Following An Autumn Rye Cover Crop

Silage corn yields were reported to be lower when planted after autumn rye (*Secale cereale* L.) cover crop vs after silage corn alone, regardless of the tillage system evaluated.[31] However, corn inhibition by an autumn rye cover crop was greater in the no-till compared to either a spring plow or spring tandem disc tillage system. Within rotation comparisons indicated that rates of growth and development and silage yields were reduced following rye in the no-till as compared to the other tillage systems. No difference was seen among tillage systems following silage corn with no rye cover crop. Reduced corn performance following autumn rye, especially when planted no-till, could not

Table 3.3. Corn Yield Response to Two Tillage Systems after a Red Clover Cover Crop (1989–1992)

	Tillage System (Yields in Mg/ha)		
Year	Moldboard		No-Till
1989	8.88		9.20
1990	7.99	*	6.63
1991	11.52		11.54
1992	6.71	*	5.40

Note: Means separated by * are different according to a LSD test at the 5% level of probability.

be attributed to depletion of soil moisture, lower soil temperatures, inadequate nitrogen availability, inferior seedbed fineness, or higher soil strengths. Raimbault et al.[31] suggested that an allelopathic interaction between autumn rye and corn occurred and that this effect was enhanced in the no-till soil due to the lack of soil mixing.

In an attempt to identify management strategies that would minimize no-till corn yield depressions following an autumn rye cover crop, timing of desiccation (1 or 14 days prior to corn planting) and various modifications to the no-till tillage system were examined.[31] Raimbault et al.[31] found that desiccation of autumn rye 14 days prior to corn planting increased silage corn yields with no-till and modified no-till tillage systems (Table 3.4). Early desiccation of rye prior to corn planting in conjunction with in-row residue removal or soil loosening or both resulted in no-till corn yields similar to those with spring moldboard plowing.

3.5. TILLAGE REQUIREMENTS FOR SMALL GRAINS

The majority of the research conducted assessing tillage requirements for small grain production has occurred on the fine sandy loam soils of Prince Edward Island[7] and sandy loam and clay soils in eastern Québec.[32-34] In Prince Edward Island, Carter[35] reported that on most fine sandy loams, bulk densities were higher on no-till treatments, which resulted in higher relative compaction values (85 to 91%) when compared to moldboard plowing (75 to 85%). A polynomial relationship was found between relative soil compaction and relative cereal yields, with maximum cereal yields occurring at about 82 to 83% relative compaction (Figure 3.2). Higher relative compaction levels in no-till treatments were associated with reduced macroporosity. Lower macropore volume in the no-till treatment was associated with high relative saturation levels (above 70% at "field capacity" moisture), indicating inadequate soil aeration for optimal root health.[36] This suggested that unfavorable relative saturation conditions due to low macropore volumes were responsible for a higher incidence of root diseases in no-till treatments. Carter et al.[37] reported

Table 3.4. Silage Corn Yields in Response to Spring Moldboard, Strip-Till, and Various Modifications of No-Till Tillage Systems when Rye was Killed Either 1 or 14 Days Prior to Corn Planting

	Time of Rye Kill (yields in Mg/ha)	
Tillage Treatment	14 Days	1 Day
Spring moldboard	13.7 ab[a]	13.5 a
No-till (ripple coulters)	12.8 b	11.6 b
No-till (disc furrowers + plow coulters)	14.4 a	13.4 a
No-till (ripple + plow coulters)	13.3 ab	11.5 b
Strip tillage	13.7 ab	12.3 b

[a] Within a column, silage corn yields followed by the same letter are not statistically different according to the protected LSD test (p = 0.05).

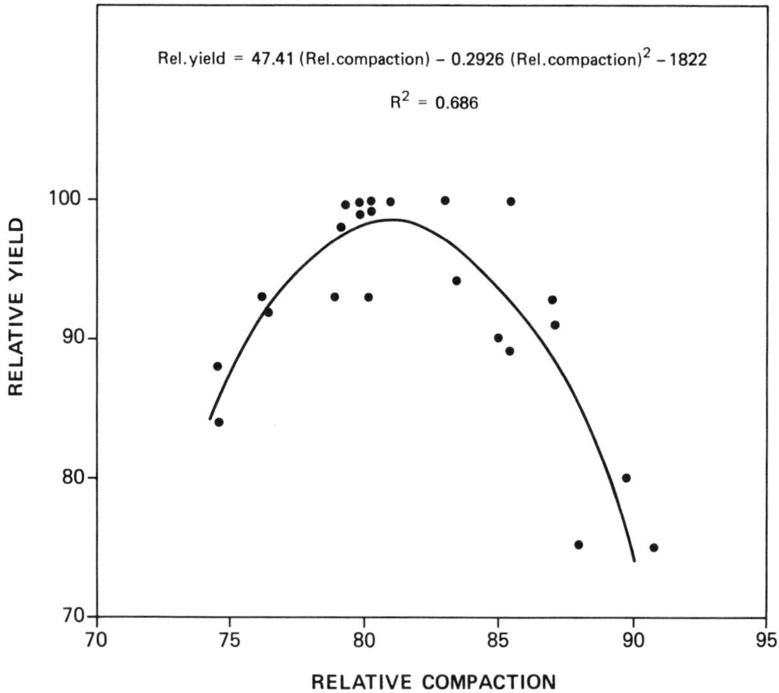

Figure 3.2. Effect of relative compaction on relative small grain yields on a fine sandy loam in Prince Edward Island. (Adapted from Carter, M. R., *Can. J. Soil Sci.*, 70, 425–433, 1990.)

that cereal yields and grain protein content were inferior on no-till treatments as compared to locations in which soil was loosened with a paraplow just prior to drilling. In this study, although little difference was observed between grain yield in the no-till and spring moldboard plow tillage systems, no-till consistently produced a lower grain protein content (Table 3.5); however, loosening soil with a paraplow increased both grain and grain protein yield. The superiority

Table 3.5. Spring Barley and Wheat Yield Response to a Spring Moldboard, Spring Paraplow, and No-Till Tillage System in Prince Edward Island

	Grain Yield[a] (Mg/ha)			Nitrogen Yield in Grain (kg/ha)		
	Barley (1983)	Wheat (1984)	Barley (1985)	Barley (1983)	Wheat (1984)	Barley (1985)
No-till	1.9	3.1	4.3	32	60	58
Paraplow	—	3.3	4.8	—	66	66
Spring moldboard	2.6	3.1	4.4	43	63	64
Standard error of the mean	0.28	0.14	0.16	2.5	1.4	0.8

[a] Grain moisture, 14.5%.

of the paraplow tillage system was attributed to its capability to create optimal macroporosity. Later studies indicated that reduced early cereal growth rates and subsequent reduced nitrogen accumulation under no-till were associated with low grain protein contents.[38]

Strategies to improve the potential of reduced tillage for small grain production in Prince Edward Island have centered on the accommodation of some tillage into the crop rotation, the use of improved seed drill openers, and the rotation of cereals with protein crops such as soybean. Studies with shallow tillage, where tillage is confined to the 10-cm depth and soil inversion is omitted, have shown that replacement of moldboard plowing with rotary harrows had no detrimental effect on cereal productivity and also reduced machinery costs and energy requirements by 25 to 48%.[38] However, long-term shallow tillage for acidic, fine sandy loams would require a periodic increase in tillage depth to remove middle topsoil compaction (below 10 cm) and provide deeper mixing of lime amendments.[39]

Replacing the conventional triple-disc seed drill with a hoe drill has improved the establishment and early growth of spring cereals and provided a degree of surface tillage that prevents the development of excessive soil compaction.[40] In addition, seed drill opener evaluation studies have demonstrated the superiority of inverted-T slot ("Baker Boot"), in comparison to V-shaped (disc) or U-shaped (hoe) slots, for seedling emergence and early growth under no-till.[41]

Other strategies to remove constraints to reduced tillage in Prince Edward Island are the adoption of barley-soybean rotations under no-till to improve the nitrogen regime and reduce root-pathogen interactions.[40] Also, shifting the tillage and seeding event to autumn (to coincide with drier soil conditions) via the use of no-till for winter wheat proved beneficial for maintenance of an optimum soil physical condition.[42] For moldboard plowed systems, reducing the intensity of secondary tillage by using a one-pass moldboard plow tillage system resulted in spring cereal yields similar to either autumn or spring moldboard plow with secondary tillage[24] and winter wheat yields equal to or superior than shallow tillage or no-till.[42] The one-pass moldboard plow tillage system consisted of a moldboard plow followed by a furrow press packer.

No difference was observed in Ontario barley yields between an autumn moldboard and an autumn chisel plow tillage system when the crops of the

Table 3.6. Number of Winter Wheat Heads (per m² and Grain Yield in Response to Moldboard Plow, Disc, and No-Till Tillage Systems Averaged Over Various Crop Sequences

	Head No. (per m²)		Grain Yield (Mg/ha)	
Treatments	1986	1987	1986	1987
No-till	508	446	5.60	2.70
Disc	488	441	5.19	2.70
Moldboard	504	429	5.67	2.83
Contrasts[a]				
MP vs no-till + disc	NS	NS	NS	NS
Disc vs zero	NS	NS	NS	NS

Source: After Yvn, T. J., *Can. J. Plant Sci.*, 71, 669, 1991.

[a] The contrast MP vs no-till + disc refers to a comparison between moldboard plow vs the mean of disc and no-till tillage systems. NS refers to contrasts that are not significant at the 5% level of significance.

preceding year were barley, barley underseeded with red clover, or corn.[23] Limited research assessing tillage requirements for wheat indicated that no difference was found in yield potential among a moldboard plow plus secondary tillage, tandem disc, or no-till tillage system.[22] In this study, wheat head number per square meter and grain yield were unaffected by tillage system (Table 3.6), regardless of the previous crop (wheat, barley, soybeans, corn, alfalfa). Sightly lower plant populations were observed in the no-till tillage system following barley and wheat, but increased tillering compensated for the lower populations. Raimbault and Vyn[23] also reported that either autumn moldboard plow or tandem disc tillage systems resulted in similar winter wheat yields following soybeans on a silt loam soil.

3.6. TILLAGE REQUIREMENTS FOR SOYBEANS

Ontario soybean yields generally decrease with conservation tillage systems such as no-till and spring discing;[43] however, the soybean yield response to tillage was dependent on soil type and the preceding year's crop(s). There was little soybean yield response to the various tillage systems on a sandy loam soil, when the preceding year's crop was either corn, wheat, or soybeans. However, on medium (i.e., loam) and finer (i.e., clay loam) textured soils, reduced soybean yields were observed in the no-till as compared to autumn moldboard plow tillage system following winter wheat and corn (Table 3.7). Yield differences among these tillage systems were not significant following soybeans. Whenever soybean yields were lower in the no-till tillage system, often a greater incidence of Rhizoctonia (*Rhizoctonia solani*) root rot damage occurred, particularly following winter wheat.

On all three soil types, soybean yields were lower in the no-till than in the autumn moldboard plow tillage system following red clover that had been

Table 3.7. Soybean Yield Response to Autumn Moldboard, Spring Disc, and No-Till Tillage Systems Following Various Crops on Loam and Clay Loam Soils Averaged Over 1990–1992 Growing Seasons

	Soil Type (Yields in Mg/ha)	
Crop/Tillage	Loam	Clay Loam
Corn		
Zero-till	3.17 b[a]	3.25 a
Disc	3.47 ab	3.11 a
Moldboard	3.61 a	3.42 a
Wheat		
Zero-till	3.06 b	2.43 c
Disc	3.38 ab	2.80 b
Moldboard	3.58 a	3.41 a
Soybeans		
Zero-till	3.34 a	3.10 a
Disc	3.35 a	3.03 a
Moldboard	3.70 a	3.27 a

Source: Adapted from Vyn, T. J. et al, *Agron. J.*, (submitted).

[a] Tillage means within a preceding year's crop followed by the same letter are not significantly different according to a protected LSD test at $p < 0.05$.

underseeded into wheat. Relatively poor no-till soybean performance following winter wheat underseeded with red clover could be partially attributed to increased slug feeding and a higher incidence of Rhizoctonia root rot in no-till as compared to autumn moldboard plow treatments. On medium- and finer-textured soils, no-till planting soybeans resulted in reduced yields relative to the autumn moldboard plow tillage system. Minimizing tillage by using only a two-pass tandem disc tillage system resulted in yields that were intermediate to no-till and autumn moldboard plow yields (Table 3.7).

Ridge planting is often suggested as an alternative to no-till on poorly drained fine-textured soils.[16] However, the adoption of ridge planting in eastern Canada is limited by the yield reductions associated with wide-row planting and the common inclusion of cereal crops in rotation.[27] Ridge planting systems resulted in yields similar to autumn moldboard plowing if the ridge tops were strip tilled just prior to planting. Autumn chisel plowing appears to be a viable alternative to autumn moldboard plowing for soybean following either corn or soybeans.[23]

In Prince Edward Island, preliminary studies on fine sandy loams indicate that no-till is an acceptable seeding method for soybean, especially where conventional seed openers are replaced by the inverted T slot opener.[41]

3.7. INFLUENCE OF TILLAGE SYSTEMS ON SOIL CONDITIONS

Vyn et al.[14] reported differences in soil physical properties among tillage systems in southern Ontario, with greater bulk densities and soil resistances

Table 3.8 Correlation Coefficients between Maize Grain Yields and Various Soil Physical Properties, Determined across Nine Tillage Systems on a Silt Loam Soil for the Years 1976–1981

	Correlation Coefficients					
Soil Properties	1976	1977	1978	1979	1980	1981
Aggregates, <5 mm	0.71*	0.45	0.84**	0.68*	0.57	0.68*
Bulk density, 5–10 cm	–0.64	–0.40	–0.28	–0.76*	–0.41	–0.14
Bulk density, 15–20 cm	–0.11	–0.39	–0.26	–0.26		
Penetrometer resistance, 5 cm	–0.54	–0.53	–0.42	–0.41	–0.26	–0.75*
Penetrometer resistance, 15 cm	–0.18	–0.67*	–0.09	–0.34	–0.22	–0.30

Source: After Vyn, T. J. et al, *Proc. 9th Conf. International Soil Tillage Research Organization,* A. Butorac (Ed.) (Osijek, Croatia, 1982).

Note: *, ** — Correlation coefficients significant at the 5 and 1% level of significance, respectively.

and lower proportions of fine aggregates (<5 mm in diameter) in no-till compared to autumn moldboard tillage systems. Similar observations were reported by later researchers.[13,15,17,18] Studies elsewhere in eastern Canada have reported differences in soil pore size distribution and associated water permeability between no-till and moldboard plowed tillage.[42,44-46]

Vyn et al.[14] reported that corn yields from nine tillage systems were correlated to the proportion of fine aggregates (<5 mm in diameter at planting time in the seedbed), bulk density, and soil resistance (Table 3.8). The proportion of fine aggregates (<5 mm in diameter) were more closely correlated with grain yield than either bulk density at the 5 to 10 cm or the 15 to 20 cm depth or to soil resistance at 5 or 15 cm. Tillage treatments with the highest proportion of fine aggregates tended to have higher yields. Later research has also observed superior corn performance in tillage systems having a greater proportion of aggregates <5 mm in diameter.[15,17,18]

3.7.1. Soil Erosion

Frequent and often intense rainfall events combined with relatively complex topography has resulted in severe erosion problems in some regions of eastern Canada. Excessive erosion from agricultural soils has been implicated as damaging wetland and lake habitat through increasing nutrient (particularly phosphorus) concentrations and/or silt deposition. Maintaining at least 20% residue cover was reported to significantly reduce the amount of erosion following an intense rainfall event.[47] Ketcheson and Underdonk[48] reported that maintaining corn residue on the soil surface of a no-till soil substantially reduced the amount of phosphorus and suspended soil particles in surface water runoff following intense rainfall events.

Rainfall simulation studies conducted on various tillage experiments have indicated that either no-till[49] (Table 3.9) or autumn chisel plow[43] tillage systems can dramatically reduce the potential for surface water runoff and soil erosion compared to the traditional practice of autumn moldboard plowing

Table 3.9 Effect of Tillage System and Various Previous Year's Crop(s) on Surface Residue Cover and the Volume of Surface Water Runoff and Amount of Soil Loss following a Simulated Rainfall Event on a Loam Soil (1991)

Previous Year's Crop, Tillage System	Residue Cover (%)	Runoff Volume (L/m^2)	Soil Loss (Mg/ha)
Corn			
No-till	67.5 a[a]	3.65 a	0.48 b
Autumn moldboard	10.2 b	6.29 a	1.27 a
Wheat/red clover			
No-till	87.8 a	2.49 b	0.15 b
Autumn moldboard	6.3 b	7.74 a	1.13 a
Soybeans			
No-till	16.3 a	8.91 a	1.32 a
Autumn moldboard	5.5 a	8.63 a	1.57 a
All crops			
No-till	57.2 a	5.02 b	0.65 b
Autumn moldboard	7.3 b	7.55 a	1.33 a

[a] Tillage subtreatments within a year's previous crop or tillage system means overall crops followed by the same letter are not significantly different at $p < 0.05$.

with spring secondary tillage. The effectiveness of a conservation tillage system in reducing soil erosion and water runoff increased proportionately with surface residue cover.

Superior soil structure associated with reduced tillage systems may also lessen the potential for surface water runoff and soil erosion. Vyn et al.[49] reported that soil loss following simulated rainfall events was negatively correlated to both the proportion of water-stable aggregates, an indicator of soil structural stability, and residue cover. They described a relationship in which percentage of residue cover and the proportion of water-stable aggregates additively explained the amount of soil eroded during simulated rainfall events (Figure 3.3). In this study, higher levels of residue cover and wet aggregate stability were associated with chisel plow as compared to moldboard plow tillage systems. Superior structural stability in reduced (i.e., no-till and chisel plow) as compared to moldboard plow tillage systems has been reported by other authors.[14,15,23,49]

3.7.2. Soil Structural Stability

Changes in soil tillage practices can affect soil structural stability, mainly through changes in soil moisture and redistribution of organic matter, which subsequently influence soil microbial activity. For example, tillage at different depths or the absence of soil inversion can cause a shift in organic matter equilibrium toward a relatively rapid accumulation of organic matter, especially labile fractions, near the soil surface.[7,37,39,50,51] In eastern Canada increased biological activity has increased aggregate stability in the surface of

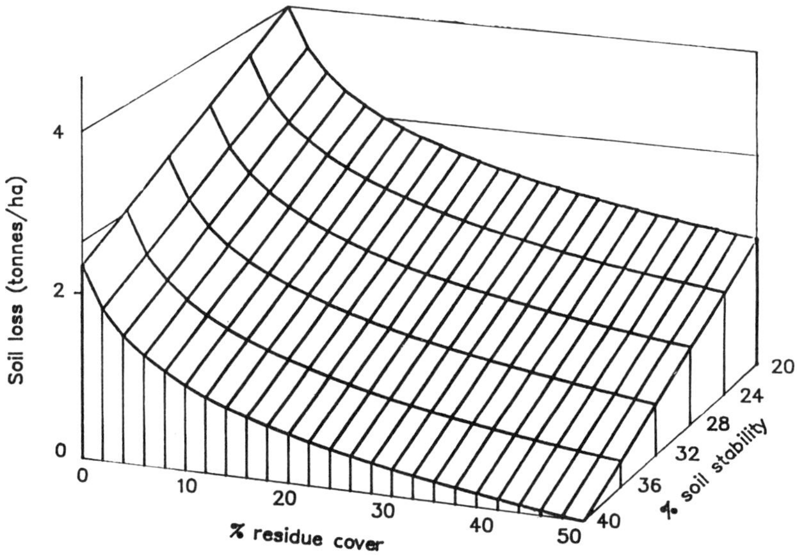

Figure 3.3. Relationship between residue cover and soil stability on the potential for soil loss measured on a long-term tillage and rotation trial.

shallow tilled,[39,52,53] chisel plow,[23,53,54] and no-till[15,39,45,53,55] topsoils. Green[56] observed that 50% of the variability in aggregate stability after long-term chisel vs moldboard plowing in a rotation experiment was due to organic matter differences near the soil surface.

In Prince Edward Island, aggregate stability has been related to both soil organic carbon and carbon fractions (e.g., microbial biomass carbon; see Figure 3.4).[53] No-till and reduced tillage increased the microbial biomass level per unit soil organic carbon above the long-term level established under grassland (Figure 3.5).[57] This change in organic matter equilibrium would indicate a shift toward organic matter accumulation over time in humid climates.[50,51,57] Such tillage-induced changes in organic matter provided the means to enhance the structural stability of inherently unstable fine sandy loams in 3 years or less.[53,55] However, the low resistance of these soils to compaction and their limited potential for regeneration of adequate macroporosity emphasizes the need for control of traffic-induced compaction if the stability benefits of improved soil structural stability are to be realized.

3.8. GENERAL SUMMARY AND CONCLUSIONS

Most of the tillage research conducted in eastern Canada has described reduced tillage systems that resulted in crop yields similar to and in some cases greater than the traditional autumn moldboard plow tillage system. However,

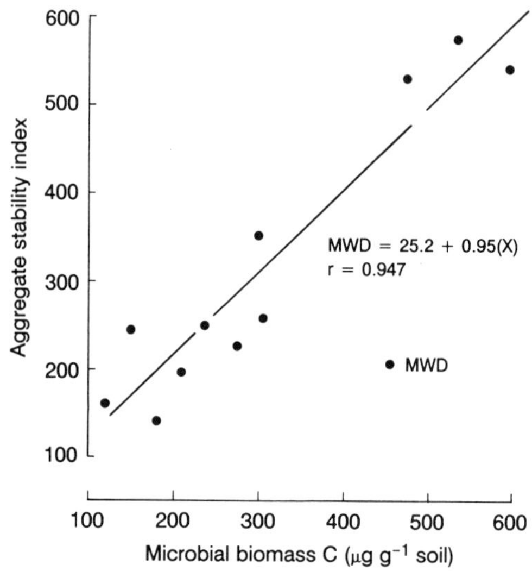

Figure 3.4. The relationship of mean weight diameter (MWD) to soil microbial biomass carbon in a Prince Edward Island fine sandy loam under different tillage systems and grass. (Adapted from Carter, M. R., *Soil Tillage Res.*, 23, 369, 1992. With permission.)

the complete absence of tillage has often resulted in significant yield reductions when compared to either reduced or moldboard plow tillage systems. Inferior crop performance in no-till systems has been attributed to either increased pest and disease problems or poorer seedbed conditions. No-till soils have been characterized as having greater bulk densities, higher soil resistances, lower macroporosity, and seedbeds that contain a greater proportion of coarse aggregates. These inferior soil conditions could inhibit root growth and/or water availability. Observations that inhibition of corn performance in no-till systems was greater in abnormally dry growing seasons, even though soil moisture contents were higher, support the hypothesis that root growth and root health is inhibited. This is further supported by the higher incidence of root diseases that is associated with wet soil conditions when barley and soybeans were planted no-till. However, no-tillage systems have not always been associated with reduced crop performance. Generally, crop performance was unaffected by tillage systems if soybeans were the preceding year's crop. Also, crop yield response to tillage has been reported to be less on coarse-textured, well-drained soils. Low levels of surface-placed residues after soybeans, and the inherent dry state of well-drained soils, result in relatively fast rates of early season soil drying. These relatively drier soil conditions may result in a soil environment which is more favorable for root growth. The ability of strip tillage systems to significantly improve corn performance relative to no-till alone indicate that tillage to improve inferior soil conditions need be limited only to a relatively small soil volume.

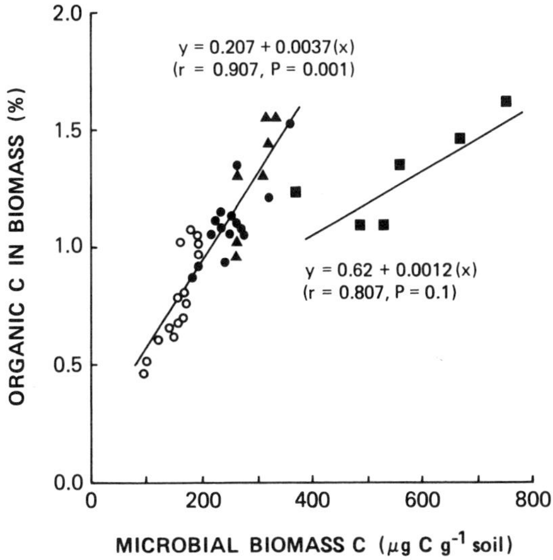

Figure 3.5. Relationship at the 0 to 5 cm soil depth between microbial biomass carbon and the organic carbon in the biomass in a range of tillage experiments and grassland sites (■) established on fine sandy loams in Prince Edward Island. The tillage treatments were moldboard plowing (○), shallow tillage and chisel plowing (▲), and direct drilling (●). (Adapted from Carter, M. R., *Biol. Fertil. Soils,* 11, 138, 1991. With permission.)

The high labor and investment costs, combined with a growing awareness of excessive soil erosion and structural deterioration associated with moldboard plow tillage systems, have prompted many eastern Canadian grain and oilseed producers to adopt reduced tillage systems. However, many other of these producers are reluctant to adopt a reduced tillage system due to reports of occasional severe yield reductions (especially on fine-textured soils) and investment costs required to purchase equipment associated with some reduced tillage systems. Adoption of reduced tillage systems has been slower in eastern Canada than in the northeast or north central United States.

The success of reducing tillage is dependent on the crop, soil type, and the preceding year's crop. The diversity of crops and soil types associated with many farming operations in eastern Canada makes selection of a single tillage system difficult. However, most of the tillage research conducted has indicated that with proper management, few situations exist in which an autumn moldboard plow tillage system was necessary to ensure agronomically acceptable crop yields.

3.9. REFERENCES

1. Roberts, P. A. " Changes in Cropping, Tillage and Land Management Practices in Southwestern Ontario for 1986 and 1991," SWEEP Report 78 (Ottawa, Ontario: Agriculture Canada, 1993).
2. Carter, M. R., and H. T. Kunelius. "Adapting Conservation Tillage in Cool, Humid Regions," *J. Soil Water Conserv.* 45:454–457 (1990).
3. Chapman, L. J., D. M. Brown and J. A. McKay. *The Climate of Southern Ontario* (Toronto, Ontario: Department of Transportation, Meteorological Branch, 1968), p. 50.
4. Brown, D. M. "Heat Units for Corn in Southern Ontario," Factsheet 78-063 (Toronto, Ontario: Ministry of Agriculture and Food, 1987).
5. Advisory Committee on Agrometeorology. "The Climate for Agriculture in Atlantic Canada," Publ. No. ACA 84-2-500, Agdex No. 070 (Truro, N.S.: N.S. Department of Agriculture, 1984), pp. 19.
6. Agriculture Canada Expert Committee on Soil Survey. "The Canadian System of Soil Classification," 2nd ed., Agriculture Canada Publ. No. 1646. (Ottawa, Ontario: Canadian Government Publication Center, 1987), pp. 164.
7. Carter, M. R., H. T. Kunelius, R. P. White and A. J. Campbell. "Development of Direct-Drilling Systems for Sandy Loam Soils in the Cool Humid Climate of Atlantic Canada," *Soil Tillage Res.* 16:371–387 (1990).
8. **Anon.,** *Soil Landscapes of Canada*, Pub. 5240/b (Ottawa, Ontario: Agriculture Canada, 1988).
9. Nowland, J. L. "The Agricultural Productivity of the Soil of the Atlantic Provinces," Research Branch, Information Division, Monogr. No. 12 (Ottawa, Ontario: Agriculture Canada, 1975), pp. 19.
10. Carter, M. R. "Physical Properties of Some Prince Edward Island Soils in Relation to Their Tillage Requirement and Suitability for Direct Drilling," *Can. J. Soil Sci.* 67:473–487 (1987).
11. Stonehouse, D. P., J. K. Baffoe and B. D. Kay. "The Impacts of Changes in Key Economic Variables on Crop Rotational Choices on Ontario Cash-Cropping Farms," *Can. J. Agric. Econ.* 35:403–420 (1987).
12. Vyn, T. J., T. B. Daynard and J. W. Ketcheson. "Research Experience with Zero-Tillage in Ontario," in *Proc. 8th Conf. International Soil Tillage Research Organization,* G. Kahnt (Ed.) (Stuttgart, Germany, 1979), pp. 27–32.
13. Stewart, G. A., D. Young, and T. J. Vyn. "The Evaluation of a Wide-Span Tractor: The Effect on Soil Properties and Crop Response," Final Report (Guelph, Ontario: Ontario Ministry of Agriculture and Food, 1991).
14. Vyn, T. J., T. B. Daynard and J. W. Ketcheson. "Effect of Reduced Tillage Systems on Soil Physical Properties and Maize Grain Yields in Ontario," in *Proc. 9th Conf. International Soil Tillage Research Organization,* A. Butorac (Ed.) (Osijek, Croatia, 1982), pp. 156–161.
15. Vyn, T. J. "Crop Sequence and Conservation Tillage Effects on Soil Structure and Maize Growth," in *Proc. 10th Conf. International Soil Tillage Research Organization* (Edinburgh: 1988), pp. 921–926.
16. Vyn, T. J., J. A. Stone and B. A. Raimbault. "Corn Development and Yield Response to Ridge Planting Systems on a Poorly Drained Soil in Southwestern Ontario," *Soil Tillage Res.* 18:207-217 (1990).

17. Vyn, T. J., and B. A. Raimbault. "Evaluation of Strip Tillage Systems for Corn Production in Ontario," *Soil Tillage Res.* 23:163-176 (1992).
18. Vyn, T. J., and B. A. Raimbault. "Long-Term Effects of Five Tillage Systems on Corn Response and Soil Structure," *Agron. J.* 85 (1993).
19. Ketcheson, J. W. "Effect of Tillage on Fertilizer Requirements for Corn on a Silt Loam Soil," *Agron. J.* 72:540-542 (1980).
20. Stone, J. A., T. J. Vyn, H. D. Martin and P. H. Groenevelt. "Ridge-Tillage and Early-Season Soil Moisture and Temperature on a Poorly Drained Soil," *Can. J. Soil Sci.* 69:181–186 (1989).
21. Janovicek, K. J., T. J. Vyn and R. P. Voroney. "No-Till Corn Response to Crop Rotation and In-Row Residue Placement," *Agron. J.* (submitted).
22. Vyn, T. J., J. C. Sutton and B. A. Raimbault. "Crop Sequence and Tillage Effects on Winter Wheat Development and Yield," *Can. J. Plant Sci.* 71:669-676 (1991).
23. Raimbault, B. A., and T. J. Vyn. "Crop Rotation and Tillage Effects on Corn Growth and Soil Structural Stability," *Agron. J.* 83:979–985 (1991).
24. Carter, M. R., R. P. White and R. G. Andrew. "Reduction of Secondary Tillage in Moldboard Plowed Systems for Silage Corn and Spring Cereals in Medium Textured Soils," *Can. J. Soil Sci.* 70:1-9 (1990).
25. Stewart, G. A., and T. J. Vyn. "Evaluation of Row Crop Planter Modifications for Corn Production," Final Report, 41, SWEEP-TED Contract No. 01686-0-0085/01-XSE (Guelph, Ontario: Agriculture Canada, 1992).
26. Vyn, T. J. "Crop Sequence and Conservation Tillage Effects on Soil Structure and Corn Production," Ph.D. thesis (Guelph, Ontario: University of Guelph, 1987).
27. Stone, J. A., T. J. Vyn and N. D. Clark. "Ridge Tillage for Corn and Soybean Production on Clay and Clay-Loam Soils in Southwestern Ontario: A Review," *Soil Tillage Res.* 18:219-230 (1990).
28. Kay, B. D., T. J. Vyn and R. W. Sheard. "Land Stewardship Cropping Systems for Corn and Soybean Production in Ontario," Final Report to Ontario Ministry of Agriculture and Food, (Guelph, Ontario, 1993), p. 165.
29. Raimbault, B. A., T. J. Vyn and M. Tollenaar. "Corn Response to Winter Rye Cover Crop Management and Spring Tillage Systems in Ontario," *Agron. J.* 82:1088–1093 (1990).
30. Janovicek, K. J. "The Role of Allelopathic Substances, Preceding Crops, and Residue Management on Corn Performance," M.Sc. thesis (Guelph, Ontario: University of Guelph, 1991).
31. Raimbault, B. A., T. J. Vyn and M. Tollenaar. "Corn Response to Rye Cover Crop, Tillage Methods and Planter Options," *Agron. J.* 83:287–290 (1991).
32. Lapierre, C. "Effet des Pratiques Culturales Réduites sur L'Efficacité de la Chaux et du Phosphore dans des Sols Agricoles du Québec," Final Report, Project 3B1-32180260-007 (Québec City, Québec: Agriculture Canada, Développement Agricole, 1991).
33. Samson, N. "Effets des Stratégies de Répression des Mauvaises Herbes sur les Systèmes de Production Céréalière Soumis à des Pratiques Culturales Réduites," Dossier 3B1-47722640-006, (Québec City, Québec: Agriculture Canada, Développement Agricole, 1991).
34. Légère, A., N. Samson, C. Lemieux and R. Rioux. "Effects of Weed Management and Reduced Tillage on Weed Populations and Barley Yields," in *Proc. EWRS Symp.* (Helsinki: 1990), pp. 111–118.

35. Carter, M. R. "Relative Measures of Bulk Density to Characterize Compaction in Tillage Studies on Fine Sandy Loams," *Can. J. Soil Sci.* 70:425-433 (1990).
36. Carter, M. R., and J. W. Johnston. "Association of Soil Macro-Porosity and Relative Saturation with Root Rot Severity of Spring Cereals," *Plant Soil* 120:149-152 (1989).
37. Carter, M. R., H. W. Johnston and J. Kimpinski. "Direct Drilling and Soil Loosening for Spring Cereals on a Fine Sandy Loam in Atlantic Canada," *Soil Tillage Res.* 12:365-384 (1988).
38. Carter, M. R. "Evaluation of Shallow Tillage for Spring Cereals on a Fine Sandy Loam. I. Growth and Yield Components, N Accumulation and Tillage Economics," *Soil Tillage Res.* 21:23–35 (1991).
39. Carter, M. R. "Evaluation of Shallow Tillage for Spring Cereals on a Fine Sandy Loam. II. Soil Physical, Chemical and Biological Properties," *Soil Tillage Res.* 21:37–52 (1991).
40. Carter, M. R., and A. J. Campbell. Unpublished results (1992).
41. Campbell, A. J. Unpublished results (1992).
42. Carter, M. R. "Characterizing the Soil Physical Condition in Reduced Tillage Systems for Winter Wheat on a Fine Sandy Loam Using Small Cores," *Can. J. Soil Sci.* 72:395–402 (1992).
43. Vyn, T. J., B. A. Raimbault and C. Roland. "Crop Rotation and Tillage Effects on Soybean Productivity, Soil Structure, and Erosion," *Agron. J.* (submitted).
44. Carter, M. R. "Temporal Variability of Soil Macroporosity in a Fine Sandy Loam under Mouldboard Ploughing and Direct Drilling," *Soil Tillage Res.* 12:37–51 (1988).
45. Gregorich, E. G., W. D. Reynolds, J. L. B. Culley, M. A. McGovern and W. E. Curnoe. "Changes in Soil Physical Properties with Depth in a Conventionally Tilled Soil After No-Tillage," *Soil Tillage Res.* 26:289–299 (1993).
46. Negi, S. C., G. S. V. Raghavan and F. Taylor. "Hydraulic Characteristics of Conventionally and Zero-Tilled Field Plots," *Soil Tillage Res.* 2:281–292 (1981).
47. Van Roestel, J. A. "Influence of Stover Placement and Tillage on Corn Grain Yield," M.Sc. thesis (Guelph, Ontario: University of Guelph, 1984).
48. Ketcheson, J. W. and J. J. Underdonk. "Effect of Corn Stover on Phosphorous in Run-Off from a Non-tilled Soil," *Agron. J.* 65:69-71 (1973).
49. Vyn, T. J., K. J. Janovicek and B. S. Green. "Influence of Tillage and Crop Rotation on Factors Affecting Soil Erosion Potential," *Can J. Soil Sci.* (submitted).
50. Angers, D. A., A. N'dayegamiye, and D. Côté. "Tillage-Induced Differences in Organic Matter of Particle Size Fractions and Microbial Biomass," *Soil Sci. Soc. Am.* 57:512–516 (1993).
51. Angers, D. A., N. Bissonette, A. Légère, and N. Samson. "Microbial and Biochemical Changes Induced by Rotation and Tillage in a Soil under Barley Production," *Can. J. Soil Sci.* 73:39–50 (1993).
52. Côté, D. " Soil Structure and Related Properties of a Silt Loam Cropped for Nine Consecutive Years with Silage Corn Under Reduced and Conventional Tillage," in *Proc. 8th ISTRO Conf.*, Vol. 1, B. D. Soane (Ed.), (Edinburgh: ISTRO, 1988), pp. 37–42.
53. Carter, M. R. "Influence of Reduced Tillage Systems on Organic Matter, Microbial Biomass, Macro-Aggregate Distribution and Stability of the Surface Soil in a Humid Climate," *Soil Tillage Res.* 23:361–372 (1992).

54. Weill, D. A., C. R. DeKimpe and E. McKyes. "Effect of Tillage Reduction and Fertilizer on Soil Macro- and Microaggregates," *Can. J. Soil Sci.* 68:489–500 (1989).
55. Angers, D. A., N. Samson and A. Legere. "Early Changes in Water-Stable Aggregation Induced by Rotation and Tillage in a Soil under Barley Production," *Can. J. Soil Sci.* 73:51–59 (1993).
56. Green, B. S. "The Effects of Long-Term Crop Rotation and Tillage on Corn Growth and Selected Soil Properties," M.Sc. thesis, (Guelph, Ontario: University of Guelph, 1992).
57. Carter, M. R. "Influence of Tillage on the Proportion of Organic Carbon and Nitrogen in the Microbial Biomass of Medium Textured Soils in a Humid Climate," *Biol. Fertil. Soils*, 11:135–139 (1991).
58. Vyn, T. J. Personal communication.

CHAPTER 4

Conservation Tillage in the Corn Belt of the United States

Rattan Lal, Terry J. Logan, Donald J. Eckert, Warren A. Dick
The Ohio State University; Columbus, Ohio

Martin J. Shipitalo
U.S. Department of Agriculture; Coshocton, Ohio

TABLE OF CONTENTS

0-87371-571-3/94/$0.00+$.50

4.1. INTRODUCTION

Conservation tillage (CT) is a generic term encompassing many different soil management practices. It can be defined as any tillage system that reduces loss of soil or water relative to conventional tillage based on soil inversion and fine seedbed. More specifically, CT refers to a seedbed preparation technique in which at least 30% of the soil surface is covered by crop residue mulch. It is an ecological approach to seedbed preparation and soil surface management.

There is a long history of using CT in the Corn Belt, and practical recommendations are based on research plots established in Ohio and elsewhere since the 1950s and 60s.[1] Principal types of CT used in the region include no-till, ridge-till, strip-till, and mulch-till or reduced-till. The ridge-till system is most often used in soils with slow internal drainage.[2-4] It is estimated that some form of CT is used on about 13 million ha, or about 46% of cropland in the Corn Belt.[5] The data in Table 4.1 show the area planted to CT in different states within the Corn Belt. The percentage of the total planted area grown with CT is 37.5 in Ohio, 39.0 in Missouri, 42.9 in Illinois, 50.8 in Indiana, and 52.7 in Iowa. Mulch tillage is the most widely used system of CT.

A principal advantage of using CT is soil erosion control. In addition, CT is adapted for convenience because it saves time and the labor involved in

Table 4.1. Area under Different Types of Conservation Tillage in the Corn Belt During 1988

Conservation	Area (1000 ha) in Different States					
Tillage	Illinois	Indiana	Iowa	Missouri	Ohio	Total
No-till	644.5	397.8	283.0	284.2	453.8	2063.30
Ridge-till	101.6	82.2	115.0	11.7	27.1	337.6
Strip-till	81.0	3.6	10.1	4.9	0.4	100.0
Mulch-till	2744.5	1766.4	3997.6	1297.2	739.7	10545.4
Reduced-till	0	6.9	0	0	159.5	166.4
Total conservation tillage	3571.6	2256.9	4405.7	1598.0	1380.5	13212.7
Total area planted	8334.8	4442.9	8363.6	4096.8	3683.4	28921.5
Percentage in conservation tillage	42.9	50.8	52.7	39.0	37.5	45.7

Note: Table recalculated from Karlen, D. L., *J. Soil Water Conserv.*, 45, 365–369, 1990.

seedbed preparation. Therefore, it is convenient to many producers, especially the growing segment of part-time farmers. The cropped area managed by CT in the region also depends on other factors, e.g. crops, soil, climate, drainage conditions, etc. CT is easier to adapt for crops that produce the residue needed to protect soil against erosion. Principal crops of this nature grown in the region are corn, small grains, and soybean. CT is usually not applicable for soils with poor or slow internal drainage. Providing surface or subsurface drainage can facilitate adoption in such soils.

Despite its popularity and widespread adoption in the Corn Belt, CT remains a debatable and controversial issue. These controversies are highlighted by differences in crop response, variable timings of farm operations affected by soil trafficability in spring, differences in effective weed control, a wide range of risks in soil degradation and environmental pollution, and differences in actual or perceived agricultural sustainability. The response of CT measured in terms of the criteria outlined above depends on several factors, including soil type, terrain characteristics, internal drainage, past land use history, climatic conditions during the growing season, and management systems used.

The objective of this chapter is to describe ecological factors that affect agronomic performance and the environmental effectiveness of CT in the United States Corn Belt. Results of crop performance and agronomic productivity are related to soil and environmental conditions with a view to establish cause-effect relationships. Agricultural sustainability of CT systems is assessed in terms of productivity and soil and environmental quality.

4.2. CHARACTERISTICS OF THE AGROECOSYSTEM

The United States Corn Belt comprises part of the north central region, including the states of Ohio, Indiana, Illinois, Missouri, and Iowa. The total land area of the region is about 58 million ha. Agronomically important

Table 4.2. Climatic Regime, Land Area, and Total Corn Production in the Corn Belt [8,178]

State	Rainfall Apr-Sept (cm)	Growing Degree Days (GDD)	Area (10^6 ha), 1991	Total Production (10^6 Mg) 1991
Ohio	52	2800–3400	1.38	8.30
Indiana	55	2700–3700	2.25	12.98
Illinois	52–55	2600–4200	4.05	29.90
Iowa	58	2600–3300	4.94	36.29
Missouri	60–65	3300–4000	0.89	5.43
Total			13.51	92.90

climatic factors of the region (shown in Table 4.2) indicate seasonal rainfall (April to September) ranging from 52 to 65 cm, a growing season length from 160 to 180 days, and growing degree days (GDD) from 2800 to 4000. Principal land uses in the Corn Belt (Table 4.3) include cropland (37.3 million ha), pasture land (10.2 million ha), and forest land (10.5 million ha). The cropland area in the region has remained practically constant over the 50-year period ending in 1987, with a total increase of only about 1.5 million ha.[6]

Accelerated soil erosion is one of the principal constraints associated with arable land use. Highly erodible soils with erosion rates exceeding twice the tolerance level (<2 tons/ha) account for 9.3 million ha of cropland. About 10 million ha of cropland has suffered productivity loss of at least 5% due to accelerated erosion in the past (Table 4.3).[7] Although less extensive, accelerated erosion is also severe on at least 1.6 millon ha of pasture land and 1.2 million ha of forest land. In addition, 5.7 million ha or about 15% of the cropland is prone to flooding, causing annual damage of about $43 million.[7] The instream damage resulting from erosion in the Corn Belt is estimated at about $800 million annually.[6]

Corn (*Zea mays* L.), the principal crop of the region, is grown on about 13.5 million ha with a total production of about 93 million Megagrams (Mg), producing an average grain yield of 6.9 Mg/ha.[8] Corn is usually grown in rotation with soybean (*Glycine max* L. Merrill). Being open row crops, land cultivated for corn and soybean production is prone to accelerated erosion with attendant risks of on- and off-site damage, including degradation of soil and environments.

4.3. ECOLOGICAL IMPACT OF CONSERVATION TILLAGE

4.3.1. Soil Physical and Hydrological Properties

Soil physical and hydrological properties play an important role in determining the effects of CT on crop performance and environments. Effects on crop performance include emergence, stand count, root development, canopy

Table 4.3. Land Use and Ecological Constraints in the Corn Belt

Land Use/Constraints	Area (10^6 ha)
Total crop land	37.25
Highly erodible	9.31
Erosion rate > 2T	9.31
Productivity loss > 5%	10.12
Total Pasture Land	10.12
Sheet and rill erosion > T	1.62
Wind erosion > T	>0.06
Total Forest Land	10.53
Sheet and rill erosion > T	1.21
Wind erosion > T	0.00
Flood-plane cropland	5.76
Non-federal wetland	0.89

Source: Recalculated from USDA, "The Second RCA Appraisal," U.S. Government Printing Office, Washington, D.C., 1989.

Note: T = Soil loss tolerance, e.g., 12.35 Mg/ha.

cover, biomass production, and yield. Effects on environment include runoff and soil erosion, leaching and macropore flow, transport of pollutants to surface water and ground-water, volatilization losses of nutrients in fertilizers and manure, and emission of radiatively active gases into the atmosphere. Important soil physical properties relevant to these effects are mechanical properties (e.g., soil bulk density, soil strength, and soil compaction), hydrologic properties (e.g., pore size distribution, soil water retention, and transmission properties), soil aeration, and soil temperature.

It is difficult to generalize the impact of CT systems on soil physical properties. The magnitude and the nature of tillage-induced alterations in these properties depend on antecedent conditions, inherent characteristics of soil, management systems, and climatic and ecological environments. Contrasting and apparently anomalous results reported in the literature are explainable when all these factors are considered. Over and above the initial state or antecedent properties, certain inherent characteristics play a major role. Important among these are soil texture, clay mineralogy, internal and surface drainage, and soil depth. In general, soils of heavy texture and with slow or poor internal drainage have a less favorable response to CT than light-textured and well-drained soils. Positive response to CT is also reported in shallow soils on undulating terrain and on those containing low-activity clays or with mixed mineralogy. Some relevant soil physical properties and processes in relation to use of CT in the Corn Belt are described below in the following.

Table 4.4. Simulated Harvest Traffic Effects on Corn Grain Yield in 1989 for Three Ohio Soils

		Corn Grain Yield (Mg/ha)	
Tillage Method	Compaction Load (Mg)	South Charleston (Silt Loam)	Hoytville (Heavy Clay)
No-till	Control	6.22	1.4
	10	8.74	0.7
	20	7.49	0.5
Plow-till	Control	9.44	4.8
	10	8.55	2.8
	20	9.02	3.4
LSD (0.05) for tillage		0.97	0.90

Source: Unpublished data of R. Lal; Lal, R. and Tanaka, H., *Soil Tillage Res.*, 2, 65–78, 1992.

Note: The axle load at South Charleston was about 9 to 18 Mg.

4.3.1.1. Soil Compaction

Compaction can be a major obstacle to the adoption of the no-till system in clayey soils, which is why ridge-tillage, chisel-tillage or mulch-tillage systems are found to be more successful than the no-till system. Soil compaction is recognized to be a major cause of low nutrient and water use efficiencies and reduced crop yields on several soils in Iowa.[9-14] Soil tilth deteriorates due to excessive traffic when soil moisture conditions cause puddling or densification.[15] An experiment on tilth analysis for five tillage systems conducted in Iowa showed that moldboard plowing and till plant systems provided comparable tilth indices and maize yield.[16] Next in decreasing order were slot plant, spring disc, and chisel plow. In Ohio and Indiana, the problem of soil compaction with CT is primarily associated with poorly drained soils,[17-19,39] and in wheel tracks.[20,21]

Simulated harvest traffic effects on soil properties and crop yields have been studied for several soils in Ohio. The data in Table 4.4 show that compaction reduced crop yield with CT only in heavy-textured soils at Hoytville, but not in medium-textured soils at South Charleston. In fact, a slight or moderate level of compaction had some positive and beneficial effects on corn yield at South Charleston, especially when rains were below average and crops suffered from drought stress.

4.3.1.2. Hydrologic Properties

The effects of tillage methods on water transmission and hydrologic properties of soils in the Corn Belt have been studied by several researchers.[22-24] Hydrologic properties are related to soil structure, soil tilth, densification, and presence of macropores.

Soil densification and deterioration in soil tilth decrease the volume of macropores. Continuous use of no-till or reduced-till systems can lead to the development of platy structure. Consequently, infiltration capacity can decrease and runoff rate and the amount can increase. The effects of densification, however, are nullified by the presence of biopores, notably those made by earthworms.[25] Continuous biopores can significantly and positively affect water transmission and infiltration capacity in no-till soil.[26] Dick et al.[27] used a dye to evaluate continuity of biopores. They observed that a large number of channels created by earthworms in long-term no-till treatments were continuous to the soil surface and increased infiltration capacity. Dick et al.[28] attributed high percolation loss from no-till as compared to plow till soil (55 vs 24%) to flow through macropores. Logsdon et al.[29] established a relationship between macroporosity and hydraulic conductivity. Despite the general trends outlined above, a high degree of spatial variability exists in the data.[30] This variability is attributed to several factors: wheel tracks, biopores, cracks, etc.

Lal and Van Doren[31] measured infiltration rate after 25 years of no-till, chisel-till, and plow-till systems of seedbed preparation. Despite the platy structure, the high infiltration rate in no-till soil was attributed to the presence of biopores. These pores were stable and continuous to a depth of about 60 cm. Earthworm activity, however, may be severely curtailed in soils prone to anaerobiosis. Methods of measurement and characterization of macropore flow have been described by Singh et al.,[32] and its importance to the transport of surface-applied chemicals in no-till soil by Shipitalo et al.[33]

4.3.1.3. Soil Structure and Moisture Retention

Through its effects on aggregation and pore size distribution, CT may affect soil moisture retention and aeration properties. In general, CT improves aggregation. For example, experiments conducted in Indiana showed that water-stable aggregation size was increased by no-till or shallow tillage as compared to moldboard plowing.[34] Mahboubi et al.[21] observed that in Ohio, the percentage of aggregation and mean weight diameter (MWD) were significantly higher in no-till as compared to chisel plowing and moldboard plowing treatments.

CT effects on soil moisture retention characteristics have also been studied.[20,35,37,38] CT generally increases soil moisture retention for light- and medium-textured soils. The moisture retention capacity of heavy-textured soils is not drastically affected by CT. Experiments conducted in Indiana showed that CT systems increased soil water retention.[39] Mahboubi et al.[21] observed that the CT system had no effect on soil moisture retention in an Ohio traffic zone. In the row zone, however, soil samples from no-till plots retained more moisture than other treatments at all suctions. In the row zone the mean difference in moisture retained in no-till over other tillage methods was 15, 16, 17, 19, 21, 19, and 20%, respectively, for saturation, 0.001, 0.003, 0.006, 0.03, and 1.5

Table 4.5. Tillage Effects on Soil Temperature at 5 cm Depth, April 1984

	Mean Daily (C°)		Hourly Summation (degree-hour)	
Tillage Method	Minimum	Maximum	All Temp	Temp > 9°C
Ridge-till	5.2	14.6	5636	3734
Beds	5.9	13.6	5601	3570
No-till	6.0	13.4	5614	3542
Plow-till	5.8	13.6	5587	3569
LSD (0.05)	0.5	0.7	72	116

Source: Recalculated from Fausey, N. R. and Lal, R., *Soil Technol.,* 2, 371–383, 1989.

MPa, and in plant-available water capacity. Differences in moisture retention were slightly more apparent at high than at low suctions.

4.3.1.4. Soil Temperature

Tillage systems' effects on soil temperature regime have been studied for several sites within the Corn Belt.[40-43] Crop response to tillage methods depends on soil temperature, especially during early spring. In the coldest part of the region (e.g., northern Minnesota), low yields using the no-till system are associated with cool soil temperatures.[44] Suboptimal soil temperatures are also considered responsible for low yields with CT on poorly drained soils in Indiana and Ohio.[17,39]

Because of the limitations of cool soil and anaerobiosis, ridge-tillage is considered an appropriate CT method for poorly drained heavily-textured soils.[3,4,45-49] Fausey and Lal[50] studied drainage and tillage effects on soil temperature on a Crosby-Kokomo soil in central Ohio (Table 4.5). Their data showed that the mean daily maximum soil temperature measured in April was highest atop the undrained ridges. Ridge-tillage and raised beds further away from the drain had higher maximum temperature than plow-till or no-till methods of seedbed preparation.

4.3.2. Soil Chemical Properties

Chemical properties of managed soils are primarily inherited but are modified over time by crop and soil management. Some properties such as mineralogy change slowly, observed only after several decades. Others such as total carbon and nitrogen content are also slow to change, but management differences may be observed in a decade or less. More dynamic properties, such as soil pH, available phosphorus, and exchangeable bases, can be greatly affected by management and changes may be seen within a year or two.

CT affects soil chemical properties primarily through concentration of chemical constituents at the surface. This occurs when fertilizer and organic wastes such as manure and sewage sludge are surface applied without incorporation and when crop residues decompose at the soil surface. The chemical constituents most stratified by CT are those such as organic carbon, which is organically bound, and those such as phosphorus, which is strongly immobilized. Mobile chemical constituents such as nitrate and chloride are not stratified by CT.

4.3.2.1. Soil pH

A general observation in soils under CT is a reduction in soil pH over a period of several years.[51,52] Most soil processes and crop management practices are acid forming. In a plowed soil, the marginal acidification produced annually is distributed throughout the plow layer and manifests itself over a period of 3 to 5 years, depending on the acid buffer capacity. In no-till systems, fertilizer and manure applications at the surface rapidly increase the acidity within the top 2 cm or so, as does residue decomposition. Blevins et al.[53] showed that the pH difference between no-till and plowed soils was about 1 to 1.5 U in the top 5 cm of the soil and 0.5 U at a depth of 5 to 15 cm. Nitrogen application to the corn crop was a major source of acidity in their study.

An important consequence of lower soil pH in the no-till soil surface is herbicide activity. Herbicides such as atrazine behave as weak acids and bases and their water solubility, soil sorption, and efficacy depend on whether they are neutral, cationic or anionic at soil pH. In the case of atrazine, it is a cation below pH 4 and reduced soil pH with no-till that has resulted in reduced efficacy,[54] probably as a consequence of increased cation adsorption by clay minerals. Published data on pesticide physicochemical data (dissociation constant, water solubility, K_{oc}, K_H) can be used to predict the effects of changes in soil chemical properties with CT on pesticide fate.

4.3.2.2. Organic Carbon and Nitrogen

A number of workers have reported increases in soil organic matter content with CT.[53,54] More recently, Havlin et al.[56] observed increases in organic carbon and nitrogen with no-till compared to plowing in the top 0 to 2.5 cm of two soils from eastern Kansas. Several crop rotations were studied. Accumulations were correlated primarily with the amount of residue added, fertilizer applications having a lesser effect. In a study of long-term tillage effects on a semiarid soil with winter wheat, organic carbon and nitrogen were higher with mulch tillage than with plowing after 44 years.[57] Both organic carbon and nitrogen increased linearly with nitrogen fertilizer addition and the nitrogen effect was similar for the different tillage systems. Dick and Daniel[52] have suggested that greater accumulations of organic carbon and nitrogen in the surface with CT are due to the concentration of plant residues at the surface and reduced biological activity in the absence of tillage.

In a unique study of tillage effects on soil organic matter composition, Stearman et al.[58] used ^{13}C nuclear magnetic resonance to characterize humic

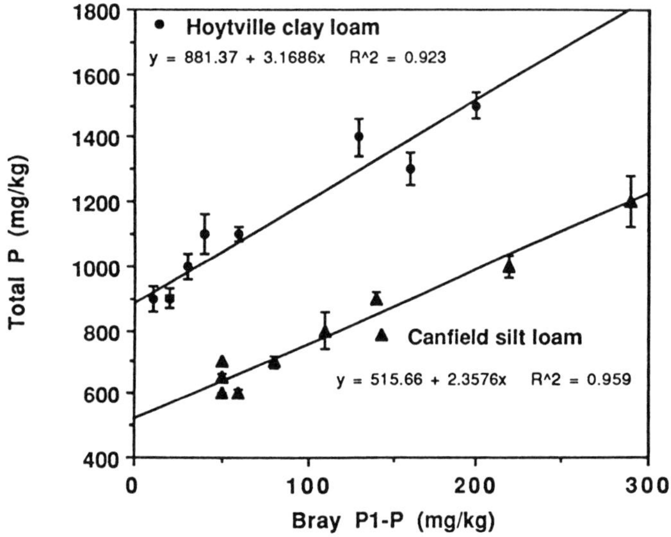

Figure 4.1. Relationship between total phosphorus and Bray P1 extractable phosphorus for two Ohio no-till soils. (Adapted from Guertal, E. A. et al, *Soil Sci. Soc. Am. J.*, 55, 410–413, 1991.)

acid extracted from soils under no-till and plow conditions for 7 years. Differences in composition were observed between the corn, soybean, and cotton (*Gossypium hirsutum* L.) crops and with depth, while only small differences were observed as a result of tillage. Soils with larger amounts of organic carbon (0 to 2 cm under no-till) had greater aliphatic to aromatic ratios, perhaps suggesting earlier stages of decomposition.

4.3.2.3. Soil Phosphorus

Phosphorus is a very immobile element in soil, the great majority of phosphorus in soil being in insoluble mineral forms and in soil humus. Most studies of phosphorus stratification in CT systems have examined changes in plant-available phosphorus rather than total phosphorus. Dick et al.[17,18,27] showed that Bray P1 available phosphorus in the top 7.5 cm of a clay-textured calcareous till soil increased from a value of 50 mg/kg with plowing to 150 mg/kg with no-till. Organic phosphorus content did not vary with tillage, however.[55] In a separate experiment with the same soil series, Guertal et al.[59] found that total phosphorus varied linearly with Bray P1 extractable phosphorus with depth in two no-till soils (Figure 4.1), one of which was the same series studied by Dick et al.[17,18,27] Using the slope of the regression line for the Hoytville soil (Figure 4.1), it is predicted that the increase in Bray P1 phosphorus of 100 mg/kg with no-till vs plowing reported by Dick[55] would result in an increase in total phosphorus of 300 mg/kg. This has great implications for erosion losses of total phosphorus: the reduction in total phosphorus losses with reduced soil loss

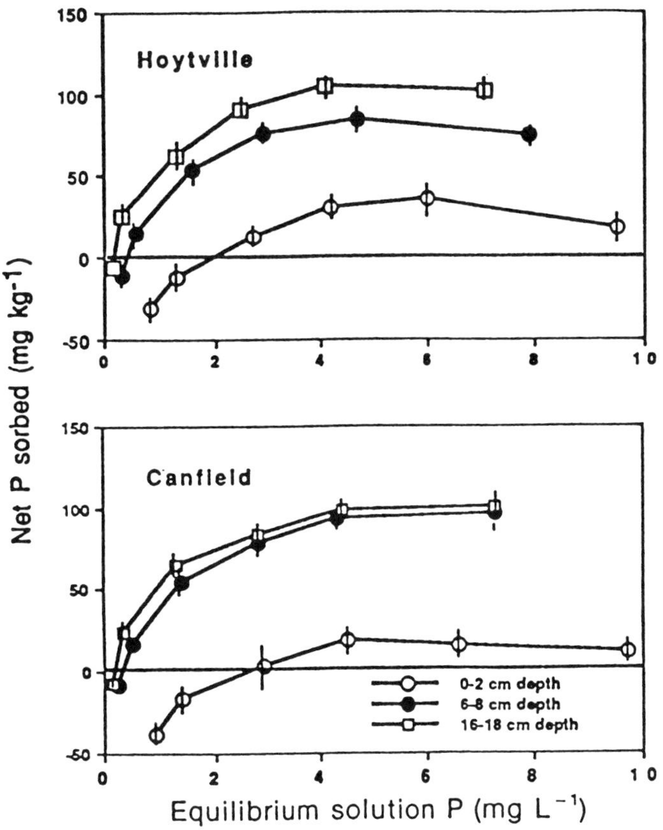

Figure 4.2. Adsorption of phosphorus by two Ohio no-till soils as a function of soil depth. (Adapted from Guertal, E. A. et al, *Soil Sci. Soc. Am. J.*, 55, 410–413, 1991.)

under no-till may be compensated by the increased total phosphorus content of the soil surface. This is supported by the findings of Logan and Adams,[60] that reductions in total phosphorus loss with no-till compared to plowing were only 89% as effective as the reductions in soil loss.

As phosphorus becomes concentrated at the soil surface with CT, the ability of the soil to buffer changes in dissolved inorganic phosphorus decrease and soil solution phosphorus at the near surface increases. This is seen in the study by Guertal et al.,[59] in which phosphorus sorption decreased and equilibrium solution phosphorus increased in the near surface (0 to 2 cm) as compared to lower depths (Figure 4.2). This explains the findings of Logan and Adams,[60] Oloya and Logan,[61] and others.[62]

These findings suggest that reducing both particulate and dissolved phosphorus in runoff will require more than the use of CT for erosion control. Phosphorus fertilizer should be applied as a band below the immediate soil surface, manure and sludge should be injected where possible, and the soil surface may have to be disturbed periodically. In a 3- to 5-year rotation, for example, a single plowing with a chisel or disc would redistribute phosphorus throughout the plow layer and over the rotation would have a minimal impact on soil erosion.

4.3.2.4. Cation Exchange Capacity and Exchangeable Bases

The observed stratification of organic matter in the near surface with CT usually results in an increase in cation exchange capacity (CEC) because of the higher specific CEC of organic matter compared to that of the clay minerals. The tillage effect on CEC would be greatest for soils low in organic matter and low in 2:1 clay minerals such as smectite or vermiculite. Twelve years under no-till produced a small but significant decrease in CEC of the near surface of a fine-textured soil from Ohio, high in organic matter and 2:1 mineralogy.[63] CEC was increased with no-till in loess- and limestone-derived soils of Kentucky with kaolinite and hydroxy interlayered vermiculite mineralogy.[65] Organic matter was somewhat lower in these soils than the Ohio soil studied by Lal et al.[63] In an Ultisol low in organic matter (<1%) and CEC (<6 $cmol_c$/kg), exchangeable potassium + calcium + magnesium increased significantly with no-till in the surface 5 cm.[64] CEC was not reported but the authors suggest that no-till increased CEC.

The exchangeable bases, calcium, magnesium, and potassium, are more mobile than phosphorus and the stratification behavior with CT seen for phosphorus is less consistently observed for these nutrients. Dick and Daniel[52] point out that the retention of these elements by soil is affected by soil acidity. As exchangeable soil acidity increases in the near surface with no-till, basic cations are displaced. This would counter any tendency for these nutrients to be stratified near the surface. Organic matter accumulation at the surface with no-till also affects the accumulation of exchangeable cations. Karathanasis and Wells[65] found that exchangeable and soluble potassium in the near surface increased two- to threefold after 6 to 16 years in no-till for loess and limestone-derived soils in western and central Kentucky. There was some tendency for ratios of exchangeable potassium to calcium + magnesium to increase with organic matter content. In contrast, no-till decreased exchangeable potassium and increased exchangeable calcium + magnesium in the surface 5 cm of a low CEC Ultisol.[64] The authors attributed this to increased organic matter content and likely increased CEC, which favors retention of divalent over monovalent cations.

4.3.3. Soil Biological and Nutritional Properties

4.3.3.1. Microbial Biomass and Activity

There is a shift in the relative concentration of microorganisms in a soil under CT.[66] Using a no-till system, plant residue is primarily localized on the soil surface, which also tends to have lower pH, and fungal growth is promoted. In plowed soils, the plant residues are distributed throughout the plowed layer and bacterial growth and activity along with residue decomposition and nutrient mineralization proceeds more rapidly. Studies have also shown that no-till tends to result in a bacterial population that has a greater percentage of anaerobic organisms.[67]

The preponderance of reports indicate that microbial populations and biomass are increased in the surface soil layer under CT.[68] In a study of soil microbial biomass carbon in fields under various number of years of continuous no-till, significant correlations were observed with years under no-till only in the subsurface soil layers, i.e., below a depth of 7.5 cm. In the surface soil layer (0 to 7.5 cm), soil organic matter concentration, but not soil microbial biomass carbon, was correlated with years under no-till. These data suggest that the subsurface layer may be a more sensitive zone for assessing biological changes caused by tillage than the surface layer.

Enzyme activities were measured in soils maintained under no-till and plowed tillage.[69] Levels at the surface were two- to sevenfold higher in the surface of a no-till soil as compared to a plowed soil. At lower depths, the plowed soil generally had higher or equal levels of enzyme activity. This increased level of enzyme activity is closely tied to organic carbon concentrations in the soil profile.

4.3.3.2. Nutrient Transformations and Management

Fertilization of most agronomic crops is generally restricted to the major nutrients of nitrogen, phosphorus and potassium and to the addition of limestone for correction of low soil pH. The effect of CT on available phosphorus and cations has been previously discussed. Acidity develops rapidly at the soil surface when CT is practiced and smaller, more frequent lime applications are generally recommended for CT systems. Subsurface soil acidity may also develop in a continuous CT soil when anhydrous ammonia is injected each year.[70] In a no-till system, numerous persistent acidic soil zones were observed scattered throughout the interrow space. When ammonia injection was localized in the same area every year, yield-limiting problems due to this acidification were considered unlikely because of the comparatively small volume of soil affected compared to the total soil volume.

Because of the greater microbial biomass concentrations and enzyme activities in a CT soil, nitrogen management strategies must be changed from that where plow tillage is practiced. For example, nitrogen fertilizer response is commonly observed to be lower under no-till than plow tillage because of nitrogen immobilization and, possibly, increased denitrication associated with no-till.[71,72] Numerous studies have shown that corn yields are often greater under plow tillage as compared to a no-till when zero or low fertilizer nitrogen rates are applied to well-drained soils.[73] When higher rates are applied, the

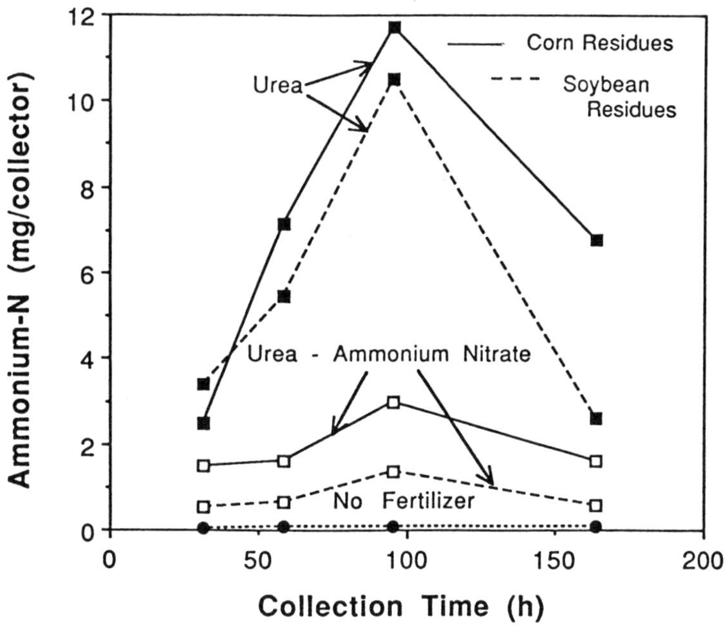

Figure 4.3. Ammonia nitrogen loss from urea and urea-ammonium nitrate fertilizers applied to a residue-covered NT soil surface.

reverse is observed and grain yields are generally equal to or greater under no-till. These findings have sometimes led to recommendations of higher nitrogen application rates for no-till than where plowing and other secondary tillage is practiced.

Probably the greatest irony is that as farmers move toward greater adaptation of CT, there is also a greater tendency to use urea or urea-based nitrogen fertilizers. Soil incorporation is a proven means of reducing ammonia losses from urea fertilizers, but this option is reduced and, in the case of no-till, completely removed when CT practices are used. Significantly greater ammonia losses occur from residue covered soils than from bare soils.[74] There is also some evidence that residue type may also affect the rate of ammonia evolution from soils fertilized with urea-based materials (Figure 4.3). Additional work needs to be conducted to confirm whether this is linked to some qualitative difference among residue types or simply that there are generally more corn than soybean residues in a field. Research has been conducted to develop urease inhibitors, but even the best inhibitors developed only decreased ammonia nitrogen losses by 50%,[74] and it is known that no-till can increase urease activity at the soil surface by a factor of two to seven times.[69]

The optimum time for nitrogen fertilizer application does not change when CT practices are used. The fertilizer nitrogen use efficiency generally increases the shorter the time between fertilizer nitrogen application and the period of active crop uptake. Both autumn moldboard plow and ridge-tillage were tested as to their effect on corn nitrogen uptake when liquid urea-ammonium nitrate was either surface applied in autumn or banded (knifed-in) between rows in the spring.[75] Compared with spring banded nitrogen, fall surface-applied nitrogen was extremely inefficient for both tillage sytems.

4.4. CONSERVATION TILLAGE AND SOIL CHEMICAL TRANSPORT

One of the concerns that may limit or slow the adoption of CT in the Corn Belt is the perception that it may increase the movement of fertilizers and pesticides to groundwater. Numerous instances of ground- and well water contamination in agricultural areas of the Corn Belt have been documented and most are apparently due to normal agricultural practices rather than to spills and misapplication.[76,77] For CT systems to be worthwhile they must reduce erosion. Usually, but not always, this is achieved by reducing runoff.[78,79] Implicitly, runoff is reduced because a greater proportion of the precipitation infiltrates. With a greater amount of water entering and moving through the soil, the potential for greater chemical movement into and through the soil is also increased. Thus, there is legitimate concern that adoption of CT may exacerbate the problem of groundwater contamination by agricultural chemicals while reducing the contamination of surface waters,[80] and some current models support this conclusion.[81]

A number of studies have been performed to determine if CT increases downward chemical movement. Most investigations have involved comparisons between no-till and moldboard plowed soils and have been laboratory or short-term field studies (Table 4.6). The results of this research generally suggest that no-till increases chemical leaching losses, with most long-term field studies documenting greater pesticide leaching in no-till than in tilled soils.[82,83] These findings, however, are not universal. For instance, Fermanich and Daniel[84] recently reported greater carbofuran movement in tilled than in no-till soil columns, whereas the leaching of chlorpyrifos was unaffected by tillage treatment. On the other hand, Clay et al.[85] reported a twofold increase in alachlor movement with no-till but little effect on bromide movement. Depending on the year studied, Kanwar et al.[86] found that no-till either reduced, increased, or had no effect on nitrate leaching from tilled and no-till field plots. Chichester and Smith[87] found that nitrate leaching losses were similar among tilled and no-till treatments. Likewise, Gaynor et al.[88] found no difference in surface or subsurface transport of atrazine and alachlor when they compared tilled, no-till, and ridge-tilled plots. Thus, while evidence exists that CT practices can increase chemical leaching losses other studies suggest that this might not always be the case depending on the nature of chemical, soil, tillage treatment, and the weather in the year investigated.

Table 4.6. Studies on the Effect of Tillage on Chemical Leaching

Tillage Treatments	Compounds	Location	Methodology	Ref
Tilled/no-till	Br, dye	New York	Laboratory lysimeters	138
Tilled/chiseled/ disced-plowed/ paraplowed/no-till	Br	Illinois	Field plots/simulated rainfall/soil samples	126
Tilled/no-till	NO_3	Ohio	Outdoor lysimeters	87
Tilled/no-till	Alachlor, Br	Minnesota	Laboratory lysimeters	85
Tilled/no-till	Carbofuran, chlorpyrifos	Wisconsin	Laboratory lysimeters	84
Till/ridge till/ no-till	Alachlor, atrazine	Ontario	Tiled field plots	88
Tilled/no-till	Br	Ohio	Outdoor column lysimeters/simulated rain	121
Tilled/no-till	Atrazine, cyanazine, metolachlor, simazine	Pennsylvania	Field plots/pan lysimeters	82
Tilled/no-till	Alachlor, atrazine, carbofuran, cyanazine	Maryland	Field plots/ shallow wells	83
Tilled/no-till	NO_3	Iowa	Field plots/simulated rainfall	180
Tilled/no-till	NO_3	Iowa	Field plots/tile flow	86
Tilled/no-till	Bacteria, Cl	Kentucky	Laboratory lysimeters	122
Tilled/no-till	Atrazine, Br	Maryland	Field plots/ infiltrometers	180
Tilled/no-till	Alachlor, atrazine, Br, carbofuran, dye, NO_3	New York	Field plots/tile flow/ wells/soil samples/ suction lysimeters	181
Tilled/no-till	Cl, NO_3	Kentucky	Field plots/pan lysimeters	182

4.4.1. Factors Promoting Infiltration Under Conservation Tillage

In order to understand how increased infiltration might lead to increased chemical movement and why there is a lack of consensus in the literature as to the effects of CT on subsurface chemical movement, the factors that promote infiltration under CT need to be examined. With CT, the number and intensity of tillage operations are reduced in order to increase the amount of crop residue remaining on the soil surface. The residue reduces evaporation and slows the movement of water off of the soil surface, thereby increasing infiltration. The increased infiltration observed with CT, however, is not directly attributable to the maintenance of a residue cover alone in all cases. Even when a comparable residue cover is maintained on tilled fields a similar reduction in runoff is not always achieved. Therefore, other factors must also contribute to the increased infiltration. Often, the bulk density is higher (and hence the total porosity of the soil is lower) when CT is practiced than when the soil is moldboard plowed.

This reduction in pore space would seemingly reduce the capacity of the soil to transmit water, yet the decrease in runoff indicates that CT can increase infiltration. Therefore, the pore space in CT soil must be more effective in transmitting water than the pore space in the plowed soil. One factor that helps to explain this paradox is that residue-free surfaces tend to crust when subjected to raindrop impact. The crust can block the entry of water to the pore space below, thereby greatly reducing infiltration. The residue cover on the CT soils reduces raindrop energy impinging on the soil surface, thereby reducing crusting. Additionally, the continued use of CT tillage practices results in increased levels of soil organic matter at the surface which in turn help to stabilize the structure and reduce the propensity of the soil to crust.

4.4.2. Characterizing Infiltration Through Macropores

A second major factor that explains how infiltration rate can increase while total porosity decreases is that CT soils may have fewer macropores but better connected pores. A plowed soil can have more macropores than CT but most are dead ended with little continuity. An increase in macroporosity attributed to a reduction in tillage intensity has been documented in a number of morphological and micromorphological studies of soil structure and porosity,[89-93] although this is not always the case.[94-96] The increase in macroporosity is usually attributable to two factors. First, the macropores created by plant roots and animal activity, which are normally disrupted by tillage, tend to persist when CT is used. Second, the presence of a residue cover can result in a more favorable environment for faunal activity and increased populations of macropore-producing arthropods and earthworms.[97] The lack of a detectable effect of tillage noted in some studies may be related to the length of time since the last tillage operation, or to physical and chemical properties of the soil that limit faunal activity or the preservation of macropores. In addition, macroporosity can vary seasonally and spatially;[95] therefore, extensive and repeated samplings may be necessary to assess the overall effect of tillage on soil porosity.

By virtue of their size, macropores can conduct a considerable volume of water through the soil. Ehlers[89] documented this in the field and noted an exponential relationship between pore size and water intake. On the other hand, based on theoretical considerations, Edwards et al.[98] demonstrated the effects of noncapillary-sized pores on infiltration. Although macropores occupy a small fraction of the soil volume, on the order of a few percent, they can account for the majority of the flow volume under some circumstances.[99-101] Thus, increased macropore continuity may compensate for the reduction in total porosity sometimes observed with CT. In fact, Roth and Joschko[102] found a highly significant negative correlation between the number of open earthworm burrows and runoff in a laboratory study, whereas Sharpley et al.[103] noted a twofold increase in runoff when they eliminated macropore-forming earthworms from pasture plots.

One consequence of water movement through macropores is that because only a small fraction of the soil may be involved in the flow processes, the velocity at which the water moves through the soil and the depth of penetration must be much greater than when the entire volume of the soil is involved in the flow process. Additionally, the amount of soil that an entrained solute encounters and the contact time with the soil must also be reduced. Under these circumstances chemical adsorption coefficients (K_d values) determined under equilibrium conditions may not be useful in predicting the extent of chemical transport.[104]

Unfortunately, subsurface movement of water and chemicals is difficult to quantify. This is especially true under conditions of macropore flow in which only a small percentage of the soil volume may contribute to movement and small amounts of solute can move well ahead of the main solute front. In laboratory studies,[33,105] and in a field lysimeter study[106] of solute transport in macropores, analyses of soil samples were insufficiently sensitive, due to errors associated with solute extraction and recovery, to detect chemical movement that was readily apparent based on analysis of percolate samples. Likewise, Bouma et al.[107] illustrated, and Shaffer et al.[108] documented, the inability of suction cup lysimeters to reliably sample the rapid flow through soil attributable to macropores. Thus, two methods commonly used to quantify chemical movement in the field, soil samples and suction cup lysimeters, may not be appropriate to fully assess solute transport in the field. Therefore, apparent inconsistences in the literature concerning the effects of CT on chemical transport may be related to problems with sampling methodology.

Recognizing the deficiencies in current methodology, increased efforts have been made to determine the consequences of macropores on water and chemical movement. One approach has been to characterize soil macroporosity using direct counts, image analysis, or computer assisted tomography.[109-111] However, not all morphologically similar macropores are equally effective in conducting water and chemicals.[33,112,113] Thus, this approach is of limited value unless all the factors that control water entry and movement through macropores can be quantified.

A second, more pragmatic approach to the problem involves improving the procedures and equipment used to obtain percolate samples in the field. Edwards et al.[114] devised a method to sample flow in individual earthworm burrows in the field. Unfortunately, only macropores ≥5 mm in diameter could be sampled in this manner and the technique is not suitable for characterizing the flow on a field scale. Pan lysimeters overcome these limitations as they can capture the flow from all pore sizes and from a relatively large, but ill-defined, area. In a field comparison, Barbee and Brown[115] concluded that pan lysimeters are superior to suction cups, especially under conditions in which preferential flow occurs. One concern with the use of pan lysimeters of a tension-free design is that the soil above them must be saturated before water from capillary-sized pores can be collected, which may cause them to shed water. Applying tension to the pan lysimeters, either mechanically or through the use of hanging water columns or wicks, may reduce this concern and improve pan collection efficiency.[116,117]

4.4.3. Water and Chemical Movement in Macropores

The amount of water transmitted in macropores can be affected by rainfall intensity and antecedent moisture content of the soil. Increased rainfall or irrigation rates have been shown to increase the relative contribution of macropores to flow.[99,118-122] The effect of antecedent moisture content on macropore flow is less certain. In general, the contribution of macropores to flow appears to increase with dry soil conditions;[33,64,123-125] however, under some conditions, dry antecedent moisture conditions may inhibit macropore flow.[119]

Chemical transport in macropores is dependent on factors that affect chemical mobility, such as placement, formulation, and reactivity with the soil, as well as the factors that influence the amount of water flowing in macropores. High intensity rainfalls yield greater water flow in macropores and greater chemical movement than low intensity rainfalls.[69,119,121,126] Rainfall timing relative to chemical application, however, can have a major influence on the amount of chemical transmitted in macropores.

Fertilizers and pesticides applied to CT soils are generally surface applied or receive only limited mechanical incorporation in order to preserve the integrity of the surface mulch. The chemicals, therefore, are not in intimate contact with the soil matrix where they either can be bypassed by the flow occurring in the macropores or are subject to adsorption. This may contribute to the observation that it is often the first few leaching events after application that results in the greatest chemical movement.[18,82,83,127,128] Conditions that promote chemical diffusion into the soil matrix can reduce the amount transmitted in leaching events. Shipitalo et al.[33] found that a light rainfall preceding a rainfall that produced macropore flow substantially reduced the movement of both adsorbed and nonabsorbed chemical tracers. Allowing time for diffusion to occur can also reduce chemical movement in macropores.[120,125,129] In addition, in the case of substances subject to decomposition or plant uptake, time can reduce the amount of chemical available for leaching. When macropore flow occurs before diffusion has occurred, chemical affinity for the soil exchange complex may not retard movement in macropores. Everts et al.[130] reported that peak concentrations of anionic and cationic tracers occurred simultaneously under conditions of macropore flow. Kladivko et al.[18] also attributed the concurrent detection of four pesticides with different adsorption coefficients in tile flow from a chiseled field to preferential flow.

Although under some circumstances chemicals can be transported through the soil with the flow in macropores regardless of their affinity for the exchange complex, reactivity with the soil will affect the total amount transmitted through the soil on a yearly basis. Using field lysimeters, Shipitalo and Edwards[106] found that disruption of macroporosity by tillage reduced the

Table 4.7. Percentage of Rainfall and Applied Br and Sr Tracers Collected in Percolate from Tilled and No-Till Column Lysimeters, 1987–1988 and 1988–1989

	Rainfall (%)	Bromide (%)	Strontium (%)
	1987–1988		
Tilled	55 a[a]	83 a	0.7 a
No-till	77 b	83 a	1.5 b
	1988–1989		
Tilled	62 a	94 a	0.7 a
No-till	84 b	91 a	2.1 b

Source: Adapted from Shipitalo, M. J. and Edwards, W. M., *Soil Sci. Soc. Am. J.*, 57, 218–223, 1993.

[a] Means in the same column, within seasons, followed by different letters are significantly different at $p \leq 0.05$.

amount of water transmitted by the soil each year of a 2-year study, but did not affect the total amount of unreactive bromide tracer leached each year (Table 4.7). Most of the applied bromide leached from the lysimeters, although the losses occurred more rapidly in the no-till than in the tilled lysimeters. On the other hand, the no-till lysimeters transmitted two to three times more reactive strontium tracer than the tilled lysimeters, although the maximum strontium losses never exceeded a small percent of that applied (Table 4.7). Most of the differences in strontium losses were the result of the first few storms after application. The more rapid leaching of bromide and the greater leaching of strontium in the no-till than in the tilled lysimeters were attributed to preferential flow in macropores. Flushing of the matrix during recharge periods accounted for the removal of most of the bromide remaining in the soil. These observations indicate that leaching losses of a chemical not retained by the soil, such as nitrate, can be slightly increased by macropore flow under some circumstances. The majority of the losses, however, were unrelated to tillage treatment and macropore flow. Therefore, in order to reduce nitrate losses, fertilizer management practices that promote efficient usage, such as soil testing, soil specific, split, and computer-controlled applications,[131] should be considered. Macropore flow, however, can substantially increase the relative leaching losses of chemicals, such as pesticides, that are normally strongly adsorbed by the soil, although the total amount transported will likely amount to only a small percentage of the application. The circumstances that enhance leaching losses in macropore flow are large, intense storms shortly after application, precisely the conditions that promote increased losses in overland flow,[78,132,133] which CT and macropore flow help to reduce. In fact, Whipkey[134] coined the term *subsurface storm flow* in reference to macropore flow because its effects on hydrology were similar to those of overland flow. The analogy to overland flow appears to extend to the factors that affect chemical transport by both mechanisms.

The effects of CT and macropores on chemical transport of strongly absorbed compounds can be reduced by allowing the chemicals to diffuse into the soil matrix or degrade before a major leaching event occurs. Weather-related factors, such as light and low intensity rainfalls, can increase diffusion, whereas increased length of time between rainfalls can result in greater degradation and allow more time for diffusion to occur. In terms of management practices, chemical applications should be postponed when heavy storms are imminent. More practical ways to reduce macropore transport might include changing the way pesticides are applied by using procedures such as point injection or by changing pesticide formulations (e.g., adjuvants[135] or controlled-release formulations[136,137]).

4.4.4. Water Quality

The main impetus for adopting CT practices in the Corn Belt is to reduce soil losses due to water erosion. In most instances, this is achieved by reducing runoff and increasing infiltration. A proportion of this increased infiltration may be due to rapid flow in macropores. The relative contribution of macropores to infiltration will be dependent on a number of factors, including soil type, slope, and management history. Macropores can contribute to flow even in plowed soils;[138] CT probably only increases their relative importance. With water moving through macropores the amount of chemical reaching the groundwater has the potential to increase. If the solute in question is one that is not retained to a significant extent by the soil matrix, the effect of macropores on transport are likely to be small. Thus, for chemicals such as nitrate it is how we manage the nutrient, not how we till the soil, that will have an overriding effect on the amount reaching the groundwater. For chemicals that are subject to strong interaction with the soil matrix, macropores may increase the amount reaching the groundwater. Whether CT and macropores contribute to increased transport in a particular field will depend on the physical and chemical properties of the soil, how and when chemicals are applied, and the weather. Heavy, intense storms shortly after surface application are likely to cause the most transport to occur. If this scenario does not occur, the effect of macropores on chemical transport are likely to be of little consequence. However, even under extreme conditions the amount of additional transport attributable to CT and macropores is likely to be small relative to the amount applied. Whether this is of concern will depend on the properties of the chemical in question. If we could accurately predict the weather, we would stand a better chance of reducing the transport of these types of compounds. Alternatively, we can adopt our management practices to further reduce the amount entering the macropores by increasing the amount that diffuses into the matrix.

Finally, we must realize that total elimination of chemical movement to groundwater is not an achievable objective under any tillage system. Therefore, we need to better define the amounts and concentrations reaching the groundwater that are tolerable. We must then balance the improvements in surface water quality attributable to the adoption of CT with the potential for increased subsurface contamination. Greater understanding of the factors affecting chemical transport and improved models will aid in this effort.

4.5. CROP RESPONSE IN CONSERVATION TILLAGE SYSTEMS

CT practices can influence soil properties differently than do tillage practices based on moldboard plowing. Such changes in the root zone may affect the development and eventual yield of several agronomic crops. Though effects on soil properties may be similar in different soils, the effects of CT on crop performance can be quite soil specific, and possibly site specific. It is important to remember that effects of tillage on crop yield usually cannot be evaluated directly without considering its effects on such factors as stand establishment, weed control, disease incidence, and fates of fertilizer nutrients. With this in mind, certain generalizations can be made.

4.5.1. Influence of Soil Water and Drainage

Most CT systems are designed to leave a covering of plant residue on the soil surface to provide protection from raindrop impact and reduce erosion potential. In some cases, this residue cover may reduce evaporation of soil water below that seen from bare surfaces. In addition, by protecting the surface from raindrop impact, residue can reduce the tendency for soils with silt-textured A horizons to form crusts, thereby allowing for greater infiltration capacity. These effects tend to increase the quantity of plant-available water in well-drained, crust-prone soils on slopes,[139,140] but may have little effect on the hydrology of noncrusting soils, poorly drained soils, or soils on level landscapes.[20,37] This differential effect on soil moisture seems to have a major effect on the response of crops to CT practices. Effects are most apparent when one compares the performance of extremely different practices such as moldboard plowing (which inverts the soil and buries all residue) and no-tillage (which causes almost no soil disturbance and leaves most residue on the surface).

4.5.1.1. Responses on Well-Drained Soils

On well-drained soils where moisture stress can cause reductions in grain yield, no-tillage practices generally produce equal or greater grain yield than moldboard plowing if equivalent plant densities and adequate weed control are obtained.[141-146] The difference has been attributed mainly to the presence of mulch cover, not lack of tillage per se, and can increase over time.[140,143,144,146,147] Van Doren et al.[146] have also indicated that maintenance of desirable soil structure, which would be destroyed by plowing, may also contribute to the yield advantage of no-tillage practices. Though numerous studies have evaluated the effects of tillage practices on soil physical properties, no positive cause and effect relationships between the effects of no-tillage on yield and on structure have emerged other than those presumed to arise due to retardation of crusting. This area remains an important one that could benefit from more research.

Continuing evaluation of corn yield on tillage-comparison plots established at Wooster, OH in the 1960s indicates that the yield advantages associated with no-tillage practices on well-drained soils persists for many years.[1] These long-term data also show that soybean yields are generally greater under no-tillage than moldboard-plow culture on such soils. Such data run counter to many farmers' fears that continuous, long-term use of no-tillage on such soils will lead to yield decline caused by such factors as increasing soil compaction, increasing insect and disease pressure, and buildup of excessive residues on the soil surface. The tendency for no-tillage practices to produce consistently equal or greater yields than moldboard plowing on well-drained soils should make them acceptable and desirable conservation practices on sloping fields that are naturally well drained. Many farmers have accepted the practice in such situations, as indicated by the data for adoption of no-tillage in Ohio counties where such fields are common.[148]

4.5.1.2. Responses on Moderately Well and Somewhat Poorly Drained Soils

On moderately well-drained and somewhat poorly drained soils, few consistent differences in corn yield have emerged in studies comparing no-tillage and moldboard plowing.[1,141,146,149] Data obtained on a tile-drained Crosby silt loam (aeric ochraqualf) at South Charleston, OH (Table 4.8) indicate that the response of corn to tillage practice varies with year, and that no-tillage may produce less corn yield in some years. Eckert[149] noted that no-tillage produced lower yields than plowing on such soils under cool, wet conditions, but greater yield in warmer, drier years. Despite a tendency for year-to-year variations in relative performance, no-tillage practices seem to produce yields comparable to moldboard plowing when averaged over longer periods of time.[1,146]

In Ohio, these soils tend to crust as do the well-drained soils described in the previous section; yet, increasing quantities of mulch do not necessarily increase yield. Rather, yields of corn produced without tillage in a relatively low-residue corn-soybean rotation are more likely to approach those obtained by plowing than those produced when corn follows corn. Van Doren et al.[146] proposed that because these soils exist on relatively flat landscapes and runoff

Table 4.8. Influence of Four Tillage Practices on Corn Yield on a Tile-Drained Crosby Silt Loam

	Corn after Corn (Mg/ha)				Corn after Soybean (Mg/ha)			
Tillage	1985	1987	1988	Avg.	1985	1987	1988	Avg.
Moldboard plow	7.8	12.9	9.9	10.2	8.3	7.9	11.5	9.2
Chisel plow	6.8	12.8	9.7	9.8	7.9	8.1	11.8	9.2
No-tillage	5.9	11.2	9.6	8.9	7.7	8.6	10.8	8.9
Ridge	7.6	11.6	9.1	9.4	7.8	8.8	9.4	8.7
LSD 0.05	1.2	1.0	NS	—[a]	NS	NS	1.1	—

Source: D. J. Eckert, unpublished data.

[a] Statistical comparison of averages not made due to differences in experimental design in different years.

can be much slower, infiltration and soil moisture may not be as affected by soil surface conditions as on better-drained soils on slopes. Therefore, the negative effects associated with mulch may dominate in some years and yield potential may be reduced in some years. The importance and balance of positive and negative effects due to mulching on these soils requires much further investigation, considering the areal extent of such soils in the region.

The data in Table 4.8 also show the effects of two other CT practices on corn yield. Chisel plowing (which leaves residue on the surface and shatters rather than inverts the soil) produced yields comparable to moldboard plowing in all comparisons, confirming observations by Griffith et al.[141] that this practice should produce comparable yields if adequate stands and weed control were obtained. Planting on undisturbed ridges produced yield equivalent to moldboard plowing in some years, but less in others. Fausey and Lal[50] also compared the relative performance of moldboard plowing, no-tillage, and ridge-planting practices on a Crosby soil at a different location and found, similarly, that practices were likely to produce relatively different yields in different years, but on the average neither ridge planting nor no-tillage produced greater or lesser yield than moldboard plowing.

4.5.1.3. Responses on Poorly Drained Soils

Early research on poorly drained soils (even with improved drainage) demonstrated that corn following corn often produced less yield when grown without tillage than with moldboard plowing,[141,146,150] but that yield differences were of lesser magnitude or nonexistent when corn followed some other crop. The tillage effects in continuous corn production tend to persist rather consistently for many years,[1,47] as do the lesser differences when corn is grown in rotation. The data in Table 4.9 illustrate these differences, showing that when corn follows corn yield generally declines as tillage intensity declines, whereas when corn follows soybean, tillage intensity has much less effect on yield.

The above-mentioned studies focused on poorly drained soils with relatively high organic matter concentrations (i.e., >3 g/kg). On a poorly drained soil with lesser organic matter, Griffith et al.[47] found that the depression in

Table 4.9. Influence of Four Tillage Practices on Corn Yield on a Tiled Drained Kokomo Silty Clay Loam

	Corn after Corn (Mg/ha)					Corn after Soybean (Mg/ha)				
Tillage	1985	1986	1987	1988	Avg.	1985	1986	1987	1988	Avg.
Moldboard plow	12.2	13.5	7.7	9.8	10.8	12.5	14.1	12.9	12.5	13.0
Chisel plow	11.4	12.4	5.8	10.1	9.9	12.4	14.8	12.1	11.5	12.7
No-Tillage	10.0	10.9	6.0	10.1	9.2	12.1	13.3	12.8	12.0	12.5
Ridge	10.8	11.1	6.9	9.4	9.6	11.9	12.0	11.4	10.5	11.4
LSD 0.05	0.7	0.8	0.9	0.5	0.3	1.2	1.1	0.8	1.1	0.4

Source: D. J. Eckert, unpublished data.

yield associated with continuous no-tillage corn production disappeared after 3 years and credited the improvement to improved soil structure, citing the data of Heard et al.[95] and Kladivko et al.[34] Dick et al.[1] also noted the tendency for this depression to disappear on a greater organic matter soil, but the time before the effect occurred was much longer. The reasons for these apparent effects deserve much further investigation, because yield depressions associated with continuous no-tillage corn production are major reasons why farmers may be unwilling to adopt CT.

Early corn growth, plant stand, and grain yield can be adversely affected by cool, wet soil conditions,[45,50,151,152] conditions that can be aggravated by the presence of residues on the soil surface. Ridge planting is an attempt to alleviate these problems by elevating the seed zone above the general land surface to allow quicker soil warming and drainage. Numerous studies have shown that planting on ridges can produce corn yields equivalent to plowing if plant densities are comparable.[3,47,50,141,142,153] Eckert[3] noted that on soils without drainage improvements, ridge planting may allow for significantly earlier planting in some years, and that in the initial years of the study corn following corn on ridges consistently produced yields equivalent to or greater than those obtained by plowing, an effect not seen in no-tillage systems. Such results indicate that ridge planting may be a preferable system for situations in which poor drainage poses a significant production constraint. The data in Tables 4.8 and 4.9 indicate, however, that on another site corn performance on ridges was erratic when compared to other practices. Eckert[3] also noted that on average, soybeans planted in row widths narrower than those achievable with ridges produced greater yield than those planted on ridges, which could cause farmers planting corn-soybean rotations to opt for other systems. Ridge planting seems to be an acceptable conservation alternative in some situations, but its degree of adoption compared to other practices continues to be rather low across the Corn Belt.[148]

4.5.2. Influence of Agronomic Factors

When the data from all studies are combined and examined, it appears that relative yields obtained with CT and plowing can be influenced by a number of factors. Soil moisture,[140] soil temperature,[141,152] and competition from unwanted vegetation[141] may all be affected by imposition of CT practices. Stand establishment and early plant development may be affected negatively by CT.[141,142,152] Moisture conservation may allow for more plant-available water.[143] Stalk rots in corn may be less prevalent in CT situations,[154,155] but *Phytophthora* root rot in soybeans may be greater.[185] The ultimate yield difference in a given year is determined by whatever factor dominates in that year. While some factors can be controlled by good management, the effects of those that cannot be controlled adequately deserve much more investigation to assess their relative importance in determining yield.

Cultural practices may have significant effects on the performance of CT practices, so much so that adoption of entire systems may be necessary to ensure success. Acidification of the surface layer of fields over time (see Eckert[156] for review) may dictate changes in liming practices or use of herbicides not sensitive to acidic soil conditions to maintain weed control. Use of *Phytophthora* root rot tolerant or resistant soybean varieties can prevent severe yield losses in this crop. Because surface broadcasting of urea-based nitrogen fertilizer materials can cause severe yield losses in corn in CT systems when compared to other materials such as anhydrous ammonia or ammonium nitrate, injection or surface banding of these materials or use of other nitrogen materials may be necessary, particularly in situations with heavy residue cover.[4,157-161] Planting dates, however, may not require change. Both Eckert[149] and Carter[94] have shown that corn planted without tillage produces more yield when planted earlier than later, and that effects of date-of-planting on yield for corn planted with and without intensive tillage are similar. The nature of the relationship between corn yield, crop rotation, and tillage intensity on some soils remains elusive. Van Doren et al.[146] proposed that increased root disease pressure in no-tillage situations may be responsible. Yakle and Cruse[9] have found that corn residues placed close to corn seed can inhibit early growth, indicating a potential chemical autotoxicity effect. However, Crookston and Kurle[162] have demonstrated that a history of corn production in a field can have more influence on corn yield than the presence of incorporated corn residues from the previous year. Some investigators have proposed that such effects could be microbiological.[163,164] In a recent study in Ohio, the authors grew corn without tillage following both corn and soybeans on a poorly drained Kokomo silty clay loam (typic argiaquoll). On some plots surface residues were switched. Results (Table 4.10) have shown trends similar to those noted by Crookston and Kurle,[162] even when residues were left on the surface. Corn residue on the surface did not affect yield; however, the history of corn the previous year did decrease yield. Such effects indicate that the soil environment created by a corn crop may be detrimental to a succeeding corn crop, and that the effect may be augmented by a lack of tillage on soils that do not benefit greatly from moisture conservation. Whether such effects are physical, microbiological, or chemical has yet to be resolved.

Table 4.10. Effects of Previous Crop and Surface Residue Type on Corn Yield on a Tile Drained Kokomo Silty Clay Loam

	Surface residue (Mg/ha)		
Previous Crop	Corn	Soybean	Avg.
Corn	5.4	6.2	5.8
Soybean	8.0[a]	9.1[a]	8.6
Avg.	6.7	7.6	

Source: A. M. Wolfe and D. J. Eckert, unpublished data.

[a] Means different at $p < 0.05$.

4.5.3. Crop Rotation

Crop rotation is one management decision that clearly affects crop yields. Classical studies have provided well-documented economic and environmental benefits of crop rotation.[165] Many factors are thought to contribute to the "rotational effect," which may be defined as the increase in yields when crops are grown, under similar conditions, in rotation compared to yields achieved under continuous monocropping. However, the most important component of this effect seems to be the benefit gained from control of insects and diseases.[166] Information on the interaction of rotation and tillage for different crops and in different climates and locations is not easily obtained because these types of experiments require studies over multiple years. Also, CT (especially no-tillage) practices have been adopted only on a wide scale in the past 25 years.

Studies in Ohio have shown that soil type, tillage, and crop rotation all interact to affect corn yields (Figures 4.4 and 4.5). As in most studies, rotating corn with another crop improved yields above that observed for continuous monoculture corn. However, the data provide a unique insight into the long-term trends of tillage and soil interactions. When considering the well-drained Wooster silt loam soil (typic fragiudalf), the corn grain yields under no-tillage have been consistently higher than yields in which spring moldboard plowing has been applied. This difference seemed to be increasing with time until the mid- to late 1980s, when several dry years caused yield depressions for all tillage treatments.

The very poorly drained Hoytville soil (mollic ochraqualf) gave different results than when no-tillage was maintained on the Wooster soil. Rotating corn with soybeans caused the no-tillage and plow yields to be similar during the initial years of the experiment. However, the yields were depressed under no-tillage in the corn-oats (*Avena sativa* L.) meadow rotation and in locations in which continuous monoculture corn was grown. These yield depressions associated with no-tillage in the corn-oats-meadow rotation are due to poorer weed control than for the other rotations during the early years of the experiment.

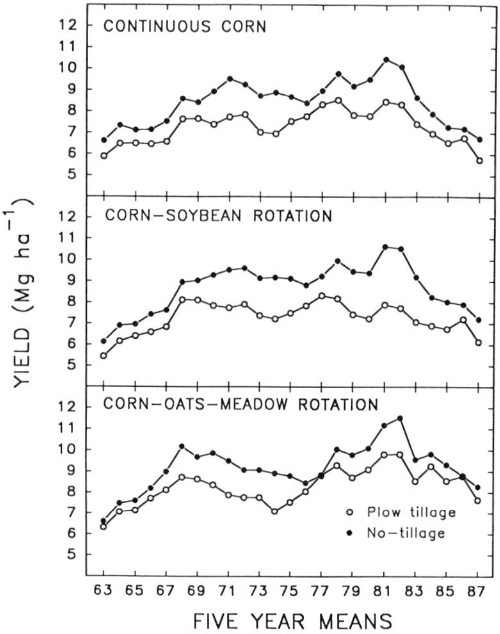

Figure 4.4. Five-year running means of corn yields, beginning with the year indicated, at Wooster, Ohio.

The continuous corn yield depressions have been attributed to a buildup of disease, although no conclusive proof of this hypothesis has been obtained.

Probably the most significant data presented in Figure 4.5 is the observation that corn grain yields under no-tillage since 1980 (after 20 years of continuous no-tillage) have again reached levels equal to or greater than levels obtained for the moldboard plow treatment. The corn grain yields associated with the corn-soybean rotation have responded especially well to the continuous application of no-tillage. The reasons for this are unclear, but the trend is quite evident. Climate, changes in soil chemical, physical and biological properties with long-term maintenance of no-tillage, and improved, more disease-resistant corn hybrids are all possible candidates that may explain the trend.

Studies in other states have reported similar findings. On a low organic matter soil under continuous NT, corn growth and yield were reduced during the first 3 years as compared to where plowing was practiced, but yields for no-tillage were equal to or better under NT for the next 4 years.[47] In contrast, corn grown in rotation yielded equally for both tillage treatments during the first 3 years and was significantly higher under no-tillage in 3 out of the next 4 years. Short- and long-term, tillage/crop rotation responses have also been reported for crops grown in other regions of the United States.[167]

Similar rotational effects have been observed for soybeans, in which the combination of a CT system (either strip-tillage or no-tillage) with a corn-soybean

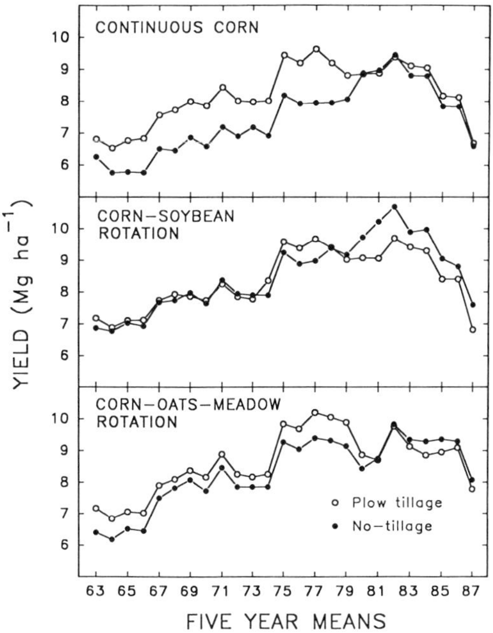

Figure 4.5. Five-year running means of corn yields, beginning with the year indicated, at Hoytville, Ohio.

rotation gave the most consistently high yields.[109] The significant tillage by rotation interaction was caused primarily by a buildup of soybean cyst nematode populations under conventional tillage and continuous soybeans.

4.5.4. Cover Crops

Cover crops have been extensively studied because of their effect on reducing soil crusting and soil erosion. Cover crops may provide additional benefits such as serving as a means of managing soil nitrogen levels, providing an effective mulch cover for increased weed control, and as a means of increasing soil organic matter. A greenhouse study indicated that corn dry weight increased 20 to 75% when legume debris was placed on the soil surface.[124] Negative allelopathic effects on corn emergence were not observed.

Field studies have shown that the type of cover crop has a significant effect on corn growth and dry matter production, with a rye (*Secale cereale* L.) cover crop resulting in lower yields than a legume cover crop.[168] Recovery of legume nitrogen by the corn crop was estimated at 40 to 45 kg of nitrogen per hectare (2-year average). The estimated fertilizer nitrogen equivalency of hairy vetch (*Vicia villosa* Roth) and bigflower vetch (*Vicia grandiflora* W. Koch *var.* Kitailbelinana) for corn production in Kentucky was 75 and 65 kg nitrogen per hectare, respectively.[169] Rye cover crop did not provide any yield increase as compared to a fallow control and corn grain yield response to applied nitrogen was greatest in a rye cover crop and minimal for a legume cover crop.

Cover crops may reduce nitrate leaching losses by taking up nitrate in autumn and then releasing it in the sping as the organic matter from the cover crop is mineralized. The particulate organic matter pool, which is composed of partially decomposed root fragments, was found to be the most important fraction influencing this internal cycling of nitrogen in Iowa soils.[170]

Ngalla and Eckert[171] evaluated inclusion of a clover cover crop in a soybean-wheat-corn rotation on a poorly drained Hoytville silty clay (mollic ochraqualf) and found that while the clover did provide nitrogen to the corn, corn yields were often less following wheat-clover than wheat alone due to plant density reductions. Eckert also noted that rye cover crops often reduced corn yield in a corn-soybean rotation on this soil, and had mixed effects on corn and soybean yields on a moderately well-drained Canfield silt loam (aquic fragiudalf), although negative responses were more common than positive. Yield reductions were attributed to plant density differences. The only case in which a rye cover crop increased yield was in a very dry season when corn followed soybeans and moisture conservation became quite important.

4.6. RESEARCH NEEDS IN CONSERVATION TILLAGE

In addition to soil-specific problems of adapting CT systems, some important researchable issues within the Corn Belt include the following:

- *Soil Structure and Tilth:* Developing systems for enhancing soil structure and tilth is a high priority. In addition, there is also a need to develop a quantitative index of soil tilth and soil structure assessment.
- *Soil Compaction:* Densification and formation of massive structure are serious problems, especially in clayey soils. Compaction is a serious problem, especially in the wheel track zone. An interdisciplinary research effort involving soil scientists, engineers, and plant scientists is needed to address the problem. Possible options include developing controlled traffic, use of deep-rooted cover crops, and ameliorative cropping systems.
- *Fate of Agricultural Chemicals:* Little is known about the fate of nitrogen and pesticides in CT systems. Evaluating transport of pollutants in relation to macropore flow remains a research priority.

Specific research needs in relation to chemical transport include the following:

1. Despite the amount of effort already expended, more long-term field studies that simultaneously quantify surface and subsurface transport are needed. In conjunction, better methods for sampling subsurface flows are needed and we must investigate how chemicals move from the vadose zone to the groundwater.

2. The acceptable levels of surface and subsurface contamination need to be determined, realizing that zero tolerances are probably not attainable for most materials.
3. Concurrent with greater knowledge of the processes affecting chemical transport, our models must be refined, taking into account the potential effects of macropore flow so that the models are usable as decision-making tools.
4. Management strategies need to be defined and refined that increase the efficiency of agricultural chemicals while minimizing both surface and subsurface losses, realizing that these strategies will differ depending on soil types, climates, etc., within the Corn Belt.
    - *Soil Guide:* Developing CT systems in relation to soil type is an important issue that needs to be addressed. A soil guide should be developed to assess tillage requirements for different soil types.
    - *Farming-by-Soil:* Risks to environmental degradation can be minimized by developing CT and fertilizer needs for all soil types within a farm. Linking tillage system with fertility requirements and automating the system will be an important step forward.
    - *Environmental Issues:* It is important to define critical limits of important pollutants in terms of health-related risks to humans, livestock, and wildlife. Determining environmental acceptibility and sustainability is critical to widespread adoption of CT.
    - *Greenhouse Gas Emissions and Carbon Sequestration:* CT may have a major impact on emissions of radiatively active gases, e.g., CO_2, NO_x, etc. It can also be a principal tool for sequestering carbon in soil. There is a need to evaluate the role of CT in greenhouse gas emissions and environmental concerns.
    - *Energy Use and Economics:* Optimizing energy use through judicious management of fertilizer, chemicals, crop residue, farm equipment, and operations is an important issue. Energy use is also intimately linked with overall economics and profitability. In the long run, use of CT depends on the profitability.

4.7. GENERAL SUMMARY AND CONCLUSIONS

Sustainable use of soil and water resources is the primary objective of agriculture. Sustainability implies profitable crop production on a continuous basis, least degradation of soils, productive capacity for future use, and minimum risks to water quality and other environmental considerations. Therefore, application of CT systems should be assessed in terms of these criteria. In fact, CT may be an important tool for sustaining the soil and water resources. Positive effects of CT on sustaining productivity are evident from the data of long-term experiments in Ohio,[1,39,172] Iowa,[172] Indiana,[39] and in the region as a whole.[38,173]

Sustainability of soil and water resources with reference to CT also depends on adopting the system to specific soil-related constraints. Some examples of appropriate adaptation are ridge-tillage and raised bed system for poorly drained soil, chisel plowing, and paraplowing for compacted soils, and growing cover crops for improving soil fertility.[174-177]

Applicability of CT systems in the Corn Belt can be improved by adopting a systems approach, a holistic approach that considers all factors of production, e.g., crop rotation, integrated pest management, residue management, nutrient cycling, fertility management, soil and water conservation, appropriate farm machinery, and environmental and socioeconomic considerations. In fact, CT may require a special set of cultural practices that may be different from those needed for conventional tillage systems. It is important to develop, adapt, and use the special set of cultural practices appropriate for CT systems. CT as a holistic and an ecological concept can be transformed into a viable package of soil and crop management by developing and using the special set of cultural practices. This special package may require a high level of management skills and a high level of science-based inputs.

CT is an integral part of an approach to adopt sustainable agricultural practices. Being a highly skilled and science-based concept, making CT an integral part of sustainable agriculture requires a multidisciplinary approach to research and development. The multidisciplinary team should involve close cooperation among soil scientists, agronomists, crop scientists, biologists, agricultural engineers, economists, and soil scientists. The rapidly increasing area in CT in the Corn Belt[184] may be partly due to successful adoption of the multidisciplinary and holistic approach.

4.8. REFERENCES

1. Dick, W. A., E. L. McCoy, W. M. Edwards and R. Lal. "Continuous Application of No-Tillage to Ohio Soils," *Agron. J.* 83:65–73 (1991).
2. Fausey, N. R. "Experience with Ridge Tillage on Slowly Permeable Soils in Ohio," *Soil Tillage Res.* 18:195–206 (1991).
3. Eckert, D. J. "Ridge Planting for Row Crops on a Poorly Drained Soil. I. Rotation and Drainage Effects," *Soil Tillage Res.* 18:181–188 (1990).
4. Eckert, D. J. "Ridge Planting for Row Crops on a Poorly Drained Soil. II. Anhydrous Ammonia and UAN Management for Maize Following Soybean," *Soil Tillage Res.* 18:189–194 (1990).
5. USDA. "Our American Land," *1987 Yearbook of Agriculture* (Washington, DC: U.S. Government Printing Office, 1987).
6. USDA. "The Second RCA Appraisal" (Washington, DC: U.S. Government Printing Office, 1987).
7. USDA. "The Second RCA Appraisal" (Washington, U.S. Government Printing Office, 1989).
8. National Corn Growers Association. *The World of Corn.* St. Louis: NCGA, 1991), p. 31.

9. Yakle, G. A., and R. M. Cruse. "Corn Plant Residue Age and Placement Effects upon Early Corn Growth," *Can. J. Plant Sci.* 63:871–877 (1983).
10. Hill, R. L., and R. M. Cruse. "Tillage Effects on Bulk Density and Soil Strength of Two Mollisols," *Soil Sci. Soc. Am. J.* 49:1270–1273 (1985).
11. Hill, R. L., R. Horton and R. M. Cruse. "Tillage Effects on Soil Water Retention and Pore Size Distribution of Two Mollisols," *Soil Sci. Soc. Am. J.* 49:1264–1270 (1985).
12. Garcia, F. R., M. Cruse and A. M. Blackmer. "Compaction and Nitrogen Placement Effect on Root Growth, Water Depletion, and Nitrogen Uptake," *Soil Sci. Soc. Am. J.* 52:792–798 (1988).
13. Benjamin, J. G., and R. M. Cruse. "Tillage Effects on Shear Strength and Bulk Density of Aggregates," *Soil Tillage Res.* 9:255–263 (1987).
14. Karlen, D. L. "Conservation Tillage Research Needs," *J. Soil Water Conserv.* 45:365–369 (1990).
15. Voorhees, W. B. "Soil Tilth Deterioration under Row Cropping in the Northern Corn Belt: Influence of Tillage and Wheel Traffic," *J. Soil Water Conserv.* 34:184–186 (1979).
16. Singh, K. K., T. S. Calvin, A. Q. Mughal, and D. C. Erbach. "A Tilth Analysis of Five Tillage Systems," 1990 No. 90-1526 (St. Joseph, MI: American Society of Agricultural Engineers, 1990).
17. Dick, W. A., D. M. Van Doren, Jr., G. B. Triplett, Jr. and J. E. Henry. "Influence of Long-Term Tillage and Rotation Combinations on Crop Yields and Selected Soil Parameters: I. Results Obtained for a Mollic Ochraqualf Soil," Res. Bull. 1180; "II. Results Obtained for a Typic Fragiudalf Soil," Res. Bull. 1181 (Wooster: Ohio Agricultural Research & Development Center, 1986).
18. Kladivko, E. J., G. E. Van Scoyoc, E. J. Monke, K. M. Oates and W. Pask. "Pesticide and Nutrient Movement into Subsurface Tile Drains on a Silt Loam Soil in Indiana," *J. Environ. Qual.* 20:264-270 (1991).
19. Larney, F. J. and E. J. Kladivko. "Soil Strength Properties under Four Tillage Systems at Three Long-Term Study Sites in Indiana," *Soil Sci. Soc. Am. J.* 53:1539–1545 (1989).
20. Lal, R., T. J. Logan and N. R. Fausey. "Long-Term Tillage and Wheel Traffic Effects on a Poorly Drained Mollic Ochraqualf in Northwest Ohio. I. Soil Physical Properties, Root Distribution and Grain Yield of Corn and Soybean," *Soil Tillage Res.* 14:341–358 (1989).
21. Mahboubi, A. A., R. Lal and N. R. Fausey. "Twenty-Eight Years of Tillage Effects on Two Soils in Ohio," *Soil Sci. Soc. Am. J.* 57:506–512 (1993).
22. Ankeny, M. D., T. C. Kasper and R. Horton. "Characterization of Tillage and Traffic Effects on Unconfined Infiltration Measurements," *Soil Sci. Soc. Am. J.* 54:837–840 (1990).
23. Datiri, B. C., and B. Lowery. "Effects of Conservation Tillage on Hydraulic Properties of a Griswold Silt Loam Soil," *Soil Tillage Res.* 21:257–271 (1991).
24. Datiri, B. C., and B. Lowery. "Tillage Effects on Rain-Generated Wetting Front Migration through a Griswold Silt Loam Soil," *Soil Tillage Res.* 21:243–256 (1991).
25. Kemper, W. D., T. J. Trout, A. Segren and M. Bullock. "Worms and Water," *J. Soil Water Conserv.* 43:401–404 (1987).

26. Edwards, W. M., and L. D. Norton. "Effect of Macropores on Infiltration into Non-Tilled Soil," *Trans. 13th Congr. Int. Soc. Soil Sci.* 5:47–48 (1986).
27. Dick, W. A., W. M. Edwards and F. Haghiri. "Water Movement through Soil to which No-Tillage Cropping Practices have been Continuously Applied," in *Proc. Agricultural Impacts on Ground Water — A Conference.* (Dublin, OH: National Water Well Association 1986), pp. 243–252.
28. Dick, W. A., R. J. Rosenberg, E. L. McCoy, W. M. Edwards and F. Haghiri. "Surface Hydrologic Response of Soils to No-Tillage," *Soil Sci. Soc. Am. J.* 53:1520–1526 (1989).
29. Logsdon, S. D., R. R. Allmaras, L. Wu, J. B. Swan and G. W. Randall. "Macroporosity and Its Relation to Saturated Hydraulic Conductivity under Different Tillage Practices," *Soil Sci. Soc. Am. J.* 54:1097–1107 (1990).
30. Mohanty, B. D., R. S. Kanwar, and R. Horton. "A Robust-Resistant Approach to Interpret Spatial Behavior of Saturated Hydraulic Conductivity of a Glacial Till Soil under No-Tillage System," *Water Resources Res.* 27:2579–2592 (1991).
31. Lal, R. and D. M. Van Doren, Jr. "Influence of 25 Years of Continuous Corn Production by 3 Tillage Methods on Water Infiltration for Two Soils in Ohio," *Soil Tillage Res.* 16:71–84 (1990).
32. Singh, P., R. S. Kanwar and M. L. Thompson. "Measurement and Characterization of Macropores by Using AUTO-CAD and Automatic Image Analysis," *J. Environ. Qual.* 20:289–294 (1991).
33. Shipitalo, M. J., W. M. Edwards, W. A. Dick and L. B. Owens. "Initial Storm Effects on Macropore Transport of Surface-Applied Chemicals in No-Till Soil," *Soil Sci. Soc. Am. J.* 54:1530–1536 (1990).
34. Kladivko, E. J., D. R. Griffith and J. V. Mannering. "Conservation Tillage Studies on a Clermont Silt Loam Soil," *Proc. Ind. Acad. Sci.* 92:441–445 (1983).
35. Kanwar, R. S. "Effect of Tillage Systems on the Variation of Soil-Water Tensions and Soil-Water Content," *Trans. Am. Soc. Agric. Eng.* 32:605–610 (1989).
36. Edwards, J. H., D. L. Thurlow and J. T. Eason. "Influence of Tillage and Crop Rotation on Yields of Corn, Soybean, and Wheat," *Agron. J.* 80:76–80 (1988).
37. Lal, R., T. J. Logan and N. R. Fausey. "Long-Term Tillage and Wheel Traffic Effects on a Poorly Drained Mollic Ochraqualf in Northwest Ohio. II. Infiltrability, Surface Runoff, Sub-Surface Flow and Sediment Transport," *Soil Tillage Res.* 14:359–373 (1989).
38. Logan, T. J., R. Lal and W. A. Dick. "Tillage Systems and Soil Properties in North America," *Soil Tillage Res.* 20:241–270 (1991).
39. Kladivko, E. J., D. R. Griffith and J. V. Mannering. "Conservation Tillage Effects on Soil Properties and Yield of Corn and Soya Beans in Indiana," *Soil Tillage Res.* 8:277–287 (1986).
40. Roseberg, R. J., and E. L. McCoy. "Time Series Analysis for Statistical Inferences in Tillage Experiments," *Soil Sci. Soc. Am. J.* 52:1771–1776 (1988).
41. Benjamin, J. G., A. D. Blaylock, H. J. Brown and R. M. Cruse. "Ridge Tillage Effects on Simulated Water and Heat Transport," *Soil Tillage Res.* 18:167–180 (1990).
42. Gupta, S. C., J. K. Radke, J. B. Swan and J. F. Moncrief. "Predicting Soil Temperatures under a Ridge-Furrow System in the U.S. Corn Belt," *Soil Tillage Res.* 18:145–165 (1990).
43. Gupta, S. C., B. Lowery, J. F. Moncrief and W. E. Larson. "Modelling Tillage Effects on Soil Physical Properties," *Soil Tillage Res.* 20:293–318 (1991).

44. Gupta, S. C., W. E. Larson and D. R. Linden. "Tillage and Surface Residue Effects on Soil Upper Boundary Temperatures," *Soil Sci. Soc. Am. J.* 47:1212–1218 (1983).
45. Fausey, N. R. "Tillage-Drainage Interaction on a Clermont Soil," *Trans. Am. Soc. Agric. Eng.* 27:403-406 (1984).
46. Fausey, N. R. "Impact of Cultural Practices on Drainage of Clay Soils," in *Drainage Design and Management. Proc. 5th National Drainage Symp.*, Chicago, December 14–15, (St. Joseph, MI: American Society of Agricultural Engineers, 1987).
47. Griffith, D. R., E. J. Kladivko, J. V. Mannering, T. D. West and S. D. Parsons. "Long-Term Tillage and Rotation Effects on Corn Growth and Yield on High and Low Organic Matter, Poorly Drained Soils," *Agron. J.* 80:599–605 (1988).
48. Griffith, D. R., S. D. Parsons and J. V. Mannering. "Mechanics and Adaptability of Ridge-Planting for Corn and Soybean," *Soil Tillage Res.* 18:113–126 (1990).
49. Pfost, D. L., Frisby, J. C. and A. L. Thompson. "Ridge-Till/No-Till Soil Compaction Study," Papers No. 90-104, (Columbia, MO: American Society of Agricultural Engineers, 1990).
50. Fausey, N. R., and R. Lal. "Drainage-Tillage Effects on Crosby-Kokomo Soil Association in Ohio. I. Effects on Stand and Corn Grain Yield," *Soil Technol.* 2:359–370; "II. Soil Temperature Regime and Infiltrability," *Soil Technol.* 2:371–383 (1989).
51. Blevins, R. L., W. W. Frye and M. S. Smith. "The Effects of Conservation Tillage on Soil Properties," *A Systems Approach to Conservation Tillage,* F. M. D'Itri (Ed)., (Chelsea, MI: Lewis Publishers, 1985), pp. 99–120.
52. Dick, W. A., and T. C. Daniel. "Soil Chemical and Biological Properties as Affected by Conservation Tillage: Environmental Implications," in *Effects of Conservation Tillage on Groundwater Quality. Nitrates and Pesticides,* T. J. Logan, J. M. Davidson, J. L. Baker and M. R. Overcash (Eds.). (Chelsea, MI: Lewis Publishers, 1987), pp. 125–147.
53. Blevins, R. L., M. S. Smith, G. W. Thomas and W. W. Frye. "Influence of Conservation Tillage on Soil Properties," *J. Soil Water Conserv.* 38:301–305 (1983).
54. Kells, J. J., C. E. Rieck, R. E. Blevins and W. M. Muir. "Atrazine Dissipation as Affected by Surface pH and Tillage," *Weed Sci.* 28:101–104 (1980).
55. Dick, W. A. "Organic Carbon, nitrogen, and Phosphorus Concentrations and pH in Soil Profiles as Affected by Tillage Intensity," *Soil Sci. Soc. Am. J.* 47:102–107 (1983).
56. Havlin, J. L., D. E. Kissel, L. D. Maddux, M. M. Claassen and J. H. Long. "Crop Rotation and Tillage Effects on Soil Organic Carbon and Nitrogen," *Soil Sci. Soc. Am. J.* 54:448–452 (1990).
57. Rasmussen, P. E., and C. R. Rohde. "Long-Term Tillage and Nitrogen Fertilization Effects on Organic Nitrogen and Carbon in a Semiarid Soil," *Soil Sci. Soc. Am. J.* 52:1114–1117 (1988).
58. Stearman, G. K., R. J. Lewis, L. J. Tortorelli and D. D. Tyler. "Characterization of Humic Acid from No-Tilled and Tilled Soils Using C-13 Nuclear Magnetic Resonance," *Soil Sci. Soc. Am. J.* 53:744–749 (1989).
59. Guertal, E. A., D. J. Eckert, S. J. Traina and T. J. Logan. "Differential Phosphorus Retention in Soil Profiles under No-Till Crop Production," *Soil Sci. Soc. Am. J.* 55:410–413 (1991).

60. Logan, T. J. and J. R. Adams. "The Effects of Conservation Tillage on Phosphate Transport From Agricultural Land," Technical Report Series, Lake Erie Wastewater Management Study (Buffalo, NY: Corps of Engineers, 1981).
61. Oloya, T. O., and T. J. Logan. "Phosphate Desorption from Soils and Sediments with Varying Levels of Extractable Phosphate," *J. Environ. Qual.* 9:526–531 (1980).
62. Sharpley, A. N., S. J. Smith, O. R. Jones, W. A. Berg and G. A. Coleman. "The Transport of Bioavailable Phosphorus in Agricultural Runoff," *J. Environ. Qual.* 21:30–35 (1992).
63. Lal, R., T. J. Logan and N. R. Fausey. "Long-Term Tillage Effects on a Mollic Ochraqualf in Northwest Ohio. III. Soil Nutrient Profile," *Soil Tillage Res.* 15:371–382 (1990).
64. Edwards, W. M., M. J. Shipitalo, W. A. Dick and L. B. Owens. "Rainfall Intensity Affects Transport of Water and Chemicals through Macropores in No-Till Soil," *Soil Sci. Soc. Am. J.* 56:52–58 (1992).
65. Karathanasis, A. D., and K. L. Wells. "Conservation Tillage Effects on the Potassium Status of Some Kentucky Soils," *Soil Sci. Soc. Am. J.* 54:800–806 (1990).
66. Hendrix, P. F., R. W. Parmalee, D. A. Crossby, D. C. Coleman, E. P. Odom and P. M. Groffman. "Detritus Food Webs in Conventional and No-Tillage Agroecoystems," *BioScience* 36:374–380 (1986).
67. Doran, J. W. "Soil Microbial and Biochemical Changes Associated with Reduced Tillage," *Soil Sci. Soc. Am. J.* 44:765–771 (1980).
68. Staley, T. E., W. M. Edwards, C. L. Scott and L. B. Owens. "Soil Microbial Biomass and Organic Component Alterations in a No-Tillage Chronosequence," *Soil Sci. Soc. Am. J.* 52:998–1005 (1988).
69. Dick, W. A. "Influence of Long-Term Tillage and Rotation Combinations on Soil Enzyme Activities," *Soil Sci. Soc. Am. J.* 48:569–574 (1984).
70. Robbins, S. G., and R. D. Voss. "Acidic Zones From Ammonia Application in Conservation Tillage Systems," *Soil Sci. Soc. Am. J.* 53:1256–1263 (1989).
71. Rice, C. W., and M. S. Smith. "Denitrification in No-Till and Plowed Soils," *Soil Sci. Soc. Am. J.* 46:1168–1173 (1982).
72. Rice, C. W., M. S. Smith and R. L. Blevins. "Soil Nitrogen Availability after Long-Term Continuous No-Tillage and Conventional Tillage Corn Production," *Soil Sci. Soc. Am. J.* 50:1206–1210 (1986).
73. Fox, R. H., and V. A. Bendel. "Nitrogen Utilization with No-Tillage." in *No-Tillage and Surface Tillage Agriculture,* M. A. Sprague and G. B. Triplett (Eds.) (New York: John Wiley & Sons, 1986), pp. 117–148.
74. Beyrouty, C. A., L. E. Sommers and D. W. Nelson. "Ammonia Volatilization from Surface-Applied Urea as Affected by Several Phosphoroamide Compounds," *Soil Sci. Soc. Am. J.* 52:1173–1178 (1988).
75. Timmons, D. R., and R. M. Cruse. "Effect of Fertilization Method and Tillage on Nitrogen-15 Recovery by Corn," *Agron. J.* 82:777–784 (1990).
76. Holden, L. R., J. A. Graham, R. W. Whitmore, W. J. Alexander, R. W. Pratt, S. K. Liddle and L. L. Piper. "Results of the National Alachlor Well Water Survey," *Environ. Sci. Technol.* 26:935–943 (1992).
77. Kross, B. C., M. I. Selim, G. R. Hallberg, D. R. Bruner and K. Cherryholmes. "Pesticide Contamination of Private Well Water, a Growing Rural Health Concern," *Environ. Int.* 18:231–241 (1992).

78. Baker, J. L. "Hydrologic Effects of Conservation Tillage and Their Importance Relative to Water Quality" in *Effects of Conservation Tillage on Groundwater Quality*, T. J. Logan, J. M. Davidson, J. L. Baker and M. R. Overcash (Eds.) (Chelsea, MI: Lewis Publishers, 1987), pp. 113–124.
79. Griffith, D. R., J. V. Mannering and J. E. Box. "Soil and Moisture Management with Reduced Tillage," in *No-Tillage and Surface-Tillage Agriculture — The Tillage Revolution*, M. A. Sprague and G. B. Triplett (Eds.) (New York: Wiley-Interscience, 1986), pp. 19–57.
80. Hinkle, M. K. "Problems with Conservation Tillage," *J. Soil Water Conserv.* 38:201–206 (1983).
81. Donigian, A. S., Jr., and R. F. Carsel. "Modeling the Impact of Conservation Tillage Practices on Pesticide Concentrations in Ground and Surface Waters," *Environ. Toxicol. Chem.* 6:241–250 (1987).
82. Hall, J. K., M. R. Murray and N. L. Hartwig. "Herbicide Leaching and Distribution in Tilled and Untilled Soil," *J. Environ. Qual.* 18:439–445 (1989).
83. Isensee, A. R., R. G. Nash and C. S. Helling. "Effect of Conventional vs. No-Tillage on Pesticide Leaching to Shallow Groundwater," *J. Environ. Qual.* 19:434–440 (1990).
84. Fermanich, K. J., and T. C. Daniel. "Pesticide Mobility and Persistence in Microlysimeter Soil Columns from a Tilled and No-Tilled Plot," *J. Environ. Qual.* 20:195–202 (1991).
85. Clay, S. A., W. C. Koskinen and P. Carlson. "Alachlor Movement through Intact Soil Columns Taken from Two Tillage Systems," *Weed Technol.* 5:485–489 (1991).
86. Kanwar, R. S., J. L. Baker and D. G. Baker. "Tillage and Split N-Fertilization Effects on Subsurface Drainage Water Quality and Crop Yields," *Trans. Am. Soc. Agric. Eng.* 31:453–461 (1988).
87. Chichester, F. W., and S. J. Smith. "Disposition of 15N-Labeled Fertilizer Nitrate Applied during Corn Culture in Field Lysimeters," *J. Environ. Qual.* 7:227–23 (1978).
88. Gaynor, J. D., D. C. MacTavish, and W. I. Findlay. "Surface and Subsurface Transport of Atrazine and Alachlor from a Brookston Clay Loam under Continuous Corn Production," *Arch. Environ. Contam. Toxicol.* 23:240–245 (1992).
89. Ehlers, W. "Observations on Earthworm Channels and Infiltration on Tilled and Untilled Loess Soil," *Soil Sci.* 119:242–248 (1975).
90. Gantzer, C. J., and G. R. Blake. "Physical Characteristics of Le Sueur Clay-Loam Soil Following No-Till and Conventional Tillage," *Agron. J.* 70:853–857 (1978).
91. Moran, C. J., A. J. Koppi, B. W. Murphy and A. B. McBratney. "Comparison of the Macropore Structure of a Sandy Loam Surface Soil Horizon Subjected to Two Tillage Treatments," *Soil Use Manage.* 4:96–102 (1988).
92. Pagliai, M., M. La Marca, and G. Lucamante. "Micromorphometric and Micromorphological Investigations of a Clay Loam Soil in Viticulture under Zero and Conventional Tillage," *J. Soil Sci.* 34:391–403 (1983).
93. Shipitalo, M. J., and R. Protz. "Comparison of Morphology and Porosity of a Soil under Conventional and Zero Tillage," *Can. J. Soil Sci.* 67:445–456 (1987).
94. Carter, M. R. "Temporal Variability of Soil Macroporosity in a Fine Sandy Loam under Mouldboard Ploughing and Direct Drilling," *Soil Tillage Res.* 12:37–51 (1988).

95. Heard, J. R., E. J. Kladivko and J. V. Mannering. "Soil Macroporosity, Hydraulic Conductivity and Air Permeability of Silty Soils under Long-Term Conservation Tillage in Indiana," *Soil Tillage Res.* 11:1–18 (1988).
96. Pikul, J. L., J. F. Zuzel and R. E. Ramig. "Effect of Tillage-Induced Soil Macroporosity on Water Infiltration," *Soil Tillage Res.* 17:153–165 (1990).
97. House, G. J., and R. W. Parmelee. "Comparison of Soil Arthropods and Earthworms from Conventional and No-Tillage Agroecosystems," *Soil Tillage Res.* 5:351–360 (1985).
98. Edwards, W. M., R. R. van der Ploeg and W. Ehlers. "A Numerical Study of the Effects of Noncapillary-Sized Pores upon Infiltration," *Soil Sci. Soc. Am. J.* 43:851–856 (1979).
99. Beven, K., and P. Germann. "Macropores and Water Flow in Soils," *Water Resource Res.* 18:1311–1325 (1982).
100. Smettem, K. R. J., and N. Collis-George. "The Influence of Cylindrical Macropores on Steady-State Infiltration in a Soil under Pasture," *J. Hydrol.* 79:107–114 (1985).
101. Watson, K. W., and R. J. Luxmoore. "Estimating Macroporosity in a Forest Watershed by Use of a Tension Infiltrometer," *Soil Sci. Soc. Am. J.* 50:578–582 (1986).
102. Roth, C. H., and M. Joschko. "A Note on the Reduction of Runoff from Crusted Soils by Earthworm Burrows and Artificial Channels," *Z. Pflanzenernahr. Bodenk.* 154:101–105 (1991).
103. Sharpley, A. N., J. K. Syers and J. A. Springett. "Effect of Surface-Casting Earthworms on the Transport of Phosphorus and Nitrogen in Surface Runoff from Pasture," *Soil Biol. Biochem.* 11:459–462 (1979).
104. Bouwer, H. B. "Research Needs in Movement of Agricultural Chemicals to Groundwater," in *Irrigation and Drainage, Proc. 1990 ASCE Conf.*, S. C. Harris (Ed.) Durango, CO, July 11 to 13, 1990, pp. 168–174.
105. Priebe, D. L., and A. M. Blackmer. "Preferential Movement of Oxygen-18-Labeled Water and Nitrogen-15-Labeled Urea through Macropores in a Nicollet Soil," *J. Environ. Qual.* 18:66–72 (1989).
106. Shipitalo, M. J., and W. M. Edwards. "Seasonal Patterns of Water and Chemical Movement in Tilled and No-Till Column Lysimeters," *Soil Sci. Soc. Am. J.* 57: 218–223 (1993).
107. Bouma, J., C. F. M. Belmans and L. W. Dekker. "Water Infiltration and Redistribution in a Silt Loam Subsoil with Vertical Worm Channels," *Soil Sci. Soc. Am. J.* 46:917–921 (1982).
108. Shaffer, K. A., D. D. Fritton and D. E. Baker. "Drainage Water Sampling in a Wet, Dual-Pore System," *J. Environ. Qual.* 8:241–246 (1979).
109. Edwards, W. M., L. D. Norton and C. E. Redmond. "Characterizing Macropores that Affect Infiltration into Nontilled Soil," *Soil Sci. Soc. Am. J.* 52:483–487 (1988).
110. Torrento, J. R., and A. Sole-Benet. "Soil Macroporosity Evaluated by a Fast Image-Analysis Technique in Differently Managed Soils," *Commun. Soil Sci. Plant Anal.* 23:1229–1244 (1992).
111. Anderson, S. H., R. L. Peyton and C. J. Gantzer. "Evaluation of Constructed and Natural Soil Macropores Using X-Ray Computed Tomography," *Geoderma* 46:13–29 (1990).

112. Ela, S. D., S. C. Gupta and W. J. Rawls. "Macropore and Surface Seal Interactions Affecting Water Infiltration into Soil," *Soil Sci. Soc. Am. J.* 56:714–72 (1992).
113. Trojan, M. D., and D. R. Linden. "Microrelief and Rainfall Effects on Water and Solute Movement in Earthworm Burrows," *Soil Sci. Soc. Am. J.* 56:727–733 (1992).
114. Edwards, W. M., M. J. Shipitalo, L. B. Owens and L. D. Norton. "Water and Nitrate Movement in Earthworm Burrows within Long-Term No-Till Corn Fields," *J. Soil Water Conserv.* 44:240–243 (1989).
115. Barbee, G. C., and K. W. Brown. "Comparison between Suction and Free-Drainage Soil Solution Samplers," *Soil Sci.* 141:149–154 (1986).
116. Boll, J., J. S. Selker, B. M. Nijssen, T. S. Steenhuis, J. Van Winkle and E. Jolles. "Water Quality Sampling Under Preferential Flow Conditions," in *Lysimeters for Evapotranspiration and Environmental Measurements, ASCE Proc.* R. G. Allen et al., (Eds.), Honolulu, 1991, pp. 290–298.
117. Holder, M., K. W. Brown, J. C. Thomas, D. Zabcik and H. E. Murray. "Capillary-Wick Unsaturated Zone Soil Pore Water Sampler," *Soil Sci. Soc. Am. J.* 55:1195–1202 (1991).
118. Edwards, J. H., C. W. Wood, D. L. Thurlow and M. E. Ruf. "Tillage and Crop Rotation Effects on Fertility Status of a Hapludult Soil," *Soil Sci. Soc. Am. J.* 56:1577–1582 (1992).
119. Coles, N., and S. Trudgill. "The Movement of Nitrate Fertiliser from the Soil Surface to Drainage Waters by Preferential Flow in Weakly Structured Soils, Slapton S. Devon," *Agric. Ecosyst. Environ.* 13:241–259 (1985).
120. Francis, G. S., K. C. Cameron, and R. A. Kemp. "A Comparison of Soil Porosity and Solute Leaching after Six Years of Direct Drilling or Conventional Cultivation," *Aust. J. Soil Res.* 26:637–649 (1988).
121. Germann, P. F., W. M. Edwards and L. B. Owens. "Profiles of Bromide and Increased Soil Moisture after Infiltration into Soils with Macropores," *Soil Sci. Soc. Am. J.* 48:237–244 (1984).
122. Smith, M. S., G. W. Thomas, R. E. White and D. Ritonga. "Transport of *Escherichia coli* through Intact and Disturbed Soil Columns," *J. Environ. Qual.* 14:87–91 (1985).
123. Everts, C. J., and R. S. Kanwar. "Estimating Preferential Flow to a Subsurface Drain with Tracers," *Trans. Am. Soc. Agric. Eng.* 33:451–457 (1990).
124. White, R. H., A. D. Worsham and U. Blum. "Allelopathic Potential of Legume Debris and Aqueous Extracts," *Weed Sci.* 37:674–679 (1989).
125. White, R. E., J. S. Dyson, Z. Gerstl and B. Yaron. "Leaching of Herbicides through Undisturbed Cores of a Structured Clay Soil," *Soil Sci. Sci. Am. J.* 50:277–283 (1986).
126. Bicki, T. J., and L. Guo. "Tillage and Simulated Intensity Effect on Bromide Movement in an Argiudoll," *Soil Sci. Soc. Am. J.* 55:794–799 (1991).
127. Bengtson, R. L., L. M. Southwick, G. H. Willis and C. E. Carter. "The Influence of Subsurface Drainage Practices on Herbicide Losses," *Trans. Am. Soc. Agric. Eng.* 33:415–418 (1990).
128. Southwick, L. M., G. H. Willis, R. L. Bengston and T. J. Lormand. "Atrazine and Metolachlor in Subsurface Drain Water in Louisiana," *J. Irrig. Drain. Eng.* 116:16–23 (1990).

129. White, R. E. "The Influence of Macropores on the Transport of Dissolved and Suspended Matter through Soil," *Adv. Soil Sci.* 3:95–120 (1985).
130. Everts, C. J., R. S. Kanwar, E. C. Alexander, Jr. and S. C. Alexander. "Comparison of Tracer Mobilities under Laboratory and Field Conditions," *J. Environ. Qual.* 18:491–498 (1989).
131. Power, J. F., and J. S. Schepers. "Nitrate Contamination of Groundwater in North America," *Agric. Ecosyst. Environ.* 26:165–187 (1989).
132. Baker, J. L., and J. M. Laflen. "Water Quality Consequences of Conservation Tillage," *J. Soil Water Conserv.* 38:186–193 (1983).
133. Wauchope, R. D. "The Pesticide Content of Surface Water Draining from Agricultural Fields — A Review," *J. Environ. Qual.* 7:459–472 (1978).
134. Whipkey, R. Z. "Theory and Mechanics of Subsurface Stormflow," in *Int. Symp. Forest Hydrol.* W. E. Sopper and H. W. Lull (Eds.) (New York: Pergamon Press, 1967), pp. 255–260.
135. Alva, A. K., and M. Singh. "Use of Adjuvants to Minimize Leaching of Herbicides in Soil," *Environ. Manage.* 15:263–267 (1991).
136. Boydston, R. A. "Controlled Release Starch Granule Formulations Reduce Herbicide Leaching in Soil Columns," *Weed Technol.* 6:317–321 (1992).
137. Fleming, G. F., L. M. Wax and F. W. Simmons. "Leachability and Efficacy of Starch-Encapsulated Atrazine," *Weed Technol.* 6:297–302 (1992).
138. Andreini, M. S., and T. S. Steenhuis. "Preferential Paths of Flow under Conventional and Conservation Tillage," *Geoderma* 46:85–102 (1990).
139. Harrold, L. L., G. B. Triplett, Jr. and R. E. Youker. "Less Soil and Water Loss from No-Tillage Corn," *Ohio Rep.* 52(2):22–23 (1967).
140. Triplett, G. B., Jr., D. M. Van Doren, Jr. and B. L. Schmidt. "Effect of Corn (*Zea mays* L.) Stover Mulch on No-Tillage Corn Yield and Water Infiltration," *Agron. J.* 236–239 (1968).
141. Griffith, D. R., J. V. Mannering, H. M. Galloway, S. D. Parsons and C. B. Richey. "Effect of Eight Tillage-Planting Systems on Soil Temperature, Percent Stand, Plant Growth, and Yield of Corn on Five Indiana Soils," *Agron. J.* 65:321–326 (1973).
142. Mock, J. J., and D. C. Erbach. "Influence of Conservation-Tillage Environments on Growth and Productivity of Corn," *Agron. J.* 69:337–340 (1977).
143. Van Doren, D. M., Jr., and G. B. Triplett, Jr. "Mechanism of Corn (*Zea mays* L.) Response to Cropping Practices Without Tillage," *Ohio Agric. Res. Dev. Ctr. Res. Circ.* 169 (1969).
144. Van Doren, D. M., Jr., and G. B. Triplett, Jr. "Mulch and Tillage Relationships in Corn Culture," *Soil Sci. Soc. Am. Proc.* 37:766–769 (1973).
145. Van Doren, D. M., Jr., G. B. Triplett, Jr. and J. E. Henry. "Long-Term Influence of Tillage, Rotation, Soil on Corn Yield," *Ohio Rep.* 60(5):80–82 (1975).
146. Van Doren, D. M., Jr., G. B. Triplett, Jr. and J. E. Henry. "Influence of Long Term Tillage, Crop Rotation, and Soil Type Combinations on Corn Yield," *Soil Sci. Soc. Am. J.* 40:100–105 (1976).
147. Triplett, G. B., Jr., D. M. Van Doren, Jr. and W. H. Johnson. "Non-Plowed, Strip Tilled Corn Culture," *Trans. Am. Soc. Agric. Eng.* 7:105–107 (1964).
148. "National Survey of Conservation Tillage Practices" (West Lafayette, IN: Conservation Technology Information Center, 1991).
149. Eckert, D. J. "Tillage × Planting Date Interactions in Corn Production," *Agron. J.* 76:580–582 (1984).

150. Graffis, D. W., M. D. McGlamery, W. O. Scott, W. R. Oschwald, R. L. Nelson, E. L. Knake, R. G. Hoeft and M. D. Thorne. *Illinois Agronomy Handbook*. Coop. Ext. Serv. Circ. 1104 (Chicago: University of Illinois, 1975).
151. Burrows, W. C., and W. E. Larson. "Effect of Amount of Mulch on Soil Temperature and Early Growth of Corn," *Agron. J.* 54:19–23 (1963).
152. Willis, W. O., W. E. Larson and D. Kirkham. "Corn Growth as Affected by Soil Temperature and Mulch," *Agron. J.* 49:323–328 (1957).
153. Eckert, D. J. "Evaluation of Ridge Planting Systems on a Poorly Drained Lake Plain Soil," *J. Soil Water Conserv.* 42:208–211 (1987).
154. Hartman, G. L., R. P. McClary, J. B. Sinclair and J. W. Hummel. "Effects of Tillage Systems on Corn Stalk Rot," *Phytopathology* (Abstr.)73:843 (1983).
155. Lipps, P. E., and I. W. Deep. "Influence of Tillage and Crop Rotation on Yield, Stalk Rot, and Recovery of *Fusarium* and *Trichoderma* spp. from Corn," *Plant Dis.* 75:828–833 (1991).
156. Eckert, D. J. "Effects of Reduced Tillage on the Distribution of Soil pH and Nutrients in Soil Profiles," *J. Fert. Iss.* 2:86–90 (1985).
157. Eckert, D. J. "UAN Managment Practices for No-Tillage Corn Production," *J. Fert. Iss.* 4:13–18 (1987).
158. Eckert, D. J., W. A. Dick and J. W. Johnson. "Response of No-Tillage Corn Grown in Corn and Soybean Residues to Several Nitrogen Fertilizer Sources," *Agron. J.* 78:231–235 (1986).
159. Griffith, D. R. "Fertilization and No-Plow Tillage," in *Proc. Indiana Plant Food and Agricultural Chemistry Conf.* (W. Lafayette, IN: Purdue University, December 17 to 18, 1974).
160. Keller, G. D., and D. B. Mengel. "Ammonia Volatilization from Nitrogen Fertilizers Surface Applied to No-Till Corn," *Soil Sci. Soc. Am. J.* 50:1060–1063 (1986).
161. Mengel, D. B., D. W. Nelson and D. M. Huber. "Placement of Nitrogen Fertilizers for No-Till and Conventional Till Corn," *Agron. J.* 74:515–518 (1982).
162. Crookston, R. K., and J. E. Kurle. "Corn Residue Effect on the Yield of Corn and Soybean Grown in Rotation," *Agron. J.* 81:229–232 (1989).
163. Johnson, N. C., P. J. Copeland, R. K. Crookston and F. L. Pfleger. "Mycorrhizae: Possible Explanation for Yield Decline with Continuous Corn and Soybean," *Agron. J.* 84:387–390 (1992).
164. Turco, R. F., M. Bischoff, D. P. Breakwell and D. R. Griffith. "Contribution of Soil-Borne Bacteria to the Rotation Effect in Corn," *Plant Soil* 122:115–120 (1990).
165. *Alternative Agriculture,* (Washington, DC: National Research Council/National Academy Press, 1989), pp. 138–140.
166. Cook, R. J. "Root health: Importance and Relationship to Farming Practices," in *Organic Farming: Current Technology and Its Role in Sustainable Agriculture.* Spec. Publ. 46, D. F. Bezdicek and J. F. Powers (Eds.) (Madison, WI: American Society of Agronomy, 1984), pp. 111–127.
167. Langdale, G. W., R. L. Wilson and R. R. Bruce. "Cropping Frequencies to Sustain Long-Term Conservation Tillage Systems," *Soil Sci. Soc. Am. J.* 54:193–198 (1990).
168. Wagger, M. G. "Cover Crop Management and Nitrogen Rate in Relation to Growth and Yield on No-Till Corn," *Agron. J.* 81:533–538 (1989).

169. Blevins, R. L., J. H. Herbek and W. W. Frye. "Legume Cover Crops as a Nitrogen Source for No-Till Corn and Grain Sorghum," *Agron. J.* 82:769–772 (1990).
170. Cambardella, C. A., and S. J. Corak. "Seasonal Dynamics of Soil Organic N and Nitrate N with and without a Cover Crop," in *Agronomy Abstracts* (Madison, WI: American Society of Agronomy, 1992), p. 251.
171. Ngalla, C. F., and D. J. Eckert. "Wheat-Red Clover Interseedings as an N Source for No-Tillage Corn," in *The Role of Legumes in Conservation Tillage Systems,* J. F. Power (Ed.) (Ankeny, IA: Soil Conservation Society of America, 1987).
172. Brown, H. J., R. M. Cruse and T. S. Colbin. "Tillage System Effects on Crop Growth and Production Costs for a Corn-Soybean Rotation," *J. Prod. Agric.* 2:273–279 (1989).
173. Allmaras, R. R., G. W. Langdale, P. W. Unger, R. H. Dowdy and D. M. Van Doren. "Adoption of Conservation Tillage and Associated Planning Systems," in *Soil Management for Sustainability*, R. Lal and F. J. Pierce (Eds.) (Ankeny, IA: Soil and Water Conservation Society, 1991), pp. 53–83.
174. Eckert, D. J. "Rye Cover Crops for No-Tillage Corn and Soybean Production," *J. Prod. Agric.* 1:207–210 (1988).
175. Eckert, D. J. "Chemical Attributes of Soils Subjected to No-Till Cropping with Rye Cover Crops," *Soil Sci. Soc. Am. J.* 55:405–409 (1991).
176. Lal, R., E. Regnier, D. J. Eckert, W. M. Edwards and R. Hammond. "Expectations of Cover Crops for Sustainable Agriculture," in *Cover Crops for Clean Water* (Ankeny, IA: Soil and Water Conservation Society, 1991) pp. 1–11.
177. Lal, R. "Tillage and Agricultural Sustainability," *Soil Tillage Res.* 20:133–146 (1991).
178. Aldrich, S. R., W. O. Scott and E. R. Long. *Modern Corn Production,* 2nd ed. (Champaign, IL: A & L Publications, 1982), p. 378.
179. Lal, R., and H. Tanaka. "Simulated Harvest Traffic Effects on Corn, Oats, and Soybean Yields in Western Ohio," *Soil Tillage Res.* 2:65–78 (1992).
180. Kanwar, R. S., J. L. Baker and J. M. Laflen. "Nitrate Movement through the Soil Profile in Relation to Tillage System and Fertilizer Application Method," *Trans. Am. Soc. Agric. Eng.* 28:1802–1807 (1985).
181. Starr, J. L., and D. E. Glotfelty. "Atrazine and Bromide Movement through a Silt Loam Soil," *J. Environ. Qual.* 19:552–558 (1990).
182. Steenhuis, T. S., W. Staubitz, M. S. Andreini, J. Surface, T. L. Richard, R. Paulsen, N. B. Pickering, J. R. Hagerman and L. D. Geohring. "Preferential Movement of Pesticides and Tracers in Agricultural Soils," *J. Irrig. Drain. Eng.* 116:50–66 (1990).
183. Tyler, D. D., and G. W. Thomas. "Lysimeter Measurements of Nitrate and Chloride Losses from Soil Under Conventional and No-Tillage Corn," *J. Environ. Qual.* 6:63–66 (1977).
184. "National Crop Residue Management Survey," (West Lafayette, IN: Conservation Technology Information Center, 1992), p. 42.
185. Schmitthenner, A. F. Unpublished data.

Humid Micro-Thermal to Humid Meso-Thermal Climates

CHAPTER 5

Reduced Cultivation and Direct Drilling for Cereals in Great Britain

Dudley G. Christian
Rothamsted Experimental Station; Harpenden, Herts, England

Bruce C. Ball
Scottish Agricultural College; Edinburgh, Scotland

TABLE OF CONTENTS

0-87371-571-3/94/$0.00+$.50

5.1. INTRODUCTION

The past four to five decades have seen a steady evolution of agriculture toward separate specializations in arable or livestock production. As a result, many arable farmers no longer needed to use mixed crop rotations such as the Norfolk four-course (grass, wheat, root crops, barley) and in arable areas cereals were grown more frequently and livestock farming was reduced. This led, during the 1970s and early 1980s, to an increasing proportion of straw becoming surplus to requirement and much was burned. During this period agriculture faced a reducing and more expensive labor force as well as increasing costs of machinery operations.[1] Thus, studies of alternative, simplified methods of tillage were topical as a means of savings in the cost of establishing crops. The introduction of new and more efficient herbicides in the 1960s made it possible for researchers to reappraise work carried out in the 1930s by Russell and Keen at Rothamsted. They concluded from a series of field experiments that the primary function of plowing was to control weeds and that to omit plowing in weed-free conditions would not result in a yield reduction.[2]

The terms *reduced cultivation* and *direct drilling* can have several meanings and require better definition. They are usually compared to the conventional system of moldboard plowing to a depth of 20 to 25 cm, followed by one or more secondary cultivations to a depth of 10 cm, and finally drilling using a lightweight drill. Direct drilling, also known as zero tillage or no-till, refers to sowing into soil left uncultivated since the previous harvest. Tillage is confined to that provided by the drill coulters, *viz.*, to the depth to which seed is to be sown. In practice, many farmers cultivate the soil surface before drilling to improve tilth and subsequent seed covering.

Reduced cultivation broadly defines methods of cultivation which do not include the use of a conventional moldboard plow. Implement working depth is normally but not necessarily shallower than conventional plowing. In this context, the term reduced cultivation applies to the component operation(s) of the system of land preparation which is individually quicker than plowing. In some reduced cultivation systems, several passes with shallow working implements operated at a high forward speed are used such that there is not necessarily a saving in time as compared to the conventional system. Shallow tillage describes cultivation of soil to depths not greater than 10 cm with cultivation implements which may include the shallow plow.

In 1981, Allen[3] reported the results of many experiments on reduced tillage systems and direct drilling carried out in the 1960s and noted that (1) direct drilling offered the prospect of time saved by avoidance of plowing, and therefore a greater area of land could be prepared and sown; (2) there was relatively little difference in yield compared to plowing where sufficient nitrogen was applied, (3) experiments showed that where crops were sown sequentially by direct drilling, yields in the second and third year were frequently heavier than in the first year. Many studies at this time identified the need to control weeds, especially grass weeds, and that drill design required further development for the narrow row-spaced crops (15 to 20 cm) used normally in Great Britain.

Reducing cultivation was seen to be of greatest potential benefit on soils difficult to cultivate, such as heavy-textured soils, and in the late 1960s and early 1970s increasing attention was given to experimenting with reduced cultivation and direct drilling on these soils.[4] However, many heavy soils, particularly those that do not shrink on drying, have low infiltration and poor drainage and are not suited to direct drilling.

The principal motivation for the use of direct drilling or reduced cultivation in Great Britain is to allow the maximum area of crop, particularly the high-yielding autumn varieties, to be sown in the limited period when the land is suitable. Alternatively, such techniques allow crop establishment where the soil type is not well suited to tillage, for example, coarse-textured or stony soils. Thus, direct drilling and reduced tillage in Great Britain differs from the concept of conservation tillage in North America which relies on about 30% of crop residues being retained at the surface. However, in Great Britain the presence of surface residues is minimized because the residues create major drilling problems via the narrow row spacing required for cereals. Such a surface mulch can also depress seedbed temperatures, which may influence winter growth. The use of such techniques primarily for the protection of soil against erosion or conservation of moisture has always been limited in Great Britain. Several farmers on sandy soils near the coasts of eastern Scotland have direct drilled as a means of limiting wind erosion, but did not persist with the technique because of weed problems.[5]

The adoption of reduced tillage techniques to cereal production has received the greatest attention from researchers in recent years and discussion in this chapter is confined to this subject area. Some work has been summarized for other crops: fodder root crops,[6] stubble brassicas,[7] oilseed rape, and root crops.[8]

5.2. CLIMATE AND AGRICULTURAL REGIONS OF GREAT BRITAIN

The British Isles are between latitudes 50° to 61°N. The climate is maritime and variable. Rain is likely in all months of the year, and compared to a continental climate, the amplitude of winter to summer temperature is small. There is a north-south temperature gradient, which affects the length of the growing season.[9,10] Rainfall varies most on an east-west division of the islands,

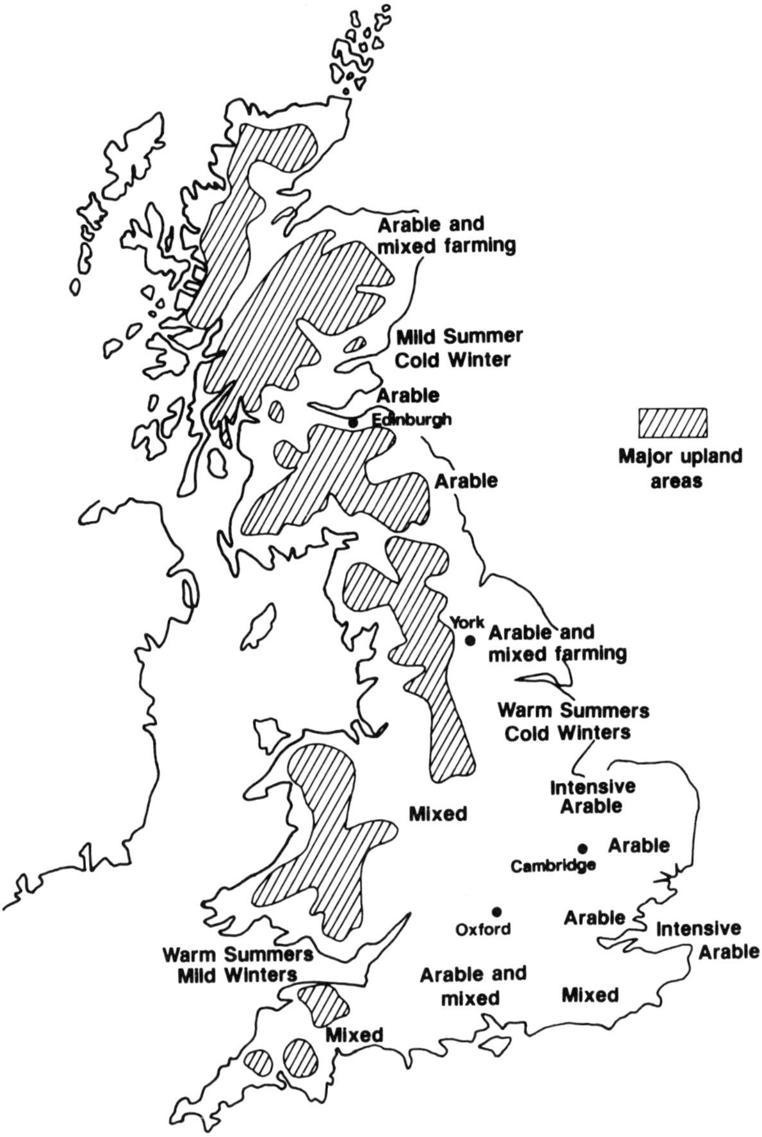

Figure 5.1. Principal farming activities in different areas of Great Britain.

with the least rainfall occurring in the eastern areas. As a consequence most arable land is found in the east and grassland farming in the west and north (Figure 5.1). Arable crops, especially cereals, are of major importance (Table 5.1).[11,12] The upland areas are unsuitable for cultivation.

Table 5.1. Land Used by Agriculture and the Proportion Occupied by the Principal Arable Crops (1000× ha)

	1970	1980	1990[a]
Area of agricultural land[b]	19,524	18,953	18,525
Arable	7,599	6,996	6,711
Wheat	2,595	1,441	2,042
Yield t/ha	4.2	5.9	6.9
Barley	5,542	2,330	1,522
Yield t/ha	3.4	4.4	5.2
Oilseed rape	10	92	397
Yield t/ha	1.9	3.3	3.1

Source: Adapted from "Annual Review of Agriculture," Cmnd. 5971 (London: Her Majesty's Stationery Office, 1975); "Annual Review of Agriculture," Cmnd. 9423 (London: Her Majesty's Stationery Office, 1985).

[a] Provisional forecast.
[b] Total area of U.K. 24 million ha approximate.

5.3. SOIL SUITABILITY FOR REDUCED TILLAGE

In Great Britain soil textural variation is considerable. It is not uncommon to have soils of different textures present within a field boundary. Soil types in southern England are predominantly rendzinas and over the rest of the country gleyic cambisols, gleyic luvisols, and entric gleysols are found in about 50% of the cereal growing areas. Entric cambisols form the main soil type in arable areas of Scotland.[8,13]

Climate and soil type are major factors influencing the choice of cultivation. The choice is made difficult by seasonal and annual variations in climate. In many regions, particularly where soils have a heavy texture, plowing was traditionally used as a means of improving topsoil drainage, thereby providing more reliable cultural management. Tillage, particularly on soils prone to compaction, is required to help retain a physical structure that is suitable for good crop growth. Tillage alters soil porosity, and can frequently result in lack of consolidation, which affects seedbed quality and reduces uniformity in the depth of seed placement. In cereals this can affect early plant development. However, studies involving the broadcasting of seed suggest that yield may not be greatly affected.[14,15] Other soil types that are in good structural condition possess a natural porosity, particularly in the surface zone, which, with careful husbandry, can be maintained. On these soils, farmers have the option to replace traditional tillage methods with reduced cultivation or direct drilling.

The relative risk of adopting reduced tillage on different soil types has been assessed.[16] Soil suitability for direct drilling is divided into three classes, taking into account soil type, climate, and experimental experience. Class 1 contains well-structured and well-drained soils on which sowing of both autumn and spring sown cereals is possible with yields similar to conventional plowing.

Table 5.2. Classification of Soil Suitability for Direct Drilling in Great Britain

Suitability Class	Soil and Climate
1: Suitable for spring and winter cereals	Stable structure, well drained, low rainfall
2: Suitable for winter cereals only	Stable structure, adequate drainage, medium rainfall
3: Unsuitable	Unstable structure, impeded drainage, high rainfall

Source: Adapted from Ball, B. C., *Proc. of EEC Workshop on Energy Saving by Reduced Tillage,* 10–12 June, 1987 (Luxembourg: Office for Official Publications of the European Communities, 1988).

Soil types include chalk and limestone soils and well-drained loams. These soils occupy about 30% of the cereal-growing area of Great Britain. Class 2 comprises soils which, with good soil and crop management, should give equal yields of winter cereals after direct drilling or traditional cultivation, but the yield of spring sown crops is likely to be reduced. Soils in this class include calcareous clays, clayey, or loamy over clayey soils that have secondary drainage. About 50% of soils in cereal areas are contained in this class. Soil types likely to be unsuitable for direct drilling form class 3. These include sandy soils with a low organic matter content, silty soils, wet clays, and alluvial types. Adoption of direct drilling on class 3 soils is considered to risk reduced yield, especially for spring sown crops.

The classification of land suitability was refined to include key soil structural and climatic requirements[17] (Table 5.2). The classification emphasizes the importance that must be placed on soil structure, porosity, and weather, particularly rainfall. Soils with resistance to compaction and a tendency to self-structure (i.e., self-mulching) are considered most suitable for reduction in tillage along with soils not suitable for tillage, such as stony soils. Because resistance to compaction is relative to soil moisture, a wider range of soil types is likely to be suitable for reduced tillage in drier areas. However, in practice, the lower rainfall allows more opportunity for cultivation and time pressures on land preparation are less. Overall, without the need to consider reduced tillage for soil conservation, the potential benefits of reduced cultivation are least in the climatic regions most suited for its use.

5.4. FACTORS AFFECTING THE ADOPTION OF NONPLOW SYSTEMS

The principal incentives to the farmer considering adopting nonplow systems are the potential saving in labor, time, and machinery costs for crop

Table 5.3. Comparison of Direct Labor and Machinery Costs (£/ha) for Plowing, Nonplowing, Reduced Cultivation, and Direct Drilling

	Plowing	Nonplowing	Reduced Cultivation	Direct Drilling
Preparation of seedbed	61	35	17	5
Applying fertilizer, drilling, harrowing, etc.	39	39	39	45
Total costs (£/ha)	100	74	56	50

Source: Adapted from Nix, J., *Tillage — What Now and What Next? Proc. SAW/MA/ADAS/ICI Conf.* (Huddersfield, U.K.: Soil and Water Management Association, Ltd., 1987), p. 16. With permission.

Table 5.4. Work Rate and Potential Area Capability of Conventional Plowing and Direct Drilling Based on a One-Man System

	Cultivation System			
	Plow, Cultivate, Drill		Spray, Direct Drill	
Soil Type/Crop	Work Rate (ha/hr)	Area Capability (ha)	Work Rate (ha/hr)	Area Capability[a] (ha)
Calcareous clay/winter wheat	0.25	88	0.99	353
Silty loam/winter wheat	0.39	132	1.09	368
Silty clay loam/spring barley	0.24	99	1.01	349

Source: Adapted from Patterson et al, *J. Agric. Eng. Res.,* 25, 1–36, 1980. With permission.

[a] Obtained by applying ADAS field efficiency factors to net work rates.

establishment. Time savings become more significant in northern Great Britain, where harvests are later, yet optimum planting dates are nearly the same as in the more favorable areas. With direct drilling such savings could be up to 50% as compared to conventional plowing and drilling (Table 5.3).[18] Savings in labor input as a result of higher work rates mean greater system capacity, with more land sown during the short periods available in the autumn and spring. Thus, it is possible to sow a greater proportion of crop at the optimum time, e.g., for autumn cropping there may be only 6 weeks available for land preparation whereas for spring cropping early sowing maximizes the growth period and reduces the risk of the land becoming unsuitable for crop germination.

As with farmers, most research workers adopt crop yield as the criteria for comparison of the success of systems. This can create an unrealistic comparison because work rates and operating costs are frequently not taken into account. In one study carried out in southeast England, the substitution of direct drilling for plowing increased the work rate approximately four times, thereby significantly increasing system capacity (Table 5.4).[19] This simplistic comparison does not take into account the full cost of producing a crop to market readiness nor the cost of buying specialist machinery. When all such costs are taken into account margins between different systems can narrow considerably, depending on individual circumstances.[17]

Despite promising results in early experiments, direct drilling has not been widely adopted by farmers in Great Britain. In 1973 it was estimated that 102,000 ha of crops were direct drilled. Of this, 30,000 ha were kale (*Brassica oleracea acephala* Dc.) and swedes (*Brassica napobrassic* L.) and 34,700 ha were cereals (3 and 0.2% in England and Scotland, respectively), 90% of which was autumn sown crops.[4] Reduced cultivation has been more widely used (12 and 1% of cereals established in England and Scotland, respectively), but unfortunately, few statistics are available to support this statement.

Farmers have been generally cautious in changing from traditional methods. The main reasons are systems reliability, straw disposal, weed control, soil drainage status and soil structure, and compaction. These factors are discussed below.

5.4.1. Systems Reliability

In the 30 years or more since experiments on direct drilling and reduced tillage began in Great Britain, much progress has been made. Compared to plowing, however, the systems are relatively new. Moldboard plowing has been used for centuries and the modern farmer is the beneficiary of much received experience. The plow inverts soil, burying residues and weeds. This distinguishes it from all other tillage implements. This is the foundation of the system's reliability. Perhaps the greatest advantage of conventional plowing systems is that they cope well with straw, weeds, compaction and drainage without any need to learn new management techniques, as required with nonplowing tillage. The provision of fresh soil on which to prepare a seedbed is valued by farmers as well as the erasure of surface compaction created during the growing and harvesting of the previous crop.

Cultivation alters the porosity in the disturbed layer sufficiently to provide a valuable drainage of surface water, particularly on heavy soils. Many of the strengths of the plowing system become weaknesses in reduced tillage systems. From an operator's point of view there is a need to obtain quickly the knowledge and skills in order to repeat success with the new methods. Although some initial experiences were favorable, others were not. This unreliability led some farmers to abandon the novel methods because of lost yields.[20]

5.4.2. Straw Disposal

Straw is not required to protect the soil from erosion in Great Britain. Straw is baled for use by animals, other uses being relatively minor.[21] Although 60% of straw is still baled,[22,23] in areas without animals the increased production of cereals in the postwar (World War II) years has led to a problem of straw surplus. Between 1977 and 1981, straw production in England and Wales rose from 10.5 to 12.6 metric tons but the amount conserved by baling remained approximately the same. Most of the surplus straw, estimated at 5 to 7 metric tons, was burned.[21] Regulations on burning were minimal during this period and, until stricter regulations were introduced after 1983, very little straw (approximately 2%) was incorporated. Reduced tillage and direct drilling appeared increasingly attractive to farmers. Experiments showed that in the absence of straw, yields of crops established after direct drilling, shallow tillage or plowing were very similar.[24] Studies in Scotland, where straw was baled and loose residues were removed by raking, also showed general success with direct drilling and reduced tillage.[15,25] When straw remains as a surface residue or is shallowly incorporated (5 to 7 cm), however, crop establishment and early growth can be severely affected. This may lead to depressed yield compared to plowing down straw.[26-30]

Fauna, e.g., slugs and carabid beetles, can increase with nonplowing systems, particularly where straw is present;[31] although these have no direct effect on soil structure, they may influence the crop. Slugs may damage seedlings, but the crop usually recovers. Carabid beetles may help control cereal aphids, thereby giving some natural control of virus diseases borne by aphids.

New regulations restricting farmers' options to burn straw have been gradually introduced and this has reduced the amount of incinerated straw burned (i.e., from 38 to 24% in 1988). Since 1992 the burning of straw and stubble has been banned in England and Wales. As a result, many farmers who opted for reduced tillage or direct drilling are returning to the use of the plow to incorporate residues (18% of straw plowed-in or cultivated in 1988).[23] In a recent survey of winter cereals grown after different methods of straw disposal, 90% of the 733 fields inspected had been plowed; in Scotland all the fields had been plowed.[32]

5.4.3. Weed Control

The spectrum of weed flora differs between direct drilling and conventional cultivation[33] because the method of tillage has a fundamental effect on the seed cycle of different species. Plowing to 20 cm will bury a large proportion of short cycle weed seeds (Figure 5.2)[34] deeply enough for any that germinate to be unlikely to reach the surface. Reduced cultivation and direct drilling retains seeds near the soil surface where they can readily germinate and increase overall populations.[35]

In recent years, new herbicides have provided the means of controlling most broadleaved weed species favored by reduced cultivation and direct drilling, but grass weeds in cereals remain a significant threat. The most important are blackgrass (*Alopecurus myosuroides* Huds.), barren brome (*Bromus sterilis* L.), meadow brome (*Bromus cummutatus* Schrad.), and meadow grasses (*Poa*

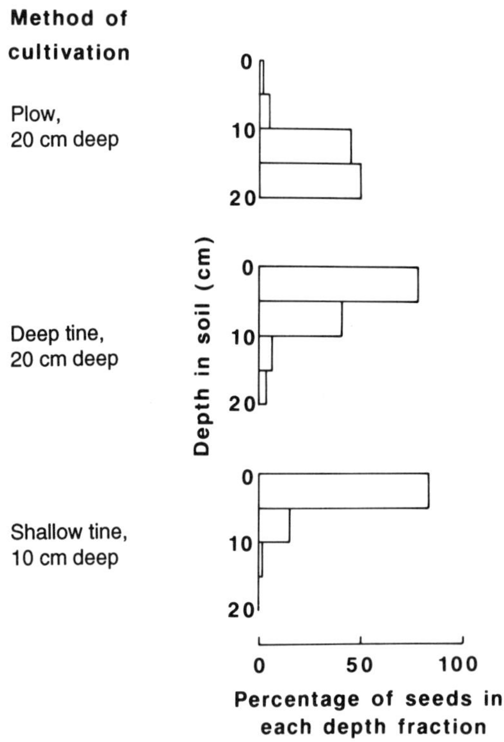

Figure 5.2. Effect of cultivation system on the incorporation of weed seeds initially placed on the soil surface before incorporation. (Adapted from Moss, S. R., *Association Naturale pour la Protection des Plantes (Commission Mauvaises Herbes-Columa) VIII ème Colloque International sur la Biologie, L'Ecologie et la Systematique des Mauvaises Herbes. Dijon.* (Paris: ANPP, 1988), p. 74. With permission.)

spp.). These weeds remain difficult to control, particularly with sequential cereal crops that are autumn sown.

The authors have referred to the change from plowing to reduced cultivation or direct drilling as causing increased competitive pressure on a cereal crop from grass weeds. The situation is made worse when residues are not burned. For example, the population of blackgrass seed, one of the most common grass weeds, builds up significantly faster under nonplowing such that a higher annual kill percentage is required to maintain a static population.[36] In the absence of burning, weed populations can build up faster.[36,37] Consequently, more emphasis must be placed on a combination of effective cultural and herbicide controls to achieve the high levels of weed control necessary. Cereal volunteers have also been found to be more numerous in nonplowing systems, persisting for several seasons after their parent crop.[38] Good husbandry can minimize the influence of weeds. Andersson[39] demonstrated that crop canopy structure was important in the control of annual broadleaved weeds and that selecting tall varieties of wheat can reduce competitive pressure from blackgrass.[40]

Herbicides have played an important role in the protection of yield. They have replaced to some extent the role of rotations in weed control and permitted sequential autumn sowing of crops. However, the problem of grass weeds in cereals with direct drilling and in nonplow systems remains to be solved by the development of new herbicides, crop cultivars, or more fundamental changes to crop husbandry. In general, increased use of herbicides is required with nonplowing systems, which makes such systems less attractive in environmental terms of reducing pesticide use.

5.4.4. Land Drainage

Great Britain's maritime climate and the high proportion of heavy-textured arable land means that drainage status plays an important role in crop production. Moisture status at sowing and during establishment is critical. Waterlogging before crop emergence can reduce the number of plants that successfully establish and postemergence waterlogging can depress crop yield, depending on the frequency and duration of waterlogging.[41] In wet conditions, direct drilling can smear the walls of the seed slot, possibly resulting in waterlogging within the seed slot itself. Even if waterlogging does not occur in the smeared slot, root breakout into the surrounding soil can be restricted during early development.[42] Davies and Cannell,[20] in their review of early experiments, cited poor performance of drills and problems of drilling as limiting the success of reduced cultivation and, in particular, direct drilling. This may have influenced farmers' attitudes and adoption of the methods. Some farmers have seen direct drilling as a quick way to complete an autumn sowing program. Late sowing by direct drilling can have a greater negative effect on yield than conventional sowing after plowing. For example, comparing drainage/cultivation interactions on a clay soil in a wet year during a long-term experiment, the grain yield of winter oats (*Avena sativa* L.) on drained land was 7.2 t/ha after plowing and 7.0 t/ha after direct drilling; however, on undrained plots the corresponding yields were 6.1 t/ha and 3.4 t/ha.[43] The 10-year total production of crops in this study is given in Table 5.5.[44] It shows that the direct drilled crops benefit significantly more from drainage than do plowed crops. Drainage also increases soil trafficability. This benefit can be critical in spring when access to the land to apply herbicides and fertilizer is required. Drainage effects combined with the greater load-bearing strength of direct drilled land results in better trafficability than after plowing and provides a greater number of potential workdays to the farmer.

Bailey[45] concluded that to minimize surface compaction and wheel ruts in clayey soil, the water table should not rise above 50 cm below the soil surface

Table 5.5. Productivity of Autumn Cereals and Oilseed Rape in Response to Drainage and Cultivation on a Clay Soil, 1981–1988

Cultivation Method	Drained	Undrained	SED	Percentage Change Due to Drainage
	(rounded values, metric ton/ha)			
Plowed	58.21	56.39		+3
Direct drilled	59.71	50.72	0.615	+18
Mean	58.96	53.39		+10

Source: Adapted from Christian, D. G., *Proc. Conf. Brimstone Experiment (October 1991)* (Harpendon, U.K.: Lawes Agricultural Trust, Rothamsted Experimental Station, 1991).

over the course of the winter. Harris et al.[46] demonstrated that this is achievable with appropriate drainage design in relation to the soil type, and showed that current drainage design specification is adequate for both direct drilled and plowed land. The risks associated with using direct drilling are mainly confined to the sowing and establishment phase, and drainage designs cannot be easily altered to accommodate this. Land drainage has suffered in recent years because of the withdrawal of grant support for artificial improvement in drainage. This has increased the dependency on conventional plowing systems to maintain topsoil drainage status.

5.4.5. Soil Structure

Maintenance of good soil structure is vital to efficient crop growth and yield. Changes in structure can be readily identified by measuring soil bulk density or strength and porosity, and in some circumstances visual inspection is all that is necessary. The influence of cultivation on soil density and moisture retention in different soil types over several years has been illustrated by extensive field experimentation.[30] Provided no physical damage occurs, the soil's density can, in the course of time, return to that prior to cultivation.[47] When the soils are dry, avoiding tillage has improved the load-bearing characteristics of a soil and changed the trafficability of the land. While this may be a desirable feature on soils that are well structured or those capable of self-structuring, on compactible soils it can increase bulk density[48] and consequently impede drainage and restrict root exploration.[49,50] Sowing seed, especially by direct drilling, can be unsatisfactory and subsequent crop growth adversely affected.

Of equal importance is the effect of tillage on soil porosity because porosity accounts for much of the difference in water retention and its distribution within the soil profile.[51] Total pore space and the volume of macropores in the topsoil decline with direct drilling as compared to plowing (Figure 5.3). The rate of change is partially governed by the previous agricultural use of the soil; however, pore continuity can be greater (Figure 5.3) in direct drilled than in plowed land, and this is important in soil hydrology.[52] Pore continuity, in addition to naturally occurring cracks and those made as a result of mole

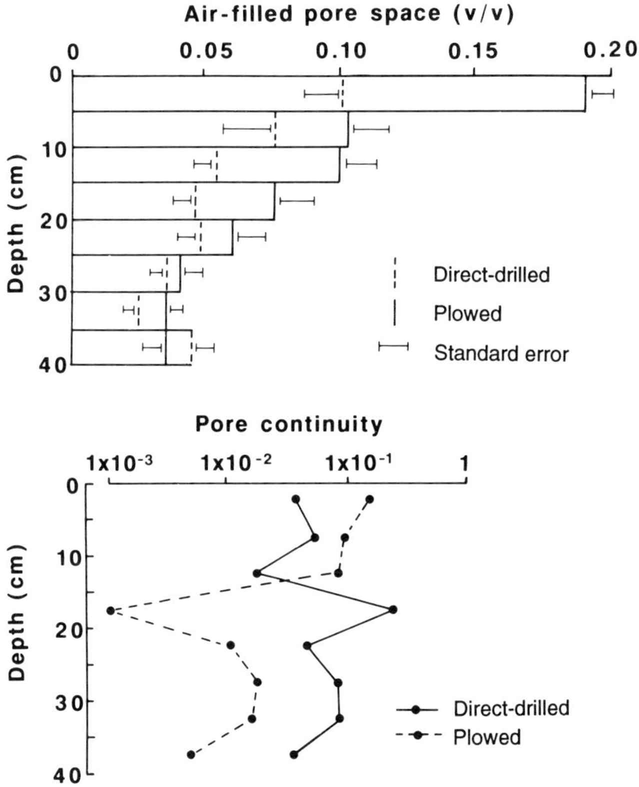

Figure 5.3. Volume of air-filled pores and pore continuity at –6 kPa water potential in a clay soil, Lawford series after direct drilling or plowing. (Adapted from Douglas, J. T. and Goss, M. J., *Soil Tillage Res.*, 10, 309 and 314, 1987. With permission of Elsevier Science Publishers, Academic Publishing Division.)

draining, assists in the rapid removal of water from the topsoil to the subsoil.[53] Channels created in the soil by earthworms also contribute to drainage. The better continuity of pores after direct drilling enables rainwater to percolate to the subsoil more rapidly than on plowed land and this alters the pattern of discharge through drains (Figure 5.4).[54] Earthworm channels are more abundant on direct drilled than on plowed land because avoidance of tillage improves the survival of earthworms, especially the deep-burrowing species.[55] The increased earthworm populations have been associated with improved aggregate stability.[56] Such improvements are particularly important where bulk densities are high, although there is little evidence that bulk densities increase after the first few years of direct drilling.[57] The overall improvement via earthworms in soil structure conditions is important for successful direct drilling, but takes several years to develop and is no substitute for the continued requirement for compaction control.

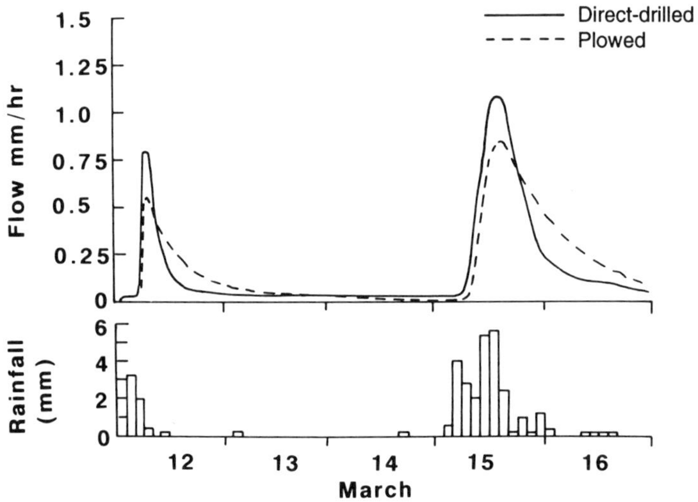

Figure 5.4. Comparison of typical drainflow response on direct drilled or plowed land in March 1982. (Adapted from Harris, G. L. and Howse, K. R., *Proc. Conf. Brimstone Experiment (October 1991)* (Harpendon, U. K.: Lawes Agricultural Trust, Rothemsted Agricultural Station, 1991).

5.4.6. Soil Compaction

Compaction can impair crop growth by restricting root exploration of the soil (high soil strength) or by reducing soil porosity (impeded drainage and aeration).[58] Soils susceptible to compaction damage contain a large proportion of clay or sand and are most susceptible when wet. Such soil and moisture conditions are common in Great Britain, particularly in Scotland. Compaction damage is most likely in times of harvest during random traffic of heavy machinery.

Under conventional tillage, such compaction damage is ameliorated by plowing. Reduced cultivation may remove near surface compaction or ruts. However, direct drilling cannot remove any such surface damage, and irregularities in surface level (including ruts) persist and may prevent sowing to uniform depth. This problem is generally worse when first adopting a reduced or zero tillage system because the strength of undisturbed soils increases with time. For example, Ball et al.[25] reported a decrease in yield of winter barley in the first 3 years of direct drilling in comparison to long-term direct drilling (Table 5.6). The greater success of the long-term direct drilled treatment was attributed to the development of a stable, protective tilth as a result of organic matter accumulation. Winter cereals were grown in this experiment and proved to be more successful than spring cereals, possibly because the root system of the winter crop is more extensive than the spring crop when soil strength increases due to drying in late spring.[59]

Table 5.6. Yields of Winter Barley and Oilseed Rape under Short- and Long-Term Direct Drilling in a Loam/Clay Loam[a]

	Short-term Direct Drilling							
Year	1	2	3	4	5	6	7	Average
Yield (t/ha)	7.33	6.51	4.53	5.42	7.51	3.72	8.48	6.21
	Long-term Direct Drilling							
Year	17	18	19	20	21	22	23	Average
Yield (t/ha)	8.15	6.58	5.52	4.96	7.28	3.67	8.82	6.43

Note: Crops in all seasons were winter barley, except years 6 and 22, which were oilseed rape.

[a] Loose straw and residues were removed by baling and raking where necessary.

Table 5.7. Problems Influencing Reduced Tillage and Possible Solutions

Problem	Possible Solutions
Surface soil condition Presence of ashes, stubble and poor structure	Reduced tillage <10 cm
Topsoil compaction	Tramlines for sowing, cultivation, and subsequent operations Low vehicle weight Low ground pressure tires Grain trailers on headlands Gantry system (zero traffic)
Perennial grass weeds	Normal plowing as part of a rotation Grow a break crop
Straw incorporation	Normal plowing as part of a rotation Shallow plowing
Lack of skill and motivation	Alert farmers to the benefits of reduced tillage Provide cash incentives

Source: Adapted from Ball, B. C., *Agric. Eng.*, 45, 2–6, 1990.

Compaction damage by agricultural operations in the first year of direct drilling was avoided completely[60] by applying an experimental zero traffic system. This resulted in the same crop yield as for plowing. Gantries, which permit almost all operations in cereal production to be made without traffic in the 6- or 12-m span between gantry wheels, are commercially available. Although these allow more frequent access to the land, their inability to accommodate harvest machinery and the high cost of purchasing and adapting the necessary machines have resulted in little interest from farmers. Simpler and less expensive methods of localizing or reducing compaction have proved more popular. These include the widespread use of trackways at regular intervals across the field for application of sprays and fertilizer between sowing and harvest. Also, compactive effort is reduced by decreasing vehicle weight and by use of large or dual tires. Such methods are generally thought to be essential for the success of reduced cultivation and direct drilling systems on soils considered to be unsuitable.[61] In the absence of compaction control, such soils require annual tillage to ameliorate compaction and erase wheelings. Some cultivations, to remove wheel marks, for example, also provide a leveling of the soil surface so that drilling will be easier.

A table summarizing the problems discussed and their solutions was given by Ball[59] and is reproduced in Table 5.7. A recent review[62] concluded that detailed guidelines for choice of reduced cultivation systems, taking into account the above five factors, should be produced. Although the advisory services had access to simple guidelines produced for conditions in England and Scotland, these have largely been ignored by growers.

5.5. ENVIRONMENTAL ISSUES IN REDUCED TILLAGE

5.5.1. Fate of Soil and Fertilizer Nitrogen

Nitrogen fertilizer, mainly in the form of nitrate, is relatively inexpensive in Great Britain and gives a good cereal yield response at high rates of application. Many arable crops take up about 200 kg/ha of nitrogen from the soil during their growth, and the quantity of nitrogen fertilizer applied to crops by the farmer reflects crop demand in relation to maximizing productivity. In recent years there has been increasing concern over the quality of natural water, especially in terms of its nitrate content. Agricultural activity has an influence on water quality. Nitrate is water soluble and is susceptible to loss by leaching. Direct drilling can help reduce the potential of agricultural practices to pollute natural water courses with nitrate.[63] Most nitrogen fertilizer is applied to cereals in the spring to coincide with the upturn in growth as temperature rises. In a well-managed fertilizer program leaching losses of fertilizer nitrogen are uncommon, except when it rains shortly after application. Denitrification rather than leaching is more likely to occur at this time,[64] and is greater on direct drilled land.[65]

The soil physical environment associated with direct drilling increases the risk of denitrification in comparison to conventional plowing. Direct drilled land has a greater moisture content, more water-filled pores (which may indicate a lower soil oxygen content), and greater bulk density than plowed land. Denitrification losses are higher in such conditions as they are also in soils that are poorly drained.[66] Although denitrification losses form only a small component of total nitrogen loss during the winter, they can be greater after reduced cultivation than after plowing.[29] Overall gaseous losses of soil nitrogen may be significant; however, such losses may be more environmentally acceptable than losses of nitrate by leaching because in compact soils nitrous oxide may be reduced to nitrogen within the soil so that relatively little is emitted as a "greenhouse" gas.[67]

The risk of the occurrence of nitrate leaching is greatest during the autumn and winter when precipitation normally exceeds evaporation. Recent studies have shown that where appropriate amounts of nitrogen fertilizer are applied to a cereal crop in the spring it is rapidly taken up by the growing plant and is unlikely to be the source of nitrate leached during the following winter. Moreover, nitrate leached in the winter is mainly derived from the mineralization of organic matter[68] and from wet and dry deposition from the atmosphere.

Cultivation increases the rate at which nitrogen is mineralized from soil organic matter.[63] Thus, avoidance of tillage will help reduce the amount of nitrogen at risk of leaching. Because reduced tillage covers a wide specification in terms of the depth and amount of soil disturbance, its effect on nitrogen losses will vary accordingly. Therefore, most studies have been confined mainly to the extreme situations provided by comparisons of plowing and direct drilling. In one study on a clay soil, nitrate leaching losses from direct drilled land were 22% less than that from plowed land.[69]

5.5.2. Soil Erosion

In Great Britain, soil erosion is less severe and apparent than in tropical and subtropical areas subject to inappropriate agricultural practice or deforestation. Erosion was largely ignored until the late 1960s, when a board of enquiry into the effect of modern farming on soil structure and fertility found evidence that soil structure was deteriorating sufficiently for erosion to occur.[70] There is evidence that water erosion increased when arable farming intensified after World War II.[71] Detailed assessments of the extent of erosion in Great Britain followed.[72-74]

Cultivation increases the risk of erosion and the soils most vulnerable are those of "light" sandy and silty textures. These soils are usually low in organic matter content and consequently have a low level of stable soil aggregates. Because they are easy to cultivate, these soils are frequently used for arable crop production. Tillage promotes the loss of organic matter, especially under intensive soil cultivation.

One study found that in England, 36% of arable land is at moderate to very high risk of erosion.[75] Direct drilling has not been adopted to any extent on such soils (many of which are sandy) because they are weakly structured and as a result unsuited to direct drilling[16] (Table 5.2). For example, the reduced volume of macropores in uncultivated soil can increase the risk of short-term runoff following a rain event. This may increase the risk of soil erosion, particularly on sloping land. A more appropriate method of tillage is tine cultivation. This helps reduce compaction in the surface layers and improve its porosity without maximizing erosion risk as would be associated with plowing.[76] Direct drilling and shallow tillage, if maintained over several years with straw burning, can have a successful outcome on some weakly structured soils,[24,77] particularly because the concentration of organic matter increases in the surface soil layer.[78] This can have positive benefits in a higher soil aggregate strength as detected by wet sieving.[79] Straw removal by baling can also cause organic matter to increase at the surface of direct drilled soil, where it may increase aggregate stability.[25] However, the farmer's commercial priorities may override any such unguaranteed long-term soil improvements associated with direct drilling.

5.6. GENERAL SUMMARY AND CONCLUSIONS

Many experiments over many years have identified the advantages of reducing the amount of cultivation used to plant crops; however, due to the complexity of soil-plant interactions the more cultivation is reduced, the greater the risk of problems that may reduce crop productivity. Many farmers used continuous reduced cultivation for many years and are only moving away from it because of straw disposal problems. Problems of weed control, straw disposal, soil compaction, and soil drainage status have prevented widespread continuous use of direct drilling in Great Britain. Farmers favor maintaining a high tillage input, especially when it includes plowing, in order to avoid the risk of encountering such problems.

The reliability of the conventional plowing system constrained adoption of direct drilling and reduced cultivation in the past and the ban on straw burning will probably lead to the virtual elimination of existing uses of these methods in the foreseeable future. The high work rates and timeliness of reduced cultivation and direct drilling can now be matched in plowing systems by use of high horsepower tractors in combination with powered cultivators and often linked with seed drills. At the same time potential problems such as surface compaction or crop residues are handled, albeit crudely. The desirable benefits of improving soil structure and drainage by eliminating tillage are less obvious to the farmer than the immediate cosmetic improvement in surface appearance created by plowing. Cultivation apparently erases much of the surface compaction created at harvest; however, long-term structural problems such as unrelieved compaction below plow depth and reduced structural stability remain.

Environmental and economic arguments point to the logic of reducing the cost of establishing crops. This must include cultivations because they are a major cost component. To control the buildup of weeds and the problem of straw associated with direct drilling and reduced tillage, a way forward may be to rotate tillage systems so the advantages of both plowing and nonplowing systems may be combined over a number of years. To prevent the loss of any long-term build-up of structural stability or natural predators, the tillage used in rotational tillage systems should be to the minimum necessary depth and to ensure minimum compaction. Complementary to rotational tillage is a more flexible crop rotation, e.g., more spring crops with reduced reliance on pesticides.

Alternatively, for autumn sowing, direct drilling oilseed rape (*Brassica* spp.) would help to control grass weeds by utilization of herbicides that cannot be used with continuous cereals. Use of such crops can reduce the likelihood of wheeling wet soil during crop establishment and can thus reduce compaction. Management decisions that relate to the choice of cultivation method for seedbed preparation could be applied with the same vigor as that asked for choice of crop and variety, for example, use of plowing in a weedy situation until infestation pressure has decreased sufficiently for reduced or shallow cultivation to take its place for a period.

For direct drilling specific uses will remain minimal, but should be encouraged on stony or very sandy soils not suitable for cultivation. Its adoption to large-scale land use crops is unlikely to occur in the foreseeable future unless difficulties of drilling at close row-spacing through surface residues are overcome.

In the future, a combination of plowed and nonplowed tillage systems should help the shift to more sustainable agricultural practices with improved soil structure by managing compaction and crop residues and with maintenance of the natural ecology of the soil.

5.7. ACKNOWLEDGMENTS

We are grateful to Dr. B. D. Soane, SCAE, Penicuik, Midlothian, for his comments. Mrs. L. Castle for preparing the figures; and Mrs. V. Payne for the typescript.

5.8. REFERENCES

1. Bullen, E. F. "How Much Cultivation?," *Span* 20:53–54 (1977).
2. Russell, E. W., and B. A. Keen. "Studies on Soil Condition. X. The Results of a 6-Year Cultivation Experiment," *J. Agric. Sci. Cambridge* 31:326–347 (1941).
3. Allen, H. P. *Direct Drilling and Reduced Cultivations* (Ipswich, U.K.: Farming Press Ltd. U.K., 1981). p. 219.
4. Allen, H. P. "ICI Plant Protection Division Experience with Direct Drilling Systems, 1961–74," *Outlook Agric.* 8:213–215 (1975).
5. Pidgeon, J. D., and J. M. Ragg. "Soil, Climatic and Management Options for Direct Drilling Cereals in Scotland," *Outlook Agric.* 10:49–55 (1979).
6. Evans, T. "New Approaches to Increasing Fodder Production. II. Swedes and Turnips in Upland Situations," *Outlook Agric.* 7(4):171–174 (1973).
7. Toosey, R. D. "New Approaches to Increasing Fodder Production. III. Stubble Catch-Cropping with Brassicas," *Outlook Agric.* 7(4):175–178 (1973).
8. Cannell, R. Q. "Reduced Tillage in North-West Europe — A Review," *Soil Tillage Res.* 5:129–177 (1985).

9. Smith, L. P. "The Agricultural Climate of England and Wales," Tech. Bull. 35 (London: Her Majesty's Stationery Office, 1984).
10. Francis, P. E. "The Climate of the Agricultural Areas of Scotland," Climatol. Memo. No. 108 (Edinburgh: Meteorological Office, 1981).
11. "Annual Review of Agriculture," Cmnd. 5971 (London: Her Majesty's Stationery Office, 1975).
12. "Annual Review of Agriculture," Cmnd. 9423 (London: Her Majesty's Stationery Office, 1985).
13. Commission of the European Communities. "Soil Map of the European Communities. Catalogue Number CD-40-84-A92-EN-D" (Luxembourg: Office for Official Publications of the European Communities, 1985).
14. Graham, J. P., and F. B. Ellis. "The Merits of Precision Drilling and Broadcasting for the Establishment of Cereal Crops in Great Britain," *ADAS Q. Rev.* 38:160–169 (1980).
15. Ball, B. C. "Cereal Production with Broadcast Seed and Reduced Tillage: A Review of Recent Experimental and Farming Experience," *J. Agric. Eng. Res.* 35:71–95 (1986).
16. Cannell, R. Q., D. B. Davies, D. Mackney and J. D. Pidgeon. "The Suitability of Soils for Sequential Direct Drilling of Combine-Harvested Crops in Great Britain: A Provisional Classification," *Outlook Agric.* 9(6):303–316 (1978).
17. Ball, B. C. "Reduced Tillage in Great Britain: Practical and Research Experience," *Proc. of EEC Workshop on Energy Saving by Reduced Tillage, 10–12, June 1987,* K. Bäumer and W. Ehlers (Eds.) (Luxembourg: Office for Official Publications of the European Communities, 1988).
18. Nix, J. "The Economics of Tillage — The Scope for Cost Savings," *In Tillage — What Now and What Next? Proc. SAWMA/ADAS/ICI Conf.* (Huddersfield, U.K.: Soil and Water Management Association Ltd., 1987), pp. 33–40.
19. Patterson, D. E., W. C. T. Chamen and C. D. Richardson, "Long-Term Experiments with Tillage Systems to Improve the Economy of Cultivations for Cereals," *J. Agric. Eng. Res.* 25:1–36 (1980).
20. Davies, D. B., and R. Q. Cannell. "Review of Experiments on Reduced Cultivation and Direct Drilling in the United Kingdom, 1957–1974," *Outlook Agric.* 8:216–220 (1975).
21. Staniforth, A. R. *Straw for Fuel, Feed and Fertilizer* (Ipswich, U.K.: Farming Press Ltd., 1982).
22. "Straw Survey 1983 England and Wales," Stats 339/83 (Guildford, U.K.: Ministry of Agriculture Fisheries and Food, 1983).
23. "Straw Survey 1988 England and Wales," Stats 25/89 (Guildford, U.K.: Ministry of Agriculture Fisheries and Food, 1989).
24. Christian, D. G., and E. T. G. Bacon. "A Long-Term Comparison of Plowing, Tine Cultivation and Direct Drilling on the Growth and Yield of Winter Cereals and Oilseed Rape on Clayey and Silty Soils," *Soil Tillage Res.* 18:311–331 (1990).
25. Ball, B. C., R. W. Lang, M. F. O'Sullivan and M. F. Franklin. "Cultivation and Nitrogen Requirements for Continuous Winter Barley on a Gleysol and a Cambisol," *Soil Tillage Res.* 13:333–352 (1989).
26. Lynch, J. M., F. B. Ellis, S. H. T. Harper and D. G. Christian. "The Effect of Straw on the Establishment and Growth of Winter Cereals," *Agric. Environ.* 5:321–328 (1980).

27. Oliphant, J. M. "The Effect of Straw and Stubble on the Yield of Winter Wheat after Cultivation or Direct Drilling," *Exp. Husb.* 38:60–68 (1982).
28. Graham, J. P., F. B. Ellis, D. G. Christian and R. Q. Cannell. "Effects of Straw Residues on the Establishment, Growth and Yield of Autumn-Sown Cereals," *J. Agric. Eng. Res.* 33:39–49 (1986).
29. Ball, B. C., and E. A. G. Robertson. "Straw Incorporation and Tillage Methods: Straw Decomposition, Denitrification and Growth and Yield of Winter Barley," *J. Agric. Eng. Res.* 46:223–243 (1990).
30. Christian, D. G., and E. T. G. Bacon "The Effects of Straw Disposal and Depth of Cultivation on the Growth, Nutrient Uptake and Yield of Winter Wheat on a Clay and Silt Soil," *Soil Use Manage.* 7:217–222 (1991).
31. Prew, R. D. "Changing Straw Disposal Practices," HGCA Rev. Art. No. 11 (London: Home-Grown Cereals Authority, 1989).
32. Cussans, G. W., F. B. Cooper, D. H. F. Davies and M. Thomas. "A Survey of the Brome Grasses: 1989," preliminary report to the British Crop Protection Council, London (1990).
33. Froud-Williams, R. J., D. S. H. Drennan and R. J. Chancellor. "Influence of Cultivation Regime on Weed Floras of Arable Cropping Systems," *J. Appl. Ecol.* 20:187–197 (1983).
34. Moss, S. R. "Influence of Cultivations on the Vertical Distribution of Weed Seeds in the Soil," in *Association Naturale pour la Protection des Plantes (Commission Mauvaises Herbes-Columa) VIII ème Colloque International sur la Biologie, L'Ecologie et la Systematique des Mauvaises Herbes, Dijon* (Paris: ANPP, 1988), pp. 71–80.
35. Froud-Williams, R. J., R. J. Chancellor and D. S. H. Drennan. "The Effects of Seed Burial and Soil Disturbance on Emergence and Survival of Arable Weeds in Relation to Minimal Cultivation," *J. Appl. Ecol.* 21:629–641 (1984).
36. Moss, S. R. "The Seed Cycle of *Alopecurus myosuroides* in Winter Cereals. A Quantitative Analysis," in *Proc. Eur. Weed Research Society Symposium. Integrated Weed Management in Cereals.* (The Netherlands: EWRS, 1990), pp. 27–35.
37. Wilson, B. J., S. R. Moss and K. J. Wright. "Long-Term Studies of Weed Populations in Winter Wheat as Affected by Straw Disposal, Tillage and Herbicide Use," in *Proc. Brighton Crop Protection Conference — Weeds* (Croydon, U.K.: BCPC Publications, 1989), pp. 131–136.
38. Christian, D. G., N. Carreck, A. J. Buck and K. Latimer, "Weed Ecology and Management," Institute of Arable Crops Res. Rep. (Harpenden, U.K.: Rothamsted Experimental Station, 1990), p. 52.
39. Andersson, B. "Influence of Crop Density and Spacing on Weed Competition and Grain Yield of Wheat and Barley," in *Proc. Eur. Weed Research Society Symposium, Economic Weed Control,* (Stutgart-Hohenheim: EWRS, 1986), pp. 121–128.
40. Moss, S. R. "The Influence of Crop Variety and Seed Rate on *Alopecurus myosuroides* Competition in Wheat and Barley," in *Proc. Brighton Crop Protection Conference — Weeds* (Croydon, U.K.: BCPC Publications, 1985), pp. 701–708.
41. Cannell, R. Q., R. K. Belford, K. Gales, R. J. Thomson and C. P. Webster. "Effects of Waterlogging and Drought on Winter Wheat and Winter Barley Grown on a Clay and a Sandy Loam Soil. I. Crop Growth and Yield," *Plant Soil* 80:53–66 (1984).

42. Prebble, R. E. "Root Penetration of Smeared Soil Surfaces," *Exp. Agric.* 6:303–308 (1970).
43. Cannell, R. Q., D. G. Christian and F. K. G. Henderson. "A Study of Mole Drainage with Simplified Cultivation for Autumn-sown Crops on a Clay Soil. IV. A Comparison of Direct-Drilling and Mouldboard Ploughing on Drained and Undrained Land on Root and Shoot Growth, Nutrient Uptake and Yield," *Soil Tillage Res.* 7:251–272 (1986).
44. Christian, D. G. "Crop Yield and Uptake of Nitrogen," in *Proc. Conf. Brimstone Experiment (October 1991)*, J. A. Catt (Ed.) (Harpenden, U.K.: Lawes Agricultural Trust, Rothamsted Experimental Station, 1991), pp. 41–53.
45. Bailey, A. D. "Drainage of Clay Soils in England and Wales," in *Proc. of the International Drainage Workshop* (Wageningen: International Institute, Land Reclamation Improvement, 1979), pp. 220–242.
46. Harris, G. L., M. J. Goss, R. J. Dowdell, K. R. Howse and P. Morgan. "A Study of Mole-Drainage with Simplified Cultivation for Autumn-Sown Crops on a Clay Soil. II. Soil Water Regimes, Water Balances and Nutrient Loss in Drain Water," *J. Agric. Sci.* 102:561–581 (1984).
47. Kuipers, H., and C. Van Ouwerkerk. "Total Pore-Space Estimations in Freshly Plowed Soil," *Neth. J. Agric. Sci.* 11:45–53 (1963).
48. Stengel, P., J. T. Douglas, J. Guérif, M. J. Goss, G. Monnier and R. Q. Cannell. "Factors Influencing the Variation of Some Properties of Soils in Relation to their Suitability for Direct Drilling," *Soil Tillage Res.* 4:35–53 (1984).
49. Ball, B. C., and M. F. O'Sullivan. "Cultivation and Nitrogen Requirements for Drilled and Broadcast Winter Barley on a Surface Water Gley (Gleysol)," *Soil Tillage Res.* 9:103–122 (1987).
50. Braim, M. A., K. Chaney and D. R. Hodgson. "Effects of Simplified Cultivation on the Growth and Yield of Spring Barley on a Sandy Loam Soil. II. Soil Physical Properties and Root Growth; Root: Shoot Relationships, Inflow Rates of Nitrogen; Water Use," *Soil Tillage Res.* 22:173–187 (1992).
51. Goss, M. J., K. R. Howse and W. Harris. "Effects of Cultivation on Soil Water Retention and Water Use by Cereals in Clay Soils," *J. Soil Sci.* 29:475–488 (1978).
52. Douglas, J. T., and M. J. Goss. "Modification of Porespace by Tillage in two Stagnogley Soils with Contrasting Management Histories," *Soil Tillage Res.* 10:303–317 (1987).
53. Goss, M. J., G. L. Harris and K. R. Howse. "Functioning of Mole Drains in a Clay Soil," *Agric. Water Manage.* 6:27–30 (1984).
54. Harris, G. L., and K. R. Howse. "Hydrology and Soil Physics Studies," in *Proc. Conf. Brimstone Experiment (October 1991),* J.A. Catt (Ed.) (Harpenden U.K.: Lawes Agricultural Trust, Rothamsted Experimental Station, 1991), pp. 15–25.
55. Barnes, B. T., and F. B. Ellis. "Effects of Different Methods of Cultivation and Direct Drilling and Disposal of Residues on Populations of Earthworms," *J. Soil Sci.* 30:669–679 (1979).
56. Gerard, B. M., and R. K. M. Hay. "The Effect on Earthworms of Plowing, Tined Cultivation, Direct Drilling and Nitrogen in a Barley Monoculture System," *J. Agric. Sci. Cambridge* 93:147–155 (1979).
57. Pidgeon, J. D., and B. D. Soane. "Soil Structure and Strength Relations Following Tillage, Zero-Tillage and Wheel Traffic in Scotland," in *Modification of Soil Structure,* W. W. Emerson, R. D. Bond and A. R. Dexter (Eds.) (Chichester: John Wiley & Sons, 1990) pp. 371–378.

58. Soane, B. D., J. W. Dickson and D. J. Campbell. "Compaction by Agricultural Vehicles: A Review. III. Incidence and Control of Compaction in Crop Production," *Soil Tillage Res.* 2:3–36 (1981).
59. Ball, B. C. "Reduced Tillage for Energy and Cost Savings with Cereals: Practical and Research Experience," *Agric. Eng.* 45:2–6 (1990).
60. Campbell, D. J., J. W. Dickson, B. C. Ball and R. Hunter. "Controlled Seedbed Traffic after Plowing or Direct Drilling under Winter Barley in Scotland, 1980–1984," *Soil Tillage Res.* 8:3–28 (1986).
61. Ball, B. C., M. F. O'Sullivan and J. K. Henshall. "Cultsave: a Computer Program to Select an Economical Tillage System for Winter Cereals," in: *Proc. 12th Conf. International Soil Tillage Research Organisation,* (Addis Abbaba: Nigeria, 1991) pp. 579–585.
62. Davies, B. D. "Reduced Cultivation for Cereals," HGCA Rev. Art. No. 5 (London: Home-Grown Cereals Authority, 1988).
63. Dowdell, R. J., and R. Q. Cannell. "Effect of Plowing and Direct Drilling on Soil Nitrate Content," *J. Soil Sci.* 26(1):51–61 (1975).
64. Dowdell, R. J. "Fate of Nitrogen to Agricultural Crops with Particular Reference to Denitrification," *Philos. Trans. R. Soc. London B.* 296:363–373 (1982).
65. Burford, J. R., R. J. Dowdell and R. Crees. "Emission of Nitrous Oxide to the Atmosphere from Direct-Drilled and Plowed Clay Soils," *J. Sci. Food. Agric.* 32:219–223 (1981).
66. Colbourn, P., and I. W. Harper. "Denitrification in Drained and Undrained Clay Soil," *J. Soil Sci.* 38:531–539 (1987).
67. Arah, J. R. M., K. A. Smith, I. J. Crichton and H. S. Li. "Nitrous Oxide Production and Denitrification in Scottish Arable Soils," *J. Soil Sci.* 42:351–367 (1991).
68. Macdonald, A. J., D. S. Powlson, P. R. Poulton and D. S. Jenkinson. "Unused Fertilizer Nitrogen in Arable Soils — Its Contribution to Nitrate Leaching," *J. Sci. Food Agric.* 46:407–419 (1989).
69. Goss, M. J., P. Colbourn, G. L. Harris and K. R. Howse. "Leaching of Nitrogen under Autumn-Sown Crops and the Effects of Tillage," in *Nitrogen Efficiency in Agricultural Soils,* D. S. Jenkinson and K. A. Smith (Eds.) (London: Elsevier, 1988), pp. 269–282.
70. Ministry of Agriculture, Fisheries and Food. "Modern Farming and the Soil," Rep. Agricultural Advisory Council on Soil Structure and Soil Fertility (London: HMSO, 1970).
71. Catt, J. A. *Soil Erosion on the Lower Greensand at Woburn Experimental Farm, Bedfordshire — Evidence, History and Causes,* M. Bell and J. Boardman, (Eds.) (Oxford: Oxbow Books, 1992), pp. 67–76.
72. Reed, A. H. "Accelerated Erosion of Arable Soils in the United Kingdom by Rainfall and Runoff," *Outlook Agric.* 10:41–48 (1979).
73. Morgan, R. P. C. "Soil Erosion and Conservation in Great Britain," *Progr. Phys. Geogr.* 4:24–47 (1980).
74. Speirs, R. B., and C. A. Frost. "The Increasing Incidence of Accelerated Soil Water Erosion on Arable Land in the East of Scotland," *Res. Dev. Agric.* 2:161–167 (1985).
75. Evans, R. "Soils at Risk of Accelerated Erosion in England and Wales," *Soil Use Manage.* 6:125–131 (1990).
76. Soane, B. D., and J. D. Pidgeon. "Tillage Requirement in Relation to Soil Physical Properties," *Soil Sci.* 119:376–384 (1975).

77. Ellis, F. B., D. G. Christian and R. Q. Cannell. "Direct Drilling, Shallow-Tine Cultivation and Plowing on a Silt Loam Soil 1974–1980," *Soil Tillage Res.* 2:115–130 (1982).
78. Fleige, H., and K. Baeumer. "Effect of Zero-Tillage on Organic Carbon and Total Nitrogen Content and their Distribution in Different N-Fractions of Loessial Soils," *Agroecosystems* 1:19–29 (1974).
79. Douglas, J. T., and M. J. Goss. "Stability and Organic Matter Content of Surface Soil Aggregates under Different Methods of Cultivation and in Grassland," *Soil Tillage Res.* 2:155–175 (1982).

CHAPTER 6

Approaches Toward Conservation Tillage in Germany

Wilfried Ehlers and Wilhelm Claupein
Institute of Agronomy and Plant Breeding, Georg August University; Goettingen, Germany

TABLE OF CONTENTS

0-87371-571-3/94/$0.00+$.50

6.1. INTRODUCTION

The plow has a long tradition in Germany. Since the Neolithic and Bronze Ages, wooden hooks were used to produce ridges and furrows in the seed furrow system. About 2000 years ago, the innovation of the moldboard plow promoted the transition to a higher yielding seedbed cultivation.[1] However, it was not until the beginning of the 11th century that the moldboard plow began to replace the hook on a larger scale, favoring the extension of cereal production during the high Middle Ages.[2]

In contrast to the tradition and art of intensive soil cultivation, the scientific approach toward conservation tillage is only 25 years old.[3] In the form of zero tillage, it became an enthusiastically supported tool to study by comparison the influence of conventional tillage (i.e., moldboard plow) on soil properties, plant reactions, and yield performance. While the scientific community demonstrated that the soil-turning action of the moldboard plow was no longer a prerequisite in crop production, the acceptance of any form of conservation tillage by German farmers remained generally low. The aim of this chapter is to outline scientific and practical experiences with conservation tillage, to summarize benefits and drawbacks, and to indicate possible ecological consequences.

6.2. CHARACTERISTICS OF THE AGROECOSYSTEM

Germany is situated in the midst of Europe. In the south it is bound by the Alps and to the north by the North and Baltic Seas. The landscape ranges from alpine mountains in southern Bavaria to medium altitude mountains in the central states and to the lowlands in the north. Due to the mild Gulfstream the climate is cool-temperate and suboceanic. The oceanic influence dominates in the northwest and gradually shifts to a more continental character in the southeast. Rain distribution is even year-round, with a slight maximum in summer. The annual precipitation may vary between 450 and 1500 mm. The lowest rainfalls are recorded within the Mainz basin at the Rhine River west of Frankfurt, in Franconia at the Main River east of Frankfurt, and in most parts of eastern Germany, especially north and west of Halle. Usually rainfall will increase with altitude. Apart from the higher mountainous regions in south and central Germany, average annual temperatures range between 7° and 10°C within agriculturally important areas. Highest temperatures are recorded from the Rhine River valley. In the south of Bavaria and in the northern lowlands the landscape was formed by the impact of glaciers and melting water. From moraine deposits developed argillic brown earths intermingled with some gley soils, whereas within the sandy areas podsols, gley, and bog soils originated with flood plain soils in between. At the North Sea, coastal marsh soils were formed. Within the mountains built from ancient rocks ranker, podzol-brown earths, and gley soils dominate, whereas the widely spread triassic formation in the central portion developed ranker, pelosols, brown earths, argillic brown earths, rendzina, and terra fusca, according to the type of parent material. Part of the central areas was covered by loess during the last Ice Age (Würm). From this eolian sediment the most fertile soils developed, such as pararendzina, chernozem, and argillic brown earth.

Germany is a small cereal-growing country. In acreage, wheat (*Triticum aestivum* L.) and barley (*Hordeum vulgare* L.) dominate rye (*Secale cereale* L.) and oat (*Avena sativa* L.) in both the former partition states. After the 1990 reunification, 42% of arable land is now grown to wheat and barley and only 13% to rye and oat.[4] Rape (*Brassica napus* L.) is grown all over the country, mainly on wheat soils, whereas sugar beet (*Beta vulgaris* L.) is concentrated on loamy-silty soils derived from loess or glacial till. Apart from local use, potato (*Solanum tuberosum* L.) production is concentrated on lighter soils in the north and east. Areas of traditional forage crops have declined while livestock husbandry has been intensified; the area grown to silage corn (*Zea mays* L.) has been extended, mainly in the cattle-feeding areas North Rhine Westphalia, Lower Saxony, and Bavaria.

Monocultures with only one arable crop in sequence are the rare exception. Usually the crops are grown in rotation, which in the last several years have been drastically simplified with only three, sometimes four, crops left in the rotation cycle. All the rotations include small cereals, and some of them are extended using sugarbeet, potato, or rape. Corn can be included usually in the rotation without the risk of increasing crop rotation diseases. The integration of nitrogen-fixing grain legumes has declined to merely minor importance because of abundant availability of low-price protein feed on the international market. On the other hand, catch-crop growing has increased to about 1.3 million ha because of soil fertility considerations.

6.3. BENEFITS AND DRAWBACKS OF MOLDBOARD PLOWING

The moldboard plow is the most widespread and important implement of primary tillage in Germany. In humid climates, the soil-turning action is highly effective in burying and thereby killing annual and perennial weeds as well as

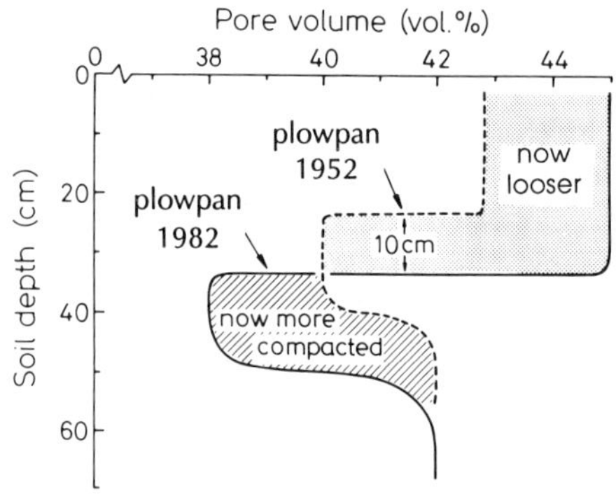

Figure 6.1. Development over 30 years: Deepening of the topsoil and increasing the plow-sole and subsoil compaction.[5]

volunteer crops. The loosening action is excellent, when used at favorable soil conditions, and the clean surface facilitates precision seeding. Leaching of nutrients such as base-forming cations is counteracted. Most important on light, sandy soils is the facility to deeply plow under organic residues, manure, or semiliquid manure mixed with straw, in order to slow the decomposition and thereby amend the water and fertility status of these soils.

Present-day changes are concerned less with the shape and the action of the plow than with the accompanying circumstances of modern farming, where the plow promotes directly and indirectly excessive pan formation and subsoil compaction. It has been demonstrated[5] that within 30 years the plowing depth has been increased by roughly 10 cm, leaving the topsoil looser and the subsoil more dense than before (Figure 6.1). The direct impact of the plow is evident from the abrupt decrease in porosity at the furrow sole. Moreover, loosening weakens the strength of the topsoil. Together with the increased load of common agricultural equipment the diminished surface bearing strength and trafficability causes an increase in depth of soil compaction under heavy machines, the more the soils are rewetted. This may happen during corn or sugar beet harvesting in a rainy autumn.

Modern agronomy facilitates soil erosion by wind and water. Reasons for this trend are well documented[6] by the following examples: plowing up permanent pasture, enlarging arable fields, destroying wind breaks, consolidating scattered holdings, switching over from farmyard manuring to immediate organic residue incorporation, controlling weeds, and simplifying crop rotations. Decreasing the acreage of forage crops such as legumes and grasses and increasing the acreage of corn (and, to a lesser extent, sugar beet) has aggravated the potential soil loss between 1960 and 1983 by 30% in the former Federal Republic of Germany (FRG) and by 60% in Bavaria.[7] The plow is not the tool to counterbalance this development toward increased erosion hazards, as clean tillage will increase the risk of continuing soil degradation.

6.4. TAXONOMY OF CONSERVATION TILLAGE SYSTEMS

Much confusion exists about the concepts of conservation tillage. In Germany the meaning of this term is far reaching and may include all methods that omit the moldboard plow. One proposal[8] emphazises the tillage implement and classifies methods of conservation tillage into those with primary tillage, using cultivators or blade implements that leave mulch at the surface, and those without primary tillage. In both cases seedbed preparation is performed with harrow, rotavator, rotary tiller, or rotary harrow. Seeding is either a separate operation or combined with seedbed preparation. The proposal also includes a method whereby primary tillage, seedbed preparation, and seeding is accomplished jointly. It seems curious that in contrast to the North American definition,[9] zero tillage is excluded from conservation tillage.[8]

Another classification scheme[10] focuses on topsoil strength and the presence of a mulch cover (Table 6.1). This scheme also classifies stubble tillage within the sequence of possible tillage operations. The added expressions of "mulch till" and "reduced till" may not completely cover the meaning they have in the United States, but are used here for further consideration.

Although the practicability of conservation tillage systems has been demonstrated during the past 25 years, the acceptance by farmers remains low. One main reason is the difficulty in management of high residue rates when straw burning is heavily restricted. Another reason is insecurity concerning weed control, when basic relief from weed propagation by burying plants and seeds with the plow is abandoned. It has been estimated[11] that in the former FRG the acreage of sequential conservation tillage comprises about 100,000 ha (i.e., 1.4% of the arable land). In the former German Democratic Republic (GDR) it is nil. For sequential conservation tillage chisel plows, cultivators and rotavators are used.[12] A combination of wing blade cultivator and rotary tiller can loosen the soil to variable depths (Figure 6.2). Noninversion cultivation, seeding, and rolling can be performed in one operation. Another machine combines cultivation and seeding via a rotavator and a specific "seeding bar." Seeds are blown by air pressure through the bar and placed atop the untilled consolidated soil, where they are covered by the lifted surface soil and organic residues[13] (Figure 6.3).

A specific mulch technique is used by some farmers to fight soil erosion in corn and sugar beet. Immediately after harvest of the preceding year's crop, the soil is moldboard plowed or cultivated by a nonturning implement. The preceding crop often is barley, which is removed from the field by the end of July. Then in August, a catch crop is sown that will be terminated in winter by frost (e.g., *Phacelia tanacetifolia* Benth., *Sinapis alba* L.) or in spring after herbicide application (e.g., *Raphanus sativus* L. var. *oleiformis* Pers.). With or without seedbed preparation (using a special dibble seeder), sugar beet or corn are seeded into the mulch. This mulching method, proposed in the early 1970s for sugar beet,[14] has been found to produce sugar yield similar to plowing.[15,16] This

Table 6.1. Tillage Systems and Allied Management Procedures, Including Frequently Used Equipment

Management	Loose-Soil-Cropping	Loose-Soil-Mulch-Cropping		Firm-Soil-Mulch-Cropping	Extreme Form of Firm-Soil-Mulch-Cropping
Procedures	Conventional Till	Low-Mulch Till	Mulch Till	Reduced Till	No-Till
Stubble tillage (<15 cm)	Stubble plow, disc harrow, rotary spade harrow, rotavator, rotary tiller, wing blade cultivator, stubble cultivator, rotary harrow, reciprocating harrow			Rotary spade, harrow, rotary tiller, rotary or reciprocating harrow, wing blade cultivator	
Primary tillage (10–40 cm)	Moldboard plow, disc plow (with crumbling roller and culti-packer)	Chisel plow, cultivator (combined with crumbling and compacting roller), wing blade cultivator	Paraplow, wing blade cultivator, deep loosener		
Seedbed preparation (<8 cm)	Combination of vibrating tine cultivator/ smoothing harrow, cage or Norwegian roller, land roller; combination of rotary or reciprocating harrow with land roller			Rotary tiller, rotary or reciprocating harrow, and land roller; in row rotavator	
Seeding	Common seed drill (row seeding, band sowing, broadcast sowing, precision sowing)		Mulch-seeders (one-, double-, triple disk seeder, put aside drill coulter, seeding bar, dibble seeder)		
Residues on soil surface (% of total)	0	<20	33–66	33–66	100
Total herbicide before seeding	Not necessary	Not necessary	Generally necessary	Generally necessary	Always necessary
Operation performance	Turning up, full depth of topsoil	Grubbing up, generally less than depth of topsoil	Loosening, predominantly in deeper layers	Stirring up, in seeding depth (not always total area)	Unavoidable soil stirring in seed row

Source: Adapted from Baeumer, K., *Integrierter Landbau,* R. Diercks and R. Heitefuß (Eds.) (Munich: BLV Verlagsgesellschaft, 1990).

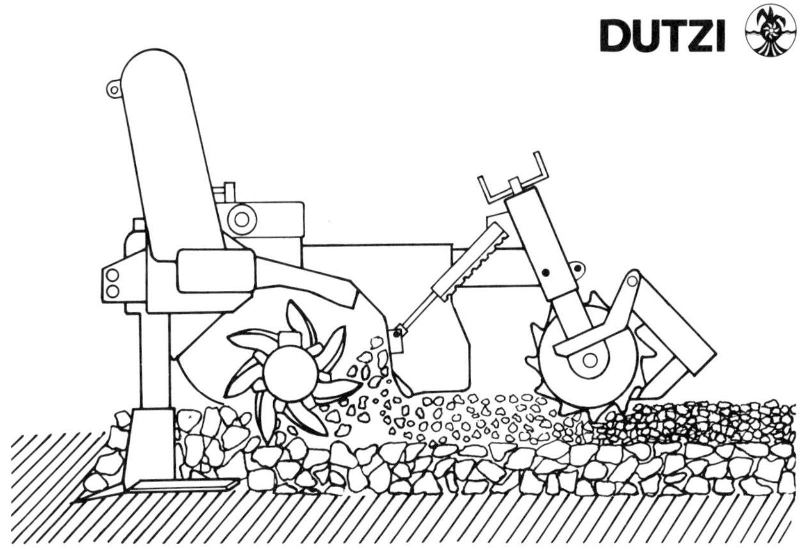

Figure 6.2. Conservation tillage with a machine, which combines wing blades (adjustable to 40 cm depth), rigid tine rotary tiller (5 to 18 cm depth), and packing roller. A seed drill can be attached to the system.

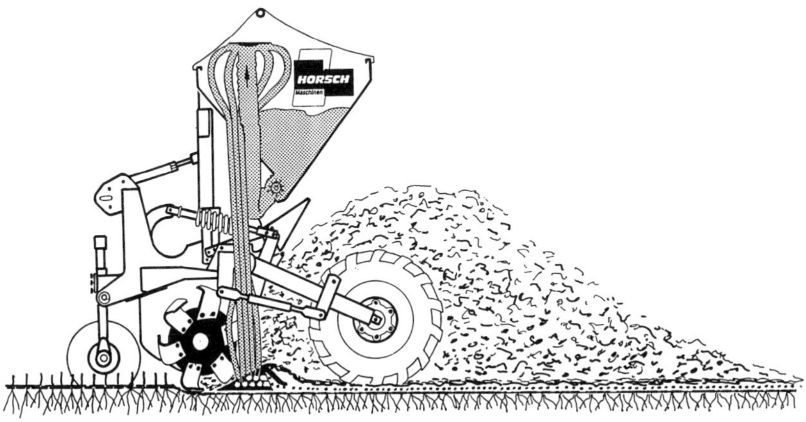

Figure 6.3. One operation of conservation tillage and seeding. The rotavator lifts soil and residues. The seeds are blown through outlets of the seeding bar on top of the untilled firm soil horizon (3 to 6 cm depth). There it is covered with the loosened soil aggregates and with the lighter organic material at the top.

technique is also promising in reducing wind erosion in the northern lowlands on sandy soils, where sequential conservation tillage is not suitable.

The acceptance of conservation tillage may be promoted by new technologies that allow loosening of compacted layers in top- and subsoil by a noninversion action. For example, the paraplow breaks up the soil without much soil surface disturbance and can be applied when conservation tillage already has been started.[17] The newly developed slit plow connects top- and subsoil by grooves, which allow water and gas transport as well as rooting.[18] This technique may present a successful strategy for starting conservation tillage on compacted soils.

6.5. SOIL RESPONSES INDUCED BY CONSERVATION TILLAGE

Omitting the turning action of moldboard plowing induces a stratification and concentration of various chemical compounds in the soil. At one of the oldest ongoing zero tillage experiments in Germany, which was started on loess-derived soil near Göttingen in 1967, organic carbon content almost doubled in the uppermost topsoil (0 to 2.5 cm) over a 20-year period, when compared to the tilled treatment. Like organic carbon, total nitrogen also accumulates in the topsoil.[19,20] This trend in time is shown in Figure 6.4. During the initial years of zero tillage some 800 to 900 kg of carbon and 70 to 90 kg of nitrogen have been enriched per year, in comparison to plowing. This period of accumulation seems to come to an end after 8 to 10 years.

As with zero till, reduced till (Table 6.1) also concentrates organic carbon (Figure 6.5),[22,23] and subsequently causes an increase in enzyme activity of soil microorganisms[17,22,23] and aggregate stability[17] (Figure 6.5). Conservation tillage causes stratification of a variety of more or less immobile plant nutrients,[22,24] but until recently it has not been proven that layering reduces nutrient acquisition by plants. Higher amounts of "available" potassium and phosphorus support the idea[24] that in a suboceanic climate the nutrient supply function of the soil may be improved by stratification. Also, protons are enriched near the soil surface by conservation tillage.[22] Therefore, lime requirement must be assessed carefully, and liming may be necessary more often but in smaller applications than with plowing.

Minimizing soil disturbance in depth may activate biotic processes, because the abundance of the mesofauna and macrofauna will increase[25,26] (Table 6.2). Under conservation tillage the effects of earthworms on soil properties are especially conspicious. The animals puncture the soil and their burrows connect the soil surface with the subsoil (Figure 6.6), allowing rapid water infiltration during heavy rainfall.[27-30] Earthworm channels also serve as a preferential growth medium of low resistance for roots.[29,31] On the other hand, the biotic channels created by the growing and expanding roots themselves will also stay preserved in a nonswelling soil, when not plowed, and will support additional

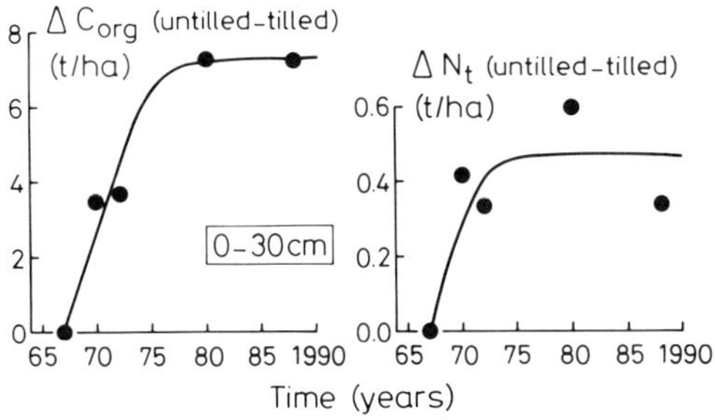

Figure 6.4. Concentration of organic carbon and total nitrogen in no-till as compared to plowed silt loam (0 to 30 cm) as a function of time. Last plowing in no-till: 1966. Data source: 1970 and 1988;[21] 1972;[19] 1980.[20]

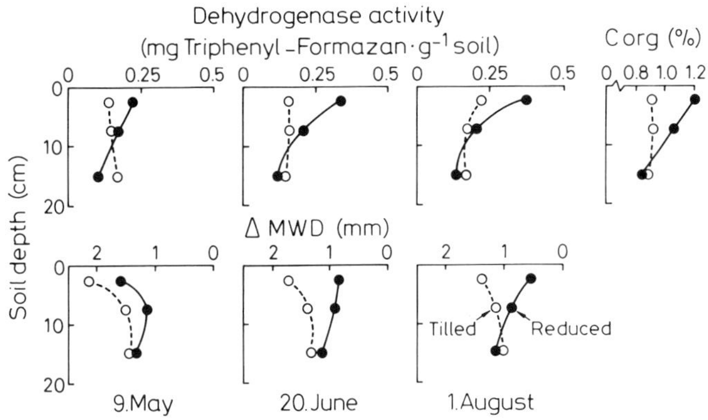

Figure 6.5. Profiles of dehydrogenase activity and aggregate stability in plowed and reduced till silt loam soil at three dates in 1985, 12 years after last plowing on reduced till. Profiles of organic carbon are also shown. The term Δ MWD means change in mean weight diameter of aggregates after wet sieving as compared to dry sieving.

root growth of subsequent crops,[31] but also water and air conductance.[32] In the tilled topsoil (0 to 10 cm) unsaturated hydraulic conductivity near saturation at 10 hPa is similar to no-till (0 to 10, 20 to 30 cm), although total porosity and porosity of 30 to 300 μm pores are enlarged by plowing (Figure 6.7). This means the continuity of 30 to 300 μm pores is higher in no-till[32] as compared to plowing. Hydraulic conductivity in the 20 to 30 cm pan layer of tilled soil with low porosity is comparatively extremely reduced.[33] This obstruction zone, which impedes both water flow and root growth, may disappear on loess-derived soil within a few years of no-till (Figure 6.8).[34] Whereas bulk density on no-till generally decreased in time, the pan layer in the plowed soil was accentuated (Figure 6.8).

Table 6.2. Effect of Tillage on Biomass and Abundance of Earthworms, and Straw Decomposition by Earthworms during a Vegetation Period (loam soils near Giessen)

Tillage	Biomass (g/m^2) (total)	Abundance of Adults ($no./m^2$) *L.t.*[a]	*A.c.*[b]	*A.r.*[c]	Straw Decomposition Rate (t/ha/year)[d]
Plow	38	2	7	4	0.5
Reduced	62	5	18	8	1.2
No-till	176	18	57	30	2.0

Source: From Friebe, B. and Henke, W., *Z. Kulturtech. Landesentwicklung,* 32, 121, 1991.

[a] *Lumbricus terrestris* L.
[b] *Allolobophora caliginosa* Savigny.
[c] *A. rosea* Savigny.
[d] Out of 6 t/ha.

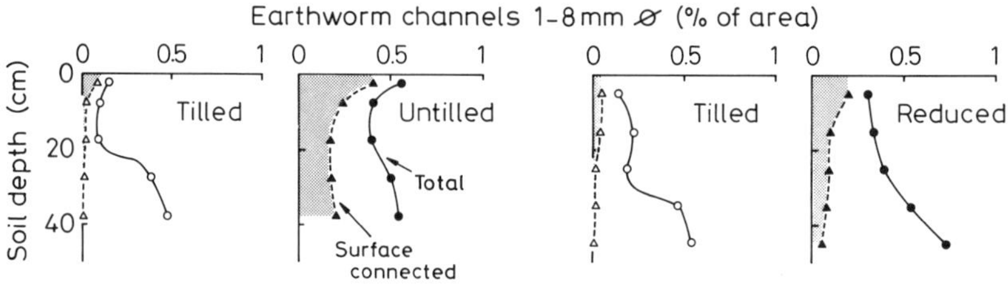

Figure 6.6. The effect of conservation tillage (no-till, reduced till) and of plowing of a silt loam on the percentage area of earthworms channels in total, and of surface-connected channels. Surface connection was ascertained by gypsum infiltration. Last plowing in no-till: 11 years; on reduced-till: 12 years before.

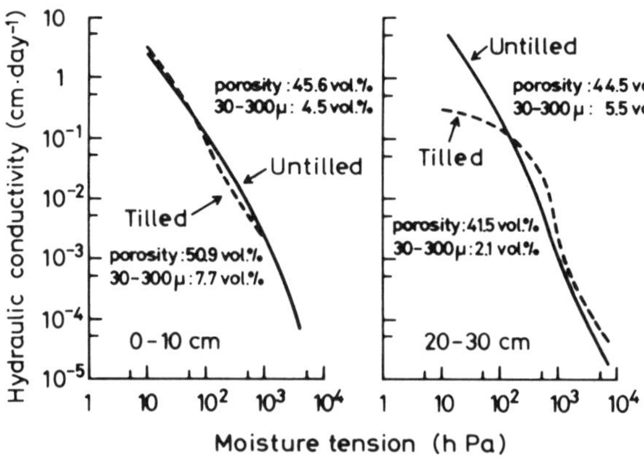

Figure 6.7. Hydraulic conductivity function in plowed and no-till silt loam. The 20 to 30 cm layer of the plowed soil is a compacted layer. Last plowing in no-till: 6 years before.

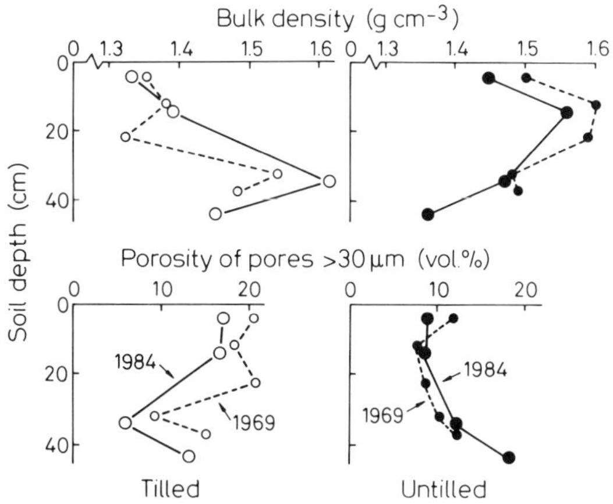

Figure 6.8. Profiles of bulk density and porosity of pores >30 µm in plowed and no-till silt loam in 2 years. Last plowing in no-till: 1966.

In those soils, in which concentration of organic matter near the soil surface gives rise to an activation of the soil macrofauna, rootability, water infiltration, and transmission as well as aeration will be optimum. Sufficient transfer functions combined with improved trafficability[35-37] due to higher bulk density (Figure 6.8) and soil strength are a miracle of many no-till soils, analogous to properties of grassland soils. Oriented preferably in a vertical direction, the biopores can support high loads without collapse.[38] On soils in which earthworms are missing, but which contain a high percentage of swelling clay minerals, vertical cracking can take over these transmission properties. Crack formation is usually favored by conservation tillage on soils with "vertic" properties.[39] However, the positive feedback of conservation tillage on soil fauna and soil properties is time dependent and cannot be expected to occur on a large scale within one growing season, during only one phase of a crop rotation.

On the above soils, compaction associated with heavy machinery may be counterbalanced by adjusting load levels, use of broad low-inflation tires, and a combination of working operations such as shallow seedbed preparation and seeding. Trafficability is improved by increased soil strength but also by relatively high moisture tension during rain.[27]

Not all the soils in Germany will show these counterbalancing actions under conservation tillage. Sandy and sandy loam soils, especially those low in organic carbon and of mixed grain size, are extremely sensitive to compacting forces which result in a reduction to root growth.[40,41] As these sandy soils lack the capability to restore structure effectively by endogenous forces,[42] they must be plowed on a regular basis. Clay soils, containing nonswelling clay minerals, may compact under agricultural vehicular traffic to such an extent that soil air content and gas diffusion will limit root and plant growth, as long as the soil is wet. When dry, however, soil strength may hinder root growth. Under these conditions, plowing will increase porosity by loosening, supported by the frost action during winter. For reasons similar to those given above, the idea of conservation tillage seems unpractical on soils with a perched water table. On those soils the plow is the only implement that will lift the topsoil above the zone of anaerobis.[39]

6.6. SUITABILITY OF SOILS FOR SEQUENTIAL CONSERVATION TILLAGE

Based on local experiences and what is known from the international literature it was concluded[39] that conservation tillage methods are feasible on those soils where positive feedback responses led to structure amelioration, stabilization, and tilth in the topsoil. The accessibility of the subsoil for roots must not be restricted by conservation tillage practices, but at best will be improved by omitting plowing. The texture of these soils range from sand and sandy loams, including humus-rich plaggen soils, to soils rich in clay with vertic properties. They belong to the orders of entisols, inceptisols, mollisols and alfisols, and their areal distribution is presented in the map of Germany in Figure 6.9. Apart from those of calcareous origin the nonforested mountainous soils have been omitted, as they seem unsuitable for sequential conservation tillage. The same is true for many sandy podzols and all soils that are subject to wetness, including poorly drained soils with a perched water table (pseudogley), gley soils, and most of the floodplain soils. Within a catena the water regime may change within short distances, making the adoption of conservation tillage even more difficult.

The success of conservation tillage with respect to soil response and crop growth depends also on climatic variables such as temperature and rainfall. Therefore, the isohyet of 700 mm annual precipitation is shown in the map as a rough borderline. The drier (and warmer) the areas are, the more suitable they seem in principle for conservation tillage. On the other hand, high annual amounts of rainfall will limit the priority of conservation tillage, for instance, on argillic brown earths from moraine deposits north of the Alps in southern Bavaria (Figure 6.9).

6.7. ADVANTAGES AND DRAWBACKS OF CONSERVATION TILLAGE

6.7.1. Nitrogen Fertilization

With conservation tillage, part of fertilizer nitrogen is immobilized, fixed, and enriched in the soil's organic matter (Figure 6.4). A high net immobilization rate is opposed by a lower net mineralization rate,[48] and therefore the nitrate-nitrogen concentration in soil solution is reduced as compared to tilled soil,[49,50] which can be best observed in the absence of root crops. This effect lowers the efficiency of nitrogen-fertilization ("Oat" in Figure 6.10), but only for a certain period of years. Thereafter, the rate of net mineralization still may be smaller in conservation tillage, but the smaller rate will be compensated for by the larger pool of organic nitrogen, so that over time the supply of soilborne nitrogen and fertilizer efficiency will approximate in both tillage systems ("Wheat" in Figure 6.10). In Figure 6.10 sequential conservation tillage with oat was performed between years 3 and 7, but with wheat it was between years 9 and 14. From this it can be concluded that the period of nitrogen accumulation will end after a period of about 10 years (Figure 6.4).

The practical implication is that the applied split applications of nitrogen fertilizer have to be balanced according to the tillage system. For instance, the first fertilizer application for winter wheat in spring during the middle to end of tillering must be stressed more in conservation than in conventional tillage. Due to a more steady net mineralization in time, the second and third applications can be lowered in the conservation system. In total, however, the fertilizer input will be more or less the same, irrespective of the tillage system, after the state of equilibrium has been achieved.

6.7.2. Soil Temperature and Water

In most years soil temperature at seeding depth in the spring is lower in the mulch layer under conservation tillage system as compared to the plowed soil. This situation is reversed in autumn. An exception to the above can occur where the slaked surface of the tilled soil has increased the albedo considerably after drying. Lower soil temperatures in the spring under conservation tillage can affect the early development of sugar beet and corn, but usually not the final yield.

Water storage in soil is improved by conservation tillage, which may be beneficial to plant growth and yield. Until recently conclusive evidence specific to Germany was still missing in this respect.[51] Minor effects are apparent, such as a moister topsoil with reduced till during dry spring months. An alternate situation may occur in the summer after cereal harvest, when soil material with higher moisture content is plowed up from deeper layers, facilitating catch crop germination on tilled soil. Under conservation tillage systems, germination may be delayed.

Altogether a move toward conservation tillage was expected after the 1990 reunification, as in large areas of eastern Germany precipitation is low (Figure 6.9), ranging between 480 and 600 mm/year. Increasing infiltration and reducing runoff results in more plant-available water and less erosion. Those matching aspects have never been stressed and comprehensively considered in the research and development of soil management systems in Germany, however.

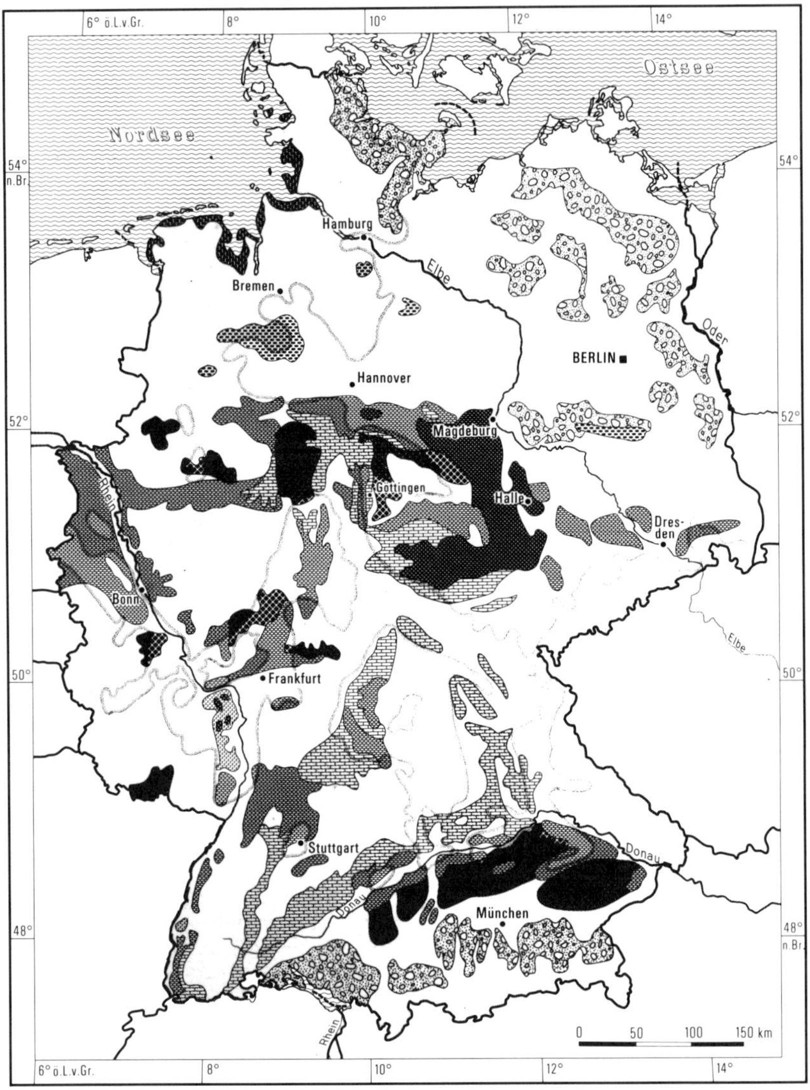

Figure 6.9. Distribution of soils in Germany, suitable for sequential conservation tillage. The isohyet of 700 mm/year precipitation (former GDR: 720 mm) is also shown. The compilation is based on References 43 to 47.

6.7.3. Crop Residues

Mulch is an essential component of any soil conservation practice. In contrast to various parts of the world, the amounts of residue in Germany are quite large. On average, winter wheat produces 6.6 t and winter barley 6.1 t/ha of grain yield. Assuming a harvest index of 0.5 results in equal yields of straw.

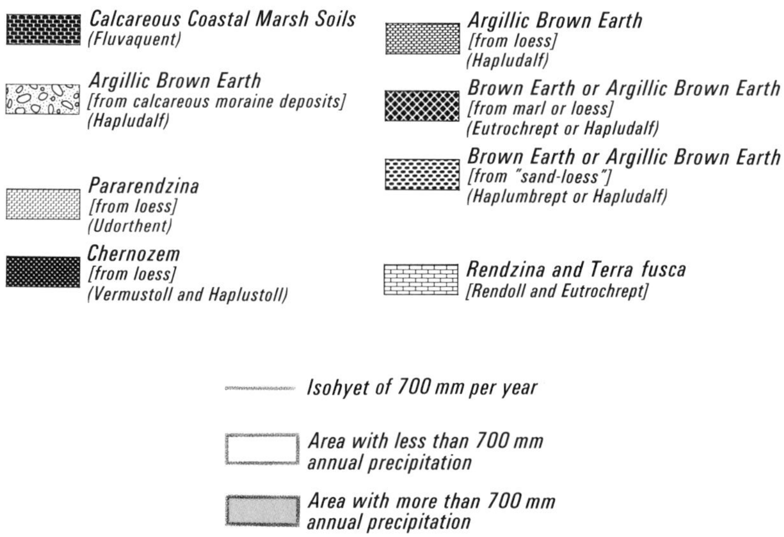

Figure 6.9. (continued)

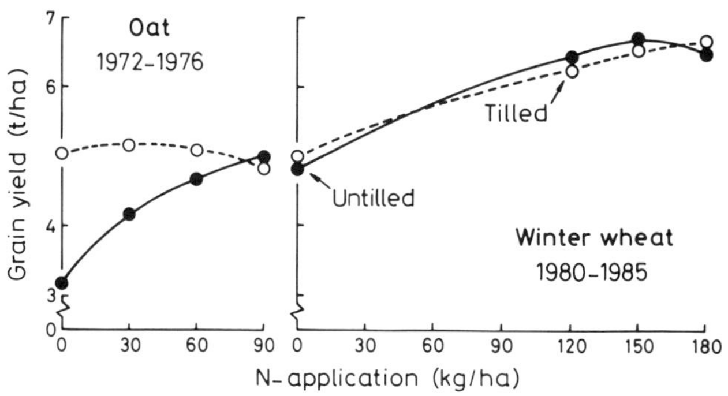

Figure 6.10. Grain yield of oat and winter wheat on silt loam as a function of applied nitrogen. Last plowing in no-till for oats in 1969, for wheat in 1971.[49]

Leading farmers are confronted with still-higher quantities, sometimes >10 t/ha. The problem is to allow the straw to decompose to such a degree that it will not negatively influence the establishment of the subsequent crop. This is more difficult to achieve as the time span between crops decreases. High amounts of uncut residues can drastically reduce the performance of the seed drill, i.e., the uniform seed placement in horizontal and vertical directions. The rate of emergence and seedling growth may also be lowered by a release of toxic compounds from straw in the initial stages of decomposition.

In conventional tillage systems the problem is solved by plowing the straw under. In conservation tillage it is important that the combine harvester cuts and evenly distributes the straw at the surface. Then the straw is mixed into the soil by PTO driven rotavating implements (Table 6.1). By choice of machine and working depth the farmer can decide within limits how much residue should be left at the surface. Microbial decomposition leads to aggregate stability (Figure 6.5) and the activity of earthworms will also contribute to straw breakdown (Table 6.2). Straw decomposition has been accelerated by splitting the straw longitudinally.[52] This kind of splitting opens the inner parts of the stalk and enlarges the surface for microbial attack.

6.7.4. Weeds and Volunteer Crops

Besides its ability to handle residues, the moldboard plow is also unsurpassed in deeply burying weed plants and their seeds. Under these circumstances, seeds of short viability may die before the next plowing shifts them back to near the soil surface. With conservation tillage systems, however, the seeds usually will be incorporated into the soil at a shallow depth, promoting weed infestation in the following crop. Weed control by mechanical and chemical means is thus absolutely necessary. If weed seed production is not prevented regularly, a large seed bank in the soil will occur over time[53] (Table 6.3). In rotations dominated by small cereals, grass weeds such as *Agropyron repens* (L.) P. Beauv., *Apera spica-venti* (L.) P. Beauv., *Alopecurus myosuroides* Huds., and more recently *Bromus sterilis* L. may proliferate because they are very well adapted to the ontogenesis of small-grain cereals. Beside chemical and mechanical means in the cereal crop these grass weeds can be controlled very effectively by the incorporation of spring sown row crops (corn, faba bean [*Vicia faba* L.], sugar beet) into the crop rotation, as these row crops can be hoed.

Successful weed control is a challenge to the farmer who adopts conservation tillage. The efficiency of some herbicides may be lowered by the concentration of organic matter near the soil surface (Figure 6.5). Furthermore, the use of some very active herbicides, including hormone weedkillers, is restricted within protected water collection areas. Overall, if weed stands can be limited, weed infestation in conservation tillage should not present a greater problem than in conventional tillage (Table 6.3).

Propagation of volunteer crops must be watched in conservation tillage. One problem is vigorous barley volunteers affecting the growth of sugar beet plants. Relief may be achieved by cultivation of the barley stubble and by growing a competitive catch crop, as both methods may control the germinating barley.[55] Another unwelcome situation is volunteer wheat in barley, which can hardly be controlled without moldboard plowing.

6.7.5. Pests

While potential weed density may be repressed by intensive soil manipulation, a similar relation between cultivation intensity and disease and pest

Table 6.3. Weed Seed Bank in Soil and Stand of Emerged Weeds in Spring with Different Tillage Systems

Primary Tillage	Seed Bank (no./m^2) (Seeds Capable of Germinating)	Emerged Weeds	
		% of Seed Bank	No./m^2
Moldboard plow	3312	3.0	99
Chisel plow with rotary harrow	4077	4.2	171
Wing blade cultivator with rotary harrow	4554	4.8	219
No-till	5637	1.8	101

Source: Adapted from Bräutigam, V., *Z. Pflanzenkr. Pflanzenschutz* SH, 12, 219, 1990. With permission.

establishment is less obvious. The epidemiology of airborne diseases such as *Erysiphe graminis* D.C. or of harmful flying insects such as aphids as vectors is influenced by tillage intensity to only a minor extent. With soilborne pests circumstances are less obvious and the occurence of pests may interact with tillage and crop rotation. A corn field attacked by European corn borer (*Ostrinia nubilalis* Hb.) will give rise to a propagation of this insect when conservation tillage and not deep conventional tillage is applied, but only in the case when corn follows corn.[56] A change in crop rotation or use of insecticides will also control this pest in conservation tillage.

Although conservation tillage leaves crop residues on the soil surface, which serve as a focus of infection, evidence indicates that cereal eyespot (*Pseudocercosporella herpotrichoides* [Fron.] Deigh.) will not be increased in winter wheat (Table 6.4) by mulching.[57,58] The promotion of soil organisms by surface mulching and the activated turnover rate of organic material seem to suppress some of the soilborne fungi.

In the very early developmental stages, sugar beet plants may be attacked by springtails (*Onychiurus* spp.), which feed on hypocotyl, cotyledons, and seminal roots, eventually killing the beet plant. In conventional tillage springtails are controlled by insecticides. In conservation tillage chemicals are needed less, as an abundant food supply of organic residues is available for the springtails. Therefore, the precision-seeded plants will be less injured (Table 6.5).[10,59]

6.7.6. Environmental Concerns

It has already been pointed out that traditional loose soil crop husbandry is the primary cause for excessive top- and subsoil compaction under heavy load. On the other hand, high soil strength and mulch at the surface will counteract the compactive forces, when the soil has been stabilized under a sequential conservation tillage method.

When conventionally tilled, a wide range of medium-textured soils show signs of surface slaking, aggregate breakdown, reduced infiltrability, and soil erosion by water during heavy rain showers. In Bavaria[7,60] the monthly

Table 6.4. Influence of Tillage on Cereal Eyespot Infestation in Winter Wheat

Soil	Tillage (infestation in %)			
	Moldboard Plow	Chisel Plow	Wing Blade Cultivator	No-Till
Hanau (silty sand)	41	27	24	15
Wernborn (silty loam)	43	46	39	13
Ossenheim (silty loam)	32	20	16	10

Source: From Tebrügge, F., *DLG-Mitteilungen,* 105, 592, 1990. With permission.

Table 6.5. Population Density of Harmful Springtails, Number of Damaged Seedlings, and Dry Weight of Sugar Beet Plants in the 6–8 Leaf Stage as Related to Tillage and Insecticide Application

Tillage	Insecticide	Density of Springtails (no./100 cm³ soil)	Damaged Seedlings (no./100 plants)	Dry matter (g/plant)
Plow	No	1.37	346	2.19
	Yes		97	2.47
Reduced	No	4.22	287	2.39
	Yes		35	2.74

Source: From Baeumer, K., *Integrierter Landbau,* R. Diercks and R. Heitefuß (Eds.) (Munich: BLV Verlagsgesellschaft, 1990). With permission.

percentage of the erosivity factor R of the universal soil-loss equation (USLE) peaks in summer from May to September (Figure 6.11), and the R factor distribution within the remaining part of Germany is assumed not to differ largely. If not protected, sheet erosion starts within the interrows, whereas tracks are the starting point of gully erosion. As the figure shows, the canopy of sugar beet, and especially of corn, does not develop coincidently with the R factor, but coverage lags behind.

Figure 6.12 indicates[7] that the cropping factor C of the USLE will increase with the area cropped to corn, but can be reduced by mulching, track loosening, and interseeding of barley. These different techniques of erosion control linked to loose soil cropping are much less effective than a technique of sequential conservation tillage, the Horsch System,[13] shown in Figure 6.3.

Although the soil conserving effects brought about by conservation tillage are evident, there is some concern with respect to herbicide application and groundwater pollution. Reduction in tillage intensity usually means an increasing need for herbicide application (i.e., mainly the number of application dates). However, pesticides applied to soils under conservation tillage are relatively immobile as are other chemicals, whereas in plowed soils pesticides and fertilizers may be washed off by surface runoff and by eroded soil, thus polluting surface watercourses. Nonpoint pollution of surface water is regularly recorded in Hesse.[61]

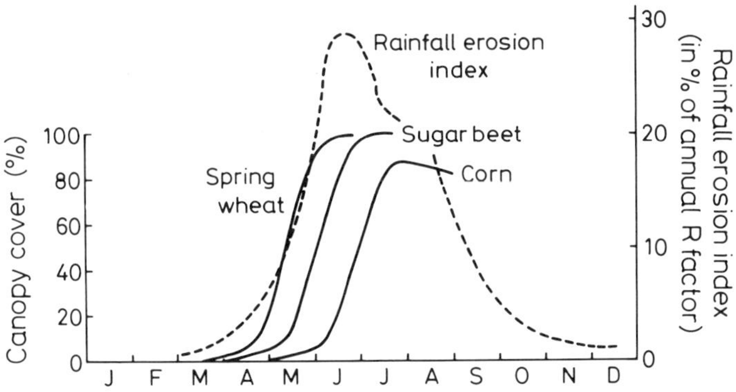

Figure 6.11. Canopy cover of three crops and monthly percentage of the average rainfall erosion index in Bavaria as a function of time. (From Schwertmann, U. and Vogl, W., *Reihe Kongreßberichte, Kongreßband 1985 Gießen* (Darmstadt: VDLUFA-Verlag, 1986). With permission.)

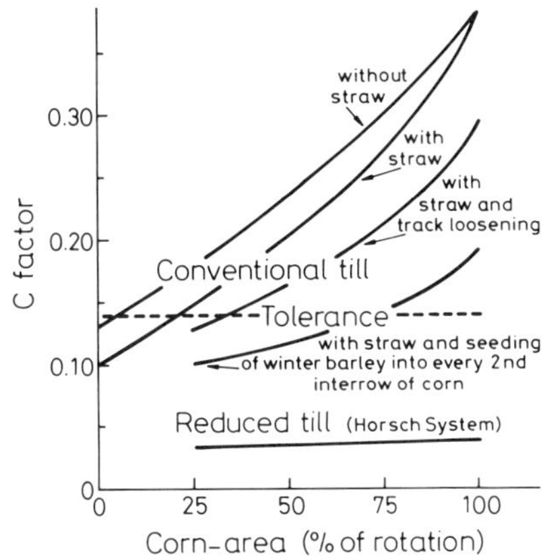

Figure 6.12. The cropping factor *C* as a function of percentage of corn in rotation and of various soil management strategies. The tolerance level is related to a 100-m 9% slope with loess soil. (From Schwertmann, U. and Vogl, W., *Reihe Kongreßberichte, Kongreßband 1985 Gießen* (Darmstadt: VDLUFA-Verlag, 1986). With permission.)

There are two opposing arguments on the impact of macrochannels on possible leaching and groundwater pollution. Sound arguments that bypass flow may either raise or diminish solute transport have not as yet presented for field situations. Both cases probably can happen, depending on circumstances.[39]

6.7.7. Crop Yields

Several investigations have shown that crop yield is not necessarily lower with conservation tillage as compared to conventional plowing, when the necessity of higher nitrogen fertilization within the initial years of sequential

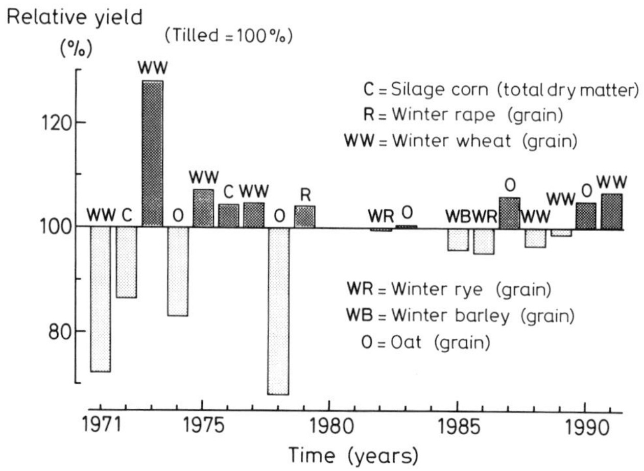

Figure 6.13. Time trend of relative yield of various crops grown on plowed and reduced till silt loam in Göttingen. The absolute yield ranged between (t/ha): C = 9.6 to 12.8; R = 2.2 to 2.3; WW = 3.1 to 8.5; WB = 5.6 to 5.9; WR = 4.5 to 4.8; and O = 3.4 to 5.4.

conservation tillage is recognized. During the course of time soil structure may improve and the farmer will learn how to manage the system. Therefore, it is not surprising that the yield of crops grown with conservation tillage may increase in time as compared to crops grown conventionally, although the reverse may happen in some cases.[62] A long-term field experiment on loessial soil indicated a positive time trend in relative yield of various crops (Figure 6.13). During a time span of 13 years, yield differences for a number of crops including small cereals, silage corn, rape, and sugar beet proved to be minimal on silt-loams between plowed and no-till treatments.[62] In another field experiment on silty sand, however, sugar beet failed with no-till but not with other conservation tillage methods.[63] Lower yields of various crops such as small cereals, rape, and faba bean were reported for no-till as compared to plowed plots from field experiments on loamy soil with signs of impeded drainage.[15]

6.8. GENERAL SUMMARY AND CONCLUSIONS

The traditional, geographic, climatic, and pedologic background of crop husbandry in Germany is described. Germany is a small cereal-growing country, in which moldboard plowing dominates. Soil compaction and erosion are environmental concerns that justify the development of conservation tillage techniques. Physical, chemical, and biological soil responses to nonplowing are characterized. Not all soils are suitable for sequential conservation tillage and a map of qualified soils within United Germany is presented. Peculiarities and difficulties are listed that deal with the management of conservation tillage systems. Fertilization, pest and weed control, residue management, as well as environmental problems with respect to higher chemical input are major points of discussion. Although extensive field experimentation as well as trials on farms have shown parity in yield, the acceptance by farmers is still low, mainly because of difficulties in residue management and concerns for weed propagation. Nevertheless, industrial enterprises have developed machines for conservation tillage with some remarkable innovation. New developments in technology raise hope that in forthcoming years conservation tillage can contribute to the goals of sustainable agriculture on a larger scale than currently available.

6.9. ACKNOWLEDGMENTS

We thank Dr. K. Priesnitz, Institute of Geography, for his kind cooperation. We are indebted to Mr. E. Höfer, Institute of Geography, for drawing the soils' map with great skill. All the other figures (except Figures 6.2 and 6.3) were prepared by Mrs. A. Bartlitz with great accuracy. We are also indebted to Dr. K. Baeumer and Dr. M. J. Goss, who reviewed the manuscript.

6.10. REFERENCES

1. Schultz-Klinken, K.-R. "Zur Geschichte der Bodenbearbeitung," *Ber. Landwirtsch.* 56:277–288 (1978).
2. Rösener, W. *Bauern im Mittelalter* (Munich: C.H. Beck, 1987), p. 335.
3. Baeumer, K. "First Experiences with Direct Drilling in Germany," *Neth. J. Agric. Sci.* 18:283–292 (1970).
4. Statistisches Bundesamt, Ed. *Statistisches Jahrbuch 1991 für das vereinte Deutschland* (Wiesbaden: Wiesbadener Graphische Betriebe GmbH, 1991).
5. Sommer, C. "Ursachen und Folgen von Bodenverdichtungen sowie Möglichkeiten zu ihrer Verminderung," *Landtechnik* 9:378–384 (1985).
6. Schröder, D. "Ursachen und Ausmaß der Erosion," *Ber. Landwirt.*, SH 205:16–27 (1991).
7. Schwertmann, U., and W. Vogl. "Landbewirtschaftung und Bodenerosion," in *Reihe Kongreßberichte, Kongreßband 1985 Gießen* (Darmstadt: VDLUFA-Verlag, 1986), pp 7–17.
8. KTBL-Arbeitsgruppe 'Bodenbearbeitung und Bestellung'. "Definition und Einordnung von Bodenbearbeitungsverfahren," KTBL-Arbeitsblatt 0236. (Lehrte, Germany: E. F. Beckmann KG, 1988).
9. Mannering, J. V., Schertz, D. L., and B. A. Julian. "Overview of Conservation Tillage," in *Effects of Conservation Tillage on Groundwater Quality,* T. J. Logan, J. M. Davidson, J. L. Baker, and M. R. Overcash (Eds.) (Chelsea, MI: Lewis Publishers, 1987), pp. 3–17.
10. Baeumer, K. "Verfahren und Wirkungen der Bodenbearbeitung," in *Integrierter Landbau,* R. Diercks and R. Heitefuß, Eds. (Munich: BLV Verlagsgesellschaft, 1990), pp. 68–87.
11. Köller, K. Personal communication.

12. Buchner, W., and K. Köller. *Integrierte Bodenbearbeitung* (Stuttgart: Eugen Ulmer, 1990), p. 127.
13. Horsch, D. "Reduzierte Bodenbearbeitung, angepaßte Saattechnik und Unkrautbekämpfung nach dem System HORSCH," in *Integrierter Landbau,* R. Diercks, and R. Heitefuß (Eds.) (Munich: BLV Verlagsgesellschaft, 1990), pp. 273–281.
14. Baeumer, K., and G. Pape. "Ergebnisse und Aussichten des Anbaus von Zuckerrüben im Ackerbausystem ohne Bodenbearbeitung," *Zucker* 25:711–718 (1972).
15. Baeumer, K. "Warum reduzierte Bearbeitung," *Arb. Dtsch. Landwirtschaftsgesell.* 179:98–113 (1984).
16. Sommer, C., Zach, M., and K. Korte. "Mit konservierender Bearbeitung mehr Bodenschutz im Zuckerrübenbau," *Zucker.* 36:58–63 (1987).
17. Ehlers, W., and K. Baeumer. "Effect of the Paraplow on Soil Properties and Plant Performance," *Proc. 11th Conf. International Soil Tillage Research Organization,* Edinburgh, (1988), pp. 637–642.
18. Steinert, P., Werner, D., and J. Lhotsky. "The Effects of Rooting on the Structure of Compacted Soil Zones after Slit Ploughing," *Sci. Agric. Bohemoslov.* 22:85–92 (1990).
19. Fleige, H., and K. Baeumer. "Effect of Zero-Tillage on Organic Carbon and Total Nitrogen Content, and Their Distribution in Different N-Fractions in Loessial Soils," *Agroecosystems* 1:19–29 (1974).
20. Lawane, G. "Mengenänderungen der organischen Bodensubstanz bei unterschiedlicher Bearbeitungsintensität," Ph.D. thesis, (Göttingen: University of Göttingen, 1984).
21. Tschirsich, C. "Die langfristige Veränderung der Phosphorgehalte und anderer Bodenkennwerte in einem Direktsaat-Fruchtfolge-Versuch," Diplom-Thesis (Göttingen: University of Göttingen, 1989).
22. Diez, Th., Kreitmayr, J., Weigelt, H., Bauchhenß, J., Beck, Th., and H. Borchert. "Einfluß langjähriger pflugloser Ackerbewirtschaftung (System HORSCH) auf Pflanzenwachstum, Wirtschaftlichkeit und Boden," *Bayer. Landwirtsch Jahrb.* 65:789–812 (1988).
23. Grocholl, J., Böhm, H., and E. Ahrens. "Einfluß unterschiedlicher Bodenbearbeitungssysteme auf die Bodenmikrobiologische Kennzahl und den Ct-Gehalt verschiedener Bodenarten und Standorte," in *Wechselwirkungen von Bodenbearbeitungssystemen auf das Ökosystem Boden,* Fachbereich Agrarwissenschaften, Ed. (Gießen, Germany, Justus-Liebig-Universität Gießen, 1989), pp. 53–62.
24. Ehlers, W., Pape, G., and W. Böhm. "Tiefenverteilung und zeitliche Änderungen der laktatlöslichen Kalium- und Phosphorgehalte während einer Vegetationsperiode in unbearbeiteten und bearbeiteten Böden," *Z. Pflanzenernaehr. Bodenkd.* 133:24–36 (1973).
25. Schwerdtle, F. "Untersuchungen zur Populationsdichte von Regenwürmern bei herkömmlicher Bearbeitung und bei 'Direktsaat'," *Z. Pflanzenkr. Pflanzenschutz* 76:635–641 (1969).
26. Friebe, B., and W. Henke. "Bodentiere und deren Strohabbauleistungen bei reduzierter Bodenbearbeitung," *Z. Kulturtech. Landesentwicklung* 32:121–126 (1991).

27. Ehlers, W., and K. Baeumer. "Soil Moisture Regime of Loessial Soils in Western Germany as Affected by Zero-Tillage Methods," *Pak. J. Sci. Ind. Res.* 17:32–39 (1974).
28. Ehlers, W. "Observations on Earthworm Channels and Infiltration on Tilled and Untilled Loess Soil," *Soil Sci.* 119:242–249 (1975).
29. Goss, M. J., Ehlers, W., Boone, F. R., White, I., and K. R. Howse. "Effects of Soil Management Practice on Soil Physical Conditions Affecting Root Growth," *J. Agric. Eng. Res.* 30:131–140 (1984).
30. Beisecker, R., Lütkemöller, D., Gäth, S., and H.-G. Frede. "Impact of Macropores on the Water and Solute Transport in Soils. I. Field Observations and Measurements," paper presented at the International Workshop on Methods of Research on Soil Structure/Soil Biota Interrelationships, Wageningen, The Netherlands, November 1991.
31. Ehlers, W., Köpke, U., Hesse, F., and W. Böhm. "Penetration Resistance and Root Growth of Oats in Tilled and Untilled Loess Soil," *Soil Tillage Res.* 3:261–275 (1983).
32. Teiwes, K. "Einfluß von Bodenbearbeitung und Fahrverkehr auf physikalische Eigenschaften schluffreicher Ackerböden," Ph.D. thesis (Göttingen: University of Göttingen, 1988).
33. Ehlers, W. "Measurement and Calculation of Hydraulic Conductivity in Horizons of Tilled and Untilled Loess-Derived Soil, Germany," *Geoderma* 19:293–306 (1977).
34. Ehlers, W. "Strukturzustand und zeitliche Änderung der Wasser- und Luftgehalte während einer Vegetationsperiode in unbearbeiteter und bearbeiteter Löß-Parabraunerde," *Z. Acker- Pflanzenbau* 137:213–232 (1973).
35. Horn, R. "Auswirkungen unterschiedlicher Bodenbearbeitung auf die mechanische Belastbarkeit von Ackerböden," *Z. Pflanzenernaehr. Bodenkd.* 149:9–18 (1986).
36. Steinkampf, H., and C. Sommer. "Zugkraftübertragung und Bodenverdichtung durch Reifen." Circ. German Agricultural Society (DLG), 1988.
37. Gruber, W., and F. Tebrügge. "Influence of Different Tillage Systems on Trafficability and Soil Compaction," paper no. 90-1090 (Columbus, OH: ASAE International Summer Meeting, June 1990).
38. Hartge, K. H., and C. Sommer. "The Effect of Geometric Patterns of Soil Structure on Compressibility of Soil," *Soil Sci.* 130:180–185 (1980).
39. Ehlers, W. "Wirkung von Bearbeitungssystemen auf gefügeabhängige Eigenschaften verschiedener Böden," *Ber. Landwirtsch., Sonderh.* 204:118–137 (1991).
40. Petelkau, H. "Auswirkungen von Schadverdichtungen auf Bodeneigenschaften und Pflanzenertrag sowie Maßnahmen zu ihrer Minderung," *Tagungsber. Akad. Landwirtschaftswiss. DDR* 227:25–34 (1984).
41. Lehfeldt, J. "Auswirkungen von Krumenbasisverdichtungen auf die Durchwurzelbarkeit sandiger und lehmiger Bodensubstrate bei Anbau verschiedener Kulturpflanzen," *Arch. Acker- Pflanzenbau Bodenkd.* 32:533–539 (1988).
42. Petelkau, H. "Ursachen, Entstehung und Prinzipien zur Einschränkung von Bodenstrukturschäden," *Tagungsber. Akad. Landwirtschaftswiss. DDR* 215:39–48 (1983).
43. Ganssen, R., and F. Hädrich. *Atlas zur Bodenkunde* (Mannheim: Bibliographisches Institut AG, 1965), p. 85.

44. Meteorologischer und Hydrologischer Dienst der DDR, Ed. *Klima-Atlas für das Gebiet der Deutschen Demokratischen Republik.* Map II 38: Mittlere Niederschlagssummen (mm) Jahr. Periode 1891–1930. Compiler E. Pelzl (Berlin: Akademie-Verlag, 1953).
45. Keller, R., and Deutsche Forschungsgemeinschaft, Eds. *Hydrologischer Atlas der Bundesrepublik Deutschland.* Map 6: Pedologie. Compiler G. Roeschmann; Map 17: Mittlere Niederschlagshöhe (mm) Jahr (1931–1960). Compiler H. Schirmer. (Boppard, Germany: Harald Boldt Verlag, 1978).
46. Roeschmann, G. *Soil Map of the Federal Republic of Germany 1:1,000,000. Legend and Explanatory Notes* (Hannover: Federal Institute for Geosciences and Natural Resources, 1986).
47. Akademie der Wissenschaften der DDR, Ed. *Atlas DDR.* Map 6: Böden. Compilers G. Haase, and R. Schmidt (Gotha: Kartographische Anstalt VEB Hermann Haack, 1985).
48. Baeumer, K., and U. Köpke. "Effects of Nitrogen Fertilization," in *Energy Saving by Reduced Soil Tillage,* Report EUR 11258, K. Baeumer, and W. Ehlers, Eds., (Luxembourg: Commission of the European Communities, 1989), pp. 145–162.
49. Claupein, W., and K. Baeumer. "Einfluß der Bodenbearbeitung auf den Stickstoffumsatz in Ackerböden," *Tagungsber. Akad. Landwirtschaftswiss.* 295:145–159 (1990).
50. Kohl, R., and T. Harrach. "Zeitliche und räumliche Variabilität der Nitratkonzentration in der Bodenlösung in einem langjährigen Bodenbearbeitungsversuch," *Z. Kulturtech. Landentwicklung* 32:80–87 (1991).
51. Ehlers, W., Khosla, B. K., Köpke, U., Stülpnagel, R., Böhm, W., and K. Baeumer. "Tillage Effects on Root Development, Water Uptake and Growth of Oats," *Soil Tillage Res.* 1:19–34 (1980/81).
52. Wieneke, F. "Strohzerkleinerung — Einrichtungen am Mähdrescher mit fasernd, spleißender Wirkung," *Landtechnik* 46:262–264 (1991).
53. Knab, W., and K. Hurle. "Einfluß der Grundbodenbearbeitung auf Ackerfuchsschwanz (*Alopecurus myosuroides* Huds.)," *Z. Pflanzenkr. Pflanzenschutz Sonderh.* 11:71–82 (1988).
54. Bräutigam, V. "Einfluß langjährig reduzierter Bodenbearbeitung auf die Unkrautentwicklung und -bekämpfung," *Z. Pflanzenkr. Pflanzenschutz Sonderh.* 12:219–227 (1990).
55. Isselstein, J., and R. Rauber. "Zwischenfrüchte zur Kontrolle von Ausfallgerste und Unkräutern vor Zuckerrüben-Mulchsaat," *Z. Pflanzenkr. Pflanzenschutz Sonderh.* 11:329–338 (1988).
56. Langenbruch, G.-A. "Einfluß der Stroh- und Bodenbearbeitung auf die Wintersterblichkeit der Maiszünslerlarven," *Nachrichtenbl. Dtsch. Pflanzenschutzdienstes* 33:86–90 (1981).
57. Tebrügge, F. "Stroheinarbeitung: Nicht bloß Resteverwertung," *DLG-Mitteilungen* 105:592–595 (1990).
58. Claupein, W. "Zur Wirkung von Fruchtfolgegestaltung und Mulchwirtschaft auf den Halmbasiskrankheitsbefall von Winterweizen," *Mitt. Ges. Pflanzenbauwiss.* 3:31–34 (1990).
59. Garbe, V. "Verunkrautung und Auftreten von Schädlingen bei unterschiedlichen Systemen der Bodenbearbeitung zu Zuckerrüben," Ph.D. thesis (Göttingen: University of Göttingen, 1987).

60. Schwertmann, U., Vogl, W., and M. Kainz. *Bodenerosion durch Wasser. Vorhersage des Abtrags und Bewertung von Gegenmaßnahmen* (Stuttgart: Eugen Ulmer, 1987), p. 64.
61. Frede, H.-G. Personal communication.
62. Baeumer, K. "Tillage Effects on Root Growth and Crop Yield," in *Agricultural Yield Potential in Continental Climates,* Proc. 16th Coll. Int. Potash Institute, Bern (1981), pp. 57–75.
63. Tebrügge, F., Gruber, W., Henke, W., Kohl, R., and H. Böhm. "Long-Term Cultural Practices Effects on the Ecologic System," paper no. 91-1009 (Albuquerque: ASAE International Summer Meeting, June 1991).

CHAPTER 7

Feasibility of Minimum Tillage Practices for Annual Cropping Systems in France

J. Massé and Denis Boisgontier
Institut Technique des Cereales et des Fourrages; Boigneville, France

Jean Marie Bodet and Jean Paul Gillet
Institut Technique des Cereales et des Fourrages; La Chappelle, St. Sauveur, France

TABLE OF CONTENTS

0-87371-571-3/94/$0.00+$.50

7.1. INTRODUCTION

The area used for agriculture in France represents about 30 million ha (almost 60% of the total area of France).[1] Annual crops, except vegetables, fruit trees, flowers, and vines, account for 13 million ha and they are cultivated every year. The principal cultivated crops are the following (in percentage of main annual crops): winter cereal crops (50%); maize (*Zea mays* L.) (grain and silage) (25%); oilseed rape (*Brassica napus* L.) and sunflower (*Helianthus annus* Mill.) (15%); protein crops (5%); other crops such as sugar beet (*Beta vulgaris* L.) and potato (*Solanum tuberosum* L.), etc. (5%).

Soil tillage takes place in many seasons (mostly from autumn to the end of spring). Moldboard plowing is mainly used, followed by one to five shallow cultivations depending on several factors such as initial and expected soil structure, weather conditions, and machines used. As economic conditions change and many new tillage techniques become available, such as direct drilling and shallow cultivation, farmers are being encouraged to consider tillage techniques that have low investment, low power consumption, and reduced labor or time. Moreover, environmental protection measures must also be considered increasingly by farmers. At the moment, the greatest concern is nitrate leaching. Many agricultural practices will have to be developed progressively according to the European Economic Communities (EEC) directive on nitrate pollution and also according to national laws.

Soil erosion is increasing in the regions in which crops are cultivated on a large scale (principally in the north of France). This is due to many factors, including: the increase of the cultivated area to the detriment of grassland, the decreasing number of hedges and ditches as fields are enlarged, the use of crop rotations largely based on crops with few residues (e.g., sugar beet and potato), and the compaction of the soil due to the use of heavy machinery.

Overall, economic limitations appear more determinative in the evolution of soil tillage than environmental issues.

7.2. DESCRIPTION OF AGROECOSYSTEM

7.2.1. Soil Types

French soils used for annual crops mainly consist of silts, calcareous, and sandy soils.[2,3] Silty soils cover a large area of France where mainly cereals are grown (the north, Bassin Parisien, Bassin Aquitain, vallée de la Saône, etc.) and also in intensive livestock regions (e.g., Brittany and Normandy). The texture of the plowed layer consists of a large proportion of silt (particles from 2 to 50 μm). Risks of soil compaction by strong rains and waterlogging depend on the extent of clay leached to the subsoil, as well as the degree of decarbonation and the type of geological formation. Well-drained silty soils are the most important soil type in the north of France and Bassin Parisien where cereals, peas, sugar beets, and other crops are grown. These soils are highly productive. Nevertheless, they are sometimes prone to water erosion, mainly when the percentage of silt is particularly high as compared to the clay content. Poorly drained silty soils can be found in different French regions; they are sensitive to waterlogging because of the accumulation of clay particles in the subsoil. Without specific surface grading, they become waterlogged during the winter. When they are drained, the time available for soil tillage increases, but the erosion risks remain significant. Silty soils upon schists, which are very common in the west of France (e.g., Brittany), are brown acidic soils. They have a silty or silt loam texture in the topsoil while the subsoil consists of impermeable schist. Drainage is often needed to increase their productivity.

Three main types of calcareous soils occur in the area mainly used for annual crops. The importance of the crack potential (mechanical disintegration) and the hardness of the calcareous rock causes the development of different types of soils. Chalk soils lie on chalk or calcareous substratum, and thus have a high calcium carbonate content. They have a high water infiltration rate due to a high porosity and a high water-holding capacity; however, these soils remain cold after winter because of their white color, and thus crops grow slowly during spring. These highly productive soils are easy to cultivate due to the availability of many field work days and low draft requirements. Soils upon limestone, the second type of calcareous soil, are found around sedimentary zones (e.g., Bassin Parisien) or old mountains. The top layer is about 20 to 50 cm deep and lies on geological substratum more or less cracked. The plowed layer has many stones which often handicaps soil tillage, particularly for plant establishment. Poor field water-holding capacity (40 to 120 mm of available water) is related to the relatively high volume of gravel and stones (20 to 50%). Marl soils are deep and have a high potential of production due to their high clay content (available water from 150 to 250 mm); however, as "heavy" soils they require much power for cultivation. These soils remain cold during spring because of their high water content. Alternate wet and dry periods allow the development of good structure in these soils. Sandy soils are characterized by an acid, brown topsoil over a sandy subsoil. Cultivation is easy, but production potential is naturally low due to poor field water-holding capacity and the poor mineral content of these soils.

Soils with a risk of waterlogging cover about 12 million ha, of which 50% can be drained to improve their productivity.[4] Currently, about 2.5 million ha are drained by tile, but soils with very low water infiltration need additional deep cultivation (by a subsoiler). In wet and nondrained soils, the number of available working days is reduced and soil cultivation as well as crop establishment are delayed. This causes lower yields and irregular crop production. In such situations, plowing is necessary and minimum tillage or direct drilling are unsuitable. Drainage, however, allows the use of better management techniques and farmers frequently change their crop and cultivation system in order to achieve higher levels of production.

Overall, certain types of soils and landscapes are major constraints to the adaptation of mechanized tillage.[5] When the slope is greater than 10%, implements or systems of cultivation must be modified (e.g., direction of plowing). Choice of tillage equipment and method must account for clay content, waterlogging, and stony soils. These factors limit the use of certain tillage techniques or increase the wear and breakage of implements (e.g., in calcareous stony soils). In very silty soils, tillage in wet conditions increases the risk of compaction; therefore, cultivation techniques must be adapted to specific soil types and conditions.

7.2.2. Climate Description

Four zones of climate can be identified in France according to the latitude, the altitude, and the distance from the sea or from the mountains (Figure 7.1). The oceanic climate type covers the west of France and produces low temperature variations and provides a good rainfall distribution. An increasing thermic gradient can be observed from the north to the south of France. The semicontinental climate type covers about 25% of France and is located in the northeast. Its main characteristics are contrasting thermic conditions between winter and summer (about 20°C difference) and a regular rainfall during all seasons. The Mediterranean climate type covers the south of France, except for the mountain regions (Alps and Massif Central). It is dry and hot in the summer and storms are produced. The mountain climate type is subject to decreasing temperature as altitudes increase and therefore is not well suited for many annual crops.

There are many microclimates and Figure 7.1 shows about 30 regional climates, determined by the National Meterological Office,[6] which are based on characteristics such as rain, temperature, solar radiation, wind, hygrometry of the air, etc. For example, rainfall reaches about 650 mm in the Mediterranean climate and increases to 1000 mm in Brittany. Solar radiation, which is about 160 kJ/m^2 during the summer in the north of France, reaches 240 kJ/m^2 in the southeast and largely determines potential evapotranspiration and the distribution of crops.

7.2.3. Cropping Systems

Crop systems are largely influenced by soils and climates, but also depend on many other factors. Crops cultivated on a large scale (cereals, oilseeds, sugar beets, potatoes, peas, etc.) cover large areas from the north to the south of France, especially in the oceanic climatic zone. Although cereals are dominant, the type of cereal varies with the region (e.g., hard wheat in the southeast or southwest, winter barley in the north, and maize for grain in the southwest). Wheat is the dominant crop (about 5 million ha), mainly as winter wheat.[7]

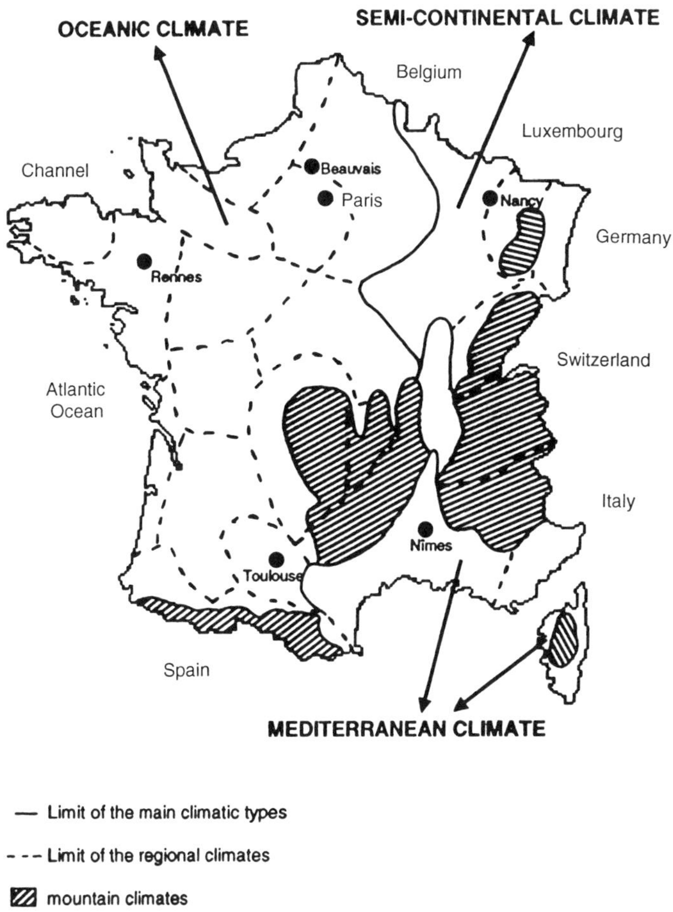

Figure 7.1. Main climatic zones in France. (From Choisnel and Payen, *Le Grand Atlas de la France Rurale,* 1989. With permission.)

Some standard rotations are as follows:

- North of France: sugar beets, potatoes, peas as break-crop, winter wheat, winter barley
- Bassin Parisien: sugar beets, peas, maize as break-crop, winter wheat
- Southwest: maize, soya, sunflower as break-crop, winter wheat, hard wheat

The number of farms was less than 1 million in 1990, with an average size of 31 ha. During the last 10 years, the number of small farms has rapidly decreased (10% reduction in the number of farms <35 ha), while large farms have increased (about 5% for farms with >100 ha). The largest farms are located in central and northern France. The number of tractors by farm averages 1.5, with a trend toward an increase in power. For example, 30% of farms own one tractor of more than 60 kW. Overall, farm productivity has increased since 1970 because of an increase in input level and greater efficiency (new varieties, new and adapted cultivation techniques, etc.). However, present economic conditions will cause a future decline in economic investment per hectare.

7.3. ADOPTION OF MINIMUM TILLAGE PRACTICES

7.3.1. Conventional Tillage Systems

Plowing is the conventional tillage system for crop establishment in France. It is usually conducted after a stubble cultivation and is followed by various surface cultivations before sowing. Stubble cultivation before plowing consists of one to three passes of an implement with tines or discs. The main objectives of stubble cultivation are to mix the stubble with the soil, to reduce the mechanical problems that stubble present, and to destroy weeds or volunteer plants of the previous crop. Stubble cultivation is now being conducted with a decreasing number of passes, and in some cases the practice has been abandoned.

Plowing is done with a moldboard plow. The disc plow is no longer used in France. Agronomists recommend a working depth of about 25 cm, but very large variations can be observed according to region, type of crops being established, or the previous crop. Generally, plowing depth varies from <20 cm (for wheat after sugar beet in the north of France) to >35 cm (for maize after wheat in the south). The plowing time depends on the soil type, the harvesting date of the previous crop, and the crop to be established. For example, for spring crops on "heavy" soil, plowing is conducted before winter, while on loam the soil is plowed just before sowing. The number of passes for surface cultivation varies from one to more than four according to the soil type and the seed bed preparation needed by the crop. For winter cereal, one or two passes are necessary, and up to four or six for sugar beet. Surface cultivation is conducted either with tine implements (rigid tine cultivators, spring tine cultivators, or harrows) or with power take-off implements (reciprocating harrow, rotary harrow, or rotary hoe cultivator). Sowing occurs either after or at the same time as the last surface cultivation. In the latter case, the seedbed is produced with combined power take-off implements.

From the 1940s until the 1980s, farmers were encouraged to practice intensive crop management. Soil cultivations became increasingly intensive. Moreover, the increase in spring crops increased the need for spring field work days. This resulted in more passes over the field with energy-consuming implements (power take-off implements). They were often used in conditions not conducive to the maintenance of soil structure, especially when heavy powerful tractors were employed. These practices increased soil erosion in regions where it was limited in the past. In the same way, the increase in vehicular traffic (for soil tillage, chemical treatments, etc.) caused soil compaction in many fields. The decreasing soil fertility, induced by the soil compaction, was compensated for by the addition of fertilizers.

Over the last decade, two aspects have emphasized the need to improve soil conservation and tillage methods. Ecological movements have placed increasing pressure on farmers to use techniques to better protect the environment, while poor economic conditions have forced farmers to reduce their costs, especially those of mechanization. One of the ways of reducing mechanization costs is the simplification of tillage.

7.3.2. Strategies to Reduce or Simplify Tillage

The approach to reduce or simplify tillage can involve a reduction in tillage events and minimum use of the moldboard plow, or a complete removal of plow and use of minimum tillage, or direct drilling.[8] Both strategies have various advantages and disadvantages.

Partial simplification consists of limiting the number of tillage passes and especially the replacement of plowing, when possible, by a reduced tillage technique. The latter may be either minimum tillage (working the soil only in the first 10 cm) or direct drilling (disturbing the soil only within the seedbed). The decision to replace plowing depends on three important aspects. First, the risk of herbicide persistence: some herbicides (e.g., atrazine, isoxaben, napropamide, etc.) can persist in the soil after the previous harvest, in particular after a dry season. To reduce the risk of phytotoxicity, a dilution through plowing may be necessary. Second, a wide range of minimum tillage implements may be required. For example, much plant residue on the soil surface may prevent minimum tillage if a farmer does not have access to a rotary cultivator. Third, the soil physical condition after the previous crop may require some tillage in the surface or at depth. Table 7.1 provides tillage solutions according to the state of the soil structure in the soil profile.

Total tillage simplification consists of replacing the plow, for every crop in the rotation, by minimum tillage or direct drilling. In most cases, a specific implement is necessary and therefore needs to be purchased. Total simplification is conducted over several years on the same area, producing cumulative effects on soil properties and fauna. The elimination of plowing can reduce working time by 30 to 50%; however, cultivation dates are often brought forward in comparison to situations in which the plow is used (i.e., drying out the soil surface is slower when soil is not plowed). When economic problems are absent, the possibility of total tillage simplification depends on three criteria. First, the quality of land drainage, which determines the level of waterlogging in winter (Table 7.2). Second, the type of crop rotation and soil

Table 7.1. Choosing a Suitable Soil Tillage Procedure in Relation to the Soil Structure Condition Remaining after the Previous Crop

Structure in Surface Soil			
Surface State	Compaction State	Structure at Soil Depth	Range of Suitable Tillage Techniques
Flat to degraded	Compact to noncompacted	Compact	Depth tillage[a]
Flat	Noncompacted	Noncompacted	Depth tillage Minimum tillage Direct drilling
Flat	Compact	Noncompacted	Depth tillage Minimum tillage
Uneven	Compact to Noncompacted	Noncompacted	Depth tillage Minimum tillage

[a] Moldboard or chisel plowing.

Table 7.2. Potential for Total Simplification of Tillage According to the Quality of the Drainage

Quality of Soil Drainage	No. of Days of Drainage after a 10–20 mm Rain	Possibility of Total Tillage Simplification
Good	< 5 days	Some rotations with risk
Medium	5–8 days	Most rotations with risk
Poor	>8 days	All rotation types at risk

type can influence the success of simplified tillage. For example, maize or sugar beet harvests are sometimes conducted under moist soil conditions and the risk of soil degradation is very great, especially in clay to silty soils. Third, the potential for adoption of total tillage simplification depends on the mechanical and physical properties of the soil which control the soil's susceptibility to degradation and its potential to recover a good structure.[9-11] Even under good drainage, crops such as grain and silage maize, sugar beet, potatoes, and silage grass can be associated with increased soil degradation, unless precautions (e.g., low pressure tires) are taken to reduce traffic-induced compaction.

7.3.3. Feasibility of Simplified Tillage Systems

A partial tillage simplification is rather easy to implement as a complete change of tillage practices is not necessary. A total simplification of soil cultivation techniques, however, brings with it the need for a complete change in farm management. This is especially true when plowing, which is a centuries-old tradition, is abandoned by the farmer and totally replaced by shallow cultivation or direct drilling. These changes do not result in a reduction of mechanization costs because high-power tractors are still required. Nevertheless, labor productivity is improved and the potential land area managed by one operator is increased.

Table 7.3. Main Characteristics Involved in Adoption of Simplified Tillage in France

Characteristic	Partial Tillage Simplification	Total Tillage Simplification
Machinery and labor	No specific requirement for implements	Generally specific requirement for implements
Adaptation to climate and soil type	*Good:* Decision to replace plowing is taken on an annual basis according to the soil conditions after the harvest of the previous crop.	*Limited:* Annual decisions replaced by medium or long-term commitments; every field must be adaptable to total simplification
Adaptation to crop rotation	*Good:* Most rotations may be adapted to crop establishment without plowing	*Limited:* For some rotations risk of degradation of soil structure is high unless precautions are taken
Technical skill	*Normal:* No special technical skill needed; any mistake can be corrected during season or the following year	*High:* Skills required for crop establishment, weed control, and maintenance of soil structure
Main advantage	Reduction of mechanization costs	Increase of labor productivity; sometimes, reduction of mechanization costs

The technical skill of the farmer becomes more important when total tillage simplification is implemented.[12] These skills are necessary to address specific changes in soil and crop management. For example, during the spring, the optimum timing for tillage is sometimes 1 to 3 days later than the optimum date for plowed fields. Furthermore, it is necessary to monitor soil structure and use tires that reduce the risk of soil compaction. Simplification of soil cultivation also increases the risk of perennial weeds. Table 7.3 give the main characteristics for both partial and total simplification of soil cultivation.

7.4. ENVIRONMENTAL ISSUES

7.4.1. Soil Erosion

In the past, soil erosion was frequent on slopes cultivated with vines or fruit trees (e.g., vallée du Rhône). However, erosion has become an increasing problem in zones of intensive production in the north of France and in silty soils because of a reduction of grass in the crop rotation and an increase in the area under cultivation. The latter has been achieved by decreasing the number of hedges, embankments, and ditches that previously limited surface water flow. The increase in soil compaction has also caused an increase in soil erosion. In the southwest, the most important erosion damage is due to strong storms on bare or partially covered soil. In contrast, in the north the surface structure breakdown of silty soils concentrates water flow, and this subsequently increases rill and gully erosion.

Approaches to control soil erosion are mainly derived from the North American experience. According to these sources, tillage plays a dominant part in protecting the soil from erosion forces. Yet, transposing these results to meet French conditions will require complementary experimentation.

7.4.2. Nitrate Leaching

Nitrate leaching can be important in zones of intensive production such as livestock areas or in cropped areas on shallow or sandy soils. Risks of nitrate leaching by rain is due to an excess of fertilizer use compared to the needs of plants and mineralization of organic matter and crop residues or manures. In order to relate the amount of fertilizers to the potential of the crop, it is necessary to know the relationship between the level of fertilizer applied and the yield obtained for each soil tillage system and to adjust fertilizer levels accordingly.[13] In addition, management of the intercrop period is important for the control of nitrate leaching.

Response curves generally show that at low nitrogen application, yields are lower in direct drilling than in plowing whereas responses become similar at optimum nitrogen. This observation seems to be due to a lack of nitrogen in the early stages of plant development (nitrogen mineralization seems lower in direct drilling than in plowing). The optimum fertilizer level is higher for direct drilling than for plowing due to low nitrogen efficiency or soil structure changes. This is especially apparent in soils with high field water capacity.[14] However, under dry conditions the reduction of evaporation by direct drilling increases crop yield at high levels of fertilization, compared to plowing (Table 7.4). Thus, it is necessary to take into account both soil tillage and weather-soil situations, which determine yield of crop and nitrogen mineralization, to determine the quantity of nitrogen to apply and to thus avoid excess of nitrates at harvest.

Nitrate leaching in the intercrop period is related to the type of previous crop. For example, crops that allow an early harvest provide a greater period for soil nitrogen mineralization in the autumn. Also, the type of crop influences the amount of nitrogen remaining in the soil after harvest and the quality of crop residue (e.g., carbon:nitrogen ratio). Nitrate leaching in the intercrop period also depends on weather conditions and levels of nitrogen fertilizer applied. The best management of the intercrop period to limit nitrate leaching is combinations of residue control and soil tillage. Figure 7.2 shows the effect of a long-term experiment for direct drilling and plowing with two straw managements on nitrate content in the soil during a long intercrop period. Direct drilling with mulch appears to be the best management to limit nitrate production in autumn and winter as compared to plowing. In this case, almost

Table 7.4. Crop Yield and Nitrogen Removed by Crop in a Comparison of Plowing and Direct Drilling for Winter Wheat

	Level of Nitrogen Fertilization (kg/ha)			
	0	150	190	230
Yields (t/ha)				
Plowing	2.65	7.23	7.06	6.85
Direct drilling	2.64	7.44	7.56	7.54
Nitrogen yield (kg/ha)				
Plowing	43	112	149	150
Direct drilling	39	125	153	182

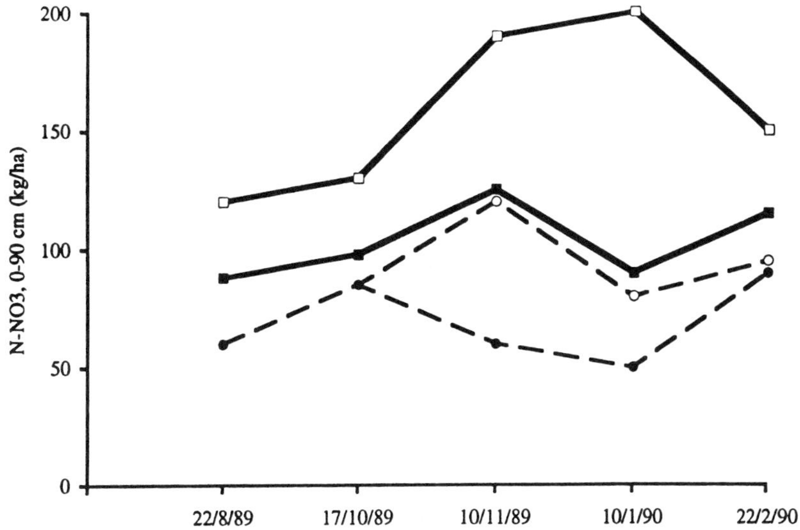

Figure 7.2. Effect of plowing with burned (□) or soil incorporated (■) straw and direct drilling with burned (○) or mulched (●) straw on soil nitrate content during the intercrop period.

200 kg/ha of nitrate is susceptible to leaching if high rainfall occurs during autumn and winter.[15]

Straws well mixed with plowing or mulched in direct drilling reduces nitrate leaching (Figure 7.2). However, the efficiency of this technique concerns only the 0 to 20 cm (or 30 cm) soil layer and in many situations, nitrate is leached below this depth. Generally, more research is needed in this area. In situations in which high nitrate contents occur in the soil after harvest, catch crops are very useful. Because of their cost, a recommended strategy has been developed by different organizations (Table 7.5) to advise farmers and other interested parties on crop residue management.[15]

Table 7.5. Recommended Techniques for Crop Residue Management to Limit Nitrate Leaching in the Intercrop Period from CORPEN[a] [15]

Source of Residue	Time of Harvest	Recommended Crop Residue Management
Cereals, oilseed rape, sunflower (Northern France)[b]	Early	Incorporate straw immediately and establish catch crop or delay until winter
Peas, soybeans, faba beans, lupins[c]	Early	Remove residues
Maize, sunflower (Southern France)[b]	Normal	Remove residues and establish catch crop (southern France) or plow in winter (Northern France).
Sugar beets[c]	Normal	Generally soil too wet to allow traffic
Grasses and catch crops	Late	Plow during winter (in Northern France) or in spring (in Southern France); can use chemicals to kill sward

[a] CORPEN = Committee for the Reduction of Pollution from Nitrates and Phosphates.
[b] High carbon:nitrogen residues.
[c] Low carbon:nitrogen residues.

7.5. GENERAL SUMMARY AND CONCLUSIONS

Because of changes in economic conditions, farmers need to reduce fixed costs. Thus, reduced tillage, which requires less power consumption and/or labor time than conventional techniques, has become more and more important in France. As soils and climate vary in different regions, soil tillage is generally adapted to permanent soil characteristics as well as to their annual variations; for example, soil structure conditions remaining after the previous crop. Thus, if tillage is well adapted to the agronomic conditions, crop yield remains generally equivalent between reduced and conventional tillage.

In the area of the environment, protection of groundwater from nitrate or pesticide pollution requires improved management of both the crop and intercrop periods. Therefore, fertilization, crop protection, weed control, and soil tillage must be improved. In this context, conservation tillage and residue management are two basic components of new agricultural practices.

7.6. REFERENCES

1. *La Statistique Aricole* (Agreste, Publication du Ministère de l'Agriculture et de la Forêt, 1992, Paris).
2. Begon, J. C., M. Berland and M. Jamagne. "Les Grands Types de Sols," in *Le Grand Atlas de la France Rurale*, J. P. de Monza, 1989, pp. 400–402.

3. Bodet, J. M. *Les Sols de l'Ouest* (Internal Document ITCF, 1991, Paris).
4. Favrot, J. C., and J. J. Herve. "Le Drainage: les Techniques," in *Le Grand Atlas de la France Rurale,* J. P. de Monza, 1989, pp. 432–434.
5. Arrouays, D., and J. P. Deffontaines. Les Contraintes de la Mécanisation, in *Le Grand Atlas de la France Rurale,* J. P. de Monza, 1989, pp. 448–450.
6. Choisnel, E., and D. Payen. "L'Agrométéorologie," in *Le Grand Atlas de la France Rurale,* J. P. de Monza, 1989, pp. 410–422.
7. Cavailhes, J. "Les Localisation des Systèmes de Production," in *Le Grand Atlas de la France Rurale,* J. P. de Monza, 1989, pp. 138–139.
8. Monnier, G., *Simplification du travail du sol* (Tiré à part des actes du Colloque du 16 Mai 1991, Paris, Perspectives Agricoles éditeur), pp. 99.
9. Stengel, P., J. T. Douglas, J. Guérif, M. J. Goss, G. Monnier and R. A. Cannell. "Factors Influencing the Variation of Some Properties of Soils in Relation to Their Suitability for Direct Drilling," *Soil Tillage Res.* 4:35–53 (1984).
10. Castillon, P., and G. Eschenbrenner. "Structure Dégradée d'un Sol Argilo-Calcaire et Comportement des Cultures d'Été," in *Les Racines* (Numéro spécial Perspectives Agricoles, 1989), pp. 64–69.
11. Massé, J., C. Colnenne, F. Tardieu and P. Crosson. "Système Racinaire du Blé et État Strutural du Sol," in *Les Racines* (Numéro spécial Perspectives Agricoles, 1989), pp. 64–69.
12. Boisgontier, D. "Qui Peut Supprimer le Labour?," in *Perspect. Agric.* 173:33–37 (1992).
13. Massé, J., and J. C. Rémy. "Profil d'Absorption des Nitrates par les Racines," in *Les Racines* (Numéro spécial Perspectives Agricoles, 1989), pp. 77–80.
14. Germon, J. C., and J. C. Taureau. *Simplification du travail du sol et transformation de l'azote* (Simplification du travail du sol, Colloque, Mai 1991, Paris, Numéro Spécial Perspectives Agricoles), pp. 58–69.
15. *Interculture,* (brochure, Committee for the Reduction of Pollution from Nitrates and Phosphates, Paris (1991).

Humid Meso-Thermal Climates

CHAPTER 8

Overcoming Constraints to Conservation Tillage in New Zealand

M. A. Choudhary and C. J. Baker
Massey University; Palmerston North, New Zealand

TABLE OF CONTENTS

0-87371-571-3/94/$0.00+$.50

8.1. INTRODUCTION

The constraints on conservation tillage in New Zealand are more philosophical than technical. New Zealand does not desperately need conservation tillage, and yet it has spawned some of the longest running and most productive research and development programs in the world. This apparent anomaly has arisen from a somewhat different approach to research than that dictated by the problems of erosion and time constraints in many of the world's major arable cropping zones. It is not that New Zealand has lagged behind. On the contrary, New Zealand has produced some of the more innovative conservation tillage technology and practices known, but New Zealand has had the luxury of time to achieve its objectives.

The country is divided across its center into two more or less equally sized main islands, one lying to the north and the other to the south of Cook Straight. Predictably, these two main islands are called North and South Islands, respectively. The native Maori name for New Zealand is *Aotearoa*, meaning "land of the long white cloud." This is how the first Polynesian Maori canoe travelers first perceived the country in their sightings from journeys across the Pacific Ocean. Such "long white clouds" are the basis of a benign temperate climate. The average annual rainfall ranges between 650 and 1500 mm in productive areas. This ensures that 40% of the land mass remains under permanent introduced pasture with intensive animal grazing year-round. Much of the rest of the land is steep and mountainous and is farmed with animals on an extensive basis.

8.2. AGROECOSYSTEM

The benign temperate climate permits year-round plant growth and/or double-cropping in most farming areas. A few farmers have attempted to grow three crops per year. The long, thin country stretches for 1800 km between latitudes 34°S and 47°S (equivalent in the Northern Hemisphere to a region stretching from Algeria to Switzerland). Mean daily maximum temperatures in summer range from 25°C in Northland to 18°C in Southland. In winter, the corresponding range is 14° to 8°C, respectively. No part of the land mass is more than 150 km from the coast, and with 2000 km of ocean between New Zealand and its nearest neighbor, Australia, it is not surprising that New Zealand's weather patterns are dominated by oceanographic events. Intercepting the dominant westerly winds that originate over the Tasman Sea (and only slightly less frequently, southerly winds originating from Antarctica) is a long range of mountains that neatly divides the country almost from end to end, bringing reliable, well-distributed (and even excessive) rainfall in the west and reliable sunshine in the east.

Only 9% of New Zealand is considered arable. Against these climatic advantages must be balanced the disadvantages of a small domestic market for farm technologies. The buying power of only 60,000 farmers represents a compact, but often uneconomic, basis for market release, especially for large and expensive machines. On the other hand, New Zealand farmers are among the most educated and technically literate farmers in the world, and the vast range of farming miniclimates within the mountainous country (ranging from 325 to 2000 mm rainfall) present an ideal variety of field test sites unequalled on such a compact and accessible basis elsewhere. Field testing is possible throughout the year.

Soils are generally recent and shallow, with limited water-holding capacity. Most soils which are extensively cropped are alluvial in origin, but there are a few examples of self-mulching, sticky clay soils. Highly abrasive volcanic pumices also abound and are ideal for accelerated wear testing of machinery. Attempts to classify soils for suitability to conservation tillage have been controversial because data are often biased by the performance of the dominant drilling technology at the time. However, within this limitation Ross and Wilson[1] classified New Zealand soils in 1985. Under this classification most of the major arable areas of New Zealand were felt to have slight or moderate limitations to the technique.

While almost every temperate arable crop known is grown in New Zealand on a limited scale, these are generally rotated regularly with permanent pastures and animal-based farming systems which serve to repair or preserve soil structure, fauna, and organic matter. The erodibility of New Zealand soils is seldom exposed to repetitive tillage practices during continuous cropping programs. Significant slumps and rill erosion occur, however, on steep-pastured hillsides due to the historical removal of native podocarp forests and their replacement by shallower rooted pasture species. Conventional wind- and water-activated erosion can be found on tilled arable soils in the Manawatu coastal belt and Hawkes Bay cropping regions of the North Island, together with the significant areas of South Canterbury and North Otago in the South Island. Not surprisingly, these areas comprise some of the more extensively tilled and cropped areas of New Zealand and provide tangible examples where conservation tillage has much to offer.

Additionally, the systematic renewal of permanent pastures in New Zealand is commonplace due to their deterioration over time from pests and climatic extremes together with imperfect pasture management, drainage, and fertility. Increasingly, farmers are turning from tillage to conservation tillage as a means of repairing or renewing such pastures.

It is ironic that with conservation tillage, the challenges presented by New Zealand agriculture have been blessings in disguise. The absence of time constraints has encouraged investigative methods and philosophies that are long term in outlook, being based on determining the biological responses to the actions of certain machine components, and then designing machines to fulfill these requirements. The results have been applicable internationally.

8.3. PASTURE RENOVATION

Forty percent (or 10.6 million ha) of the area of New Zealand is under grassland.[2] Most of this is under permanent pasture species. Annual pastures are used mainly as break-crops for short-term grazing between successive summer crops (arable or fodder). Many of the pastures based on permanent species deteriorate over time because of damage from drought, climatic extremes, overgrazing, and pests in conjunction with imperfect management and fertility, all leading to an ingress of weeds and a lowering of productivity. Because New Zealand's export income is heavily dependent on animal products, pasture production (quality and quantity) is of vital economic importance. Early technology to improve the productivity of the hill country involved clearing vegetation from forested land by hand cutting and burning, some herbicide spraying of scrub weeds, topdressing with phosphatic fertilizers, surface seeding of largely clover species, paddock fencing, and prudent stock management. Later, as the clovers improved soil fertility, grass species were introduced to balance the sward and further increase productivity.

Surface reseeding of poor pastures on steep land by aerial methods and resowing of pastures into crop rotations on flatter land using cultivation have become common practice in temperate New Zealand. Cultivation methods of sowing pastures are expensive, and potentially harmful to soil fauna (fauna increases soil erosion). The use of conservation tillage, especially overdrilling, has increased in recent years, especially in the more intensive and high-producing pastoral areas. As a result, it is estimated that currently 250,000 ha of pasturelands are resown with this method every year in New Zealand.

The New Zealand pattern of using a pasture-crop rotation on arable land means that the renewal or restoration of pasture without cultivation is an integral part of the total low energy input system. Such a system allows soil moisture conservation, nitrogen buildup and release, and organic matter accumulation from pasture root breakdown. Pasture renovation with conservation methods (especially no-tillage) also allows versatility and flexibility in the farming system, making it a valuable farm management tool. The benefits of the system allow reduced costs, reduced time for the establishment operation, and consequently lower lost-production costs, reduced delay until first and subsequent grazings, improved stock and vehicle trafficability, and less weed infestation, which may be partly and indirectly a function of the more intensive grazing management.[3]

No-tillage allows the reduction of fallow periods (traditionally 4 to 6 weeks) between pasture and spring sown crops. The availability of an increasing range of herbicides has brought about the possibility, in suitable soils and cropping patterns, of replacing conventional tillage and fallowing practices with presowing chemical weed control and direct one-pass seeding. The advent of glyphosate in particular has had a major impact on pasture renovation programs in New Zealand. These features, combined with appropriate equipment availability, have allowed a marked reduction in tillage requirements to obtain profitable and sustained crop and pasture production under increasingly intensive management.

8.4. SEED DRILLING TECHNOLOGY

8.4.1. Residue Management

Excessive cultivation has been regarded as the major cause of soil degradation internationally. To a more limited extent, such is also the case in New Zealand. In contrast, conservation tillage systems allow the retention of surface residues and reduce soil losses by erosion, improve soil structure,[4,5] enhance soil moisture retention,[6] and increase organic matter exchange and nutrient levels. The nutrients added to the soil in organic residues have the advantage of being released gradually and are less sensitive to leaching, volatilization, or fixation.

An unwelcome biological factor is the possible allelopathic effects from phytotoxins of decomposing crop residue. Where residue-covered soils reduce surface temperatures in temperate climates, this may adversely affect seed germination of temperature-sensitive species[7] and reduce root development from direct residue contact.[8] These effects were first reported in the cold damp conditions of England and eastern Washington State (United States), but have also been found to a limited extent in the more moderate ambient conditions in New Zealand. However, New Zealand workers also found that where the sown seeds can be deliberately separated from the residue by seed drill openers, even by only 10 mm, most phytotoxic effects can be avoided.[9]

The quantity and distribution patterns of plant residue on the soil surface have a major influence on the design of soil-engaging tools for proper residue management. One of New Zealand's most important contributions to conservation tillage has been to demonstrate that rather than remain a constraint by causing blockages to soil-engaging machines, correctly managed crop residues create a favorable seed microenvironment.[10] The main reason that drills using double- or triple-rolling disc openers (V-shaped slots) have persisted in practice is that these openers handle surface residue reasonably well without substantial blockages. In the minds of most design engineers, unfortunately, this mechanical expediency has so far outweighed shortcomings in their biological performance. Their most common failing is a tendency to "hairpin" surface residue rather than cut through it,[11] causing reduced germination in relatively dry soils due to low seed-soil contact[12] together with reduced germination in wet soils because of the phytotoxic contact between seeds and decomposing residue referred to earlier.[8]

Mechanical expediency has dictated that residues are sometimes chopped up, swept aside, or vertically pushed into the slot to facilitate the passage of the opener. The solution to the problem of trading off residue-handling capacity against biological advantages lies in prioritizing the biological tolerance of the openers.

In New Zealand a number of types of drill openers have been tried for drilling cereals in 150 mm rows through stubble or residue with varying degrees of success. In common with experience elsewhere, disc-type openers (double, triple, concave, and angled flat disc) all handle residue well, but leave much to be desired insofar as seed placement and microenvironment are concerned. Hoe-type openers with rigid shanks have been found to be more susceptible to residue blockages than disc-type openers. A winged opener developed in New Zealand (Cross Slot™) has been shown to successfully seed through a wide range of residue conditions while maintaining a virtually complete soil and residue cover over the seed and avoiding seed-residue contact.[13,14] The residue-handling capability of the Cross Slot™ opener is a key feature rarely found on other seed drills. The opener can handle successfully a wide range of both flat, detached residue or standing stubbles, and provides the seed with an ideal microenvironment for germination.

Residues left after crop harvest are a cost-effective means of reducing environmental pollution by runoff, nutrient loss, sedimentation, and soil erosion. Research and development of soil openers must therefore utilize rather than destroy this important natural resource.

8.4.2. Plant Establishment

The main purpose of tillage has been described as providing seedbeds with adequate soil moisture, aeration, consolidation, and freedom from weeds to encourage consistently optimum responses from seeds and seedlings during germination and emergence. Of these, only weed control is a function not performed by well-designed no-tillage drills. Opportunities to increase crop production in conservation tillage have been constrained significantly by the problems associated with adequate and timely seedling establishment using one-pass seed drills instead of multi-pass tillage. The most fundamental aim of any conservation tillage practice should be to achieve 100% seedling emergence regardless of prevaling soil, weather, or crop residue conditions. In New Zealand, scientists have advanced toward that goal with demonstratable success.

For conservation tillage to succeed as a management tool, the biological risks of stand establishment and growth must be less or no worse than those of conventional tillage.[15] A considerable volume of data suggests that field plant establishment is associated more with opener design (Figure 8.1) and the effect

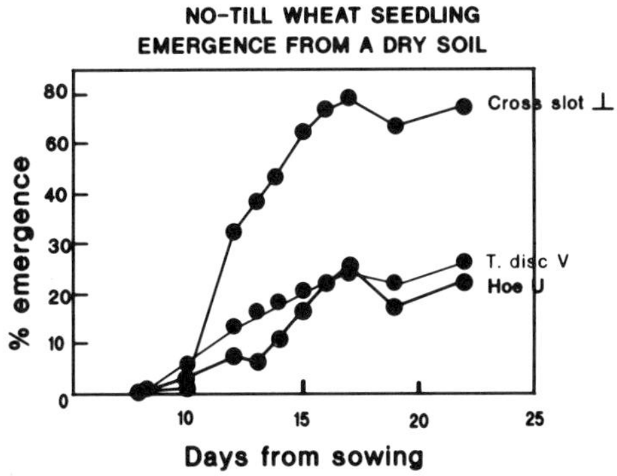

Figure 8.1. Effects on seedling emergence of selected no-tillage opener designs. (Taken from Baker, C. J., *J. Agric. Eng. Res.*, 32, 133–145, 1976.)

this has on soil physical parameters than any other single factor.[16-20] The influence of opener design on soil cover appears to be the key factor. Cover was categorized by Baker[16] as

Class 1: Negligible loose soil
Class 2: Complete loose soil
Class 3: Intermittent residue or mulch
Class 4: Complete residue or mulch

Baker showed a strong relationship between cover and seedling emergence. Choudhary and Baker[21,22] later attempted to categorize opener designs according to their biological performance by measuring their in-groove microenvironments (Figure 8.2). While such efforts were successful in relatively dry soils, only partial success was achieved in relating crop emergence patterns to the in-groove physical microenvironments under favorable climatic conditions. Chemical microenvironments in favorable soil conditions may be more limiting than physical microenvironments.

Research at Massey University, Palmerston North, clearly demonstrated that in dry soils vapor-phase soil water has a major effect on the emergence and subsurface survival of the crop seedlings. In-groove soil humidity itself is a function of the nature and amount of groove cover. When the in-groove humidity decreases too rapidly after drilling because of low diffusion resistance through the groove cover, seeds germinate, but many seedlings die before emergence.[22] In wet soils, failure of seeds to germinate or emerge has been shown to be associated with low soil oxygen diffusion rates, which are themselves related to opener design, particularly their ability to manipulate residue in, over, or adjacent to the slot.[9] Earthworm activity and numbers are also affected by opener design and residues, and in turn they have a direct impact on the oxygen diffusion rates in and around the slot.[23]

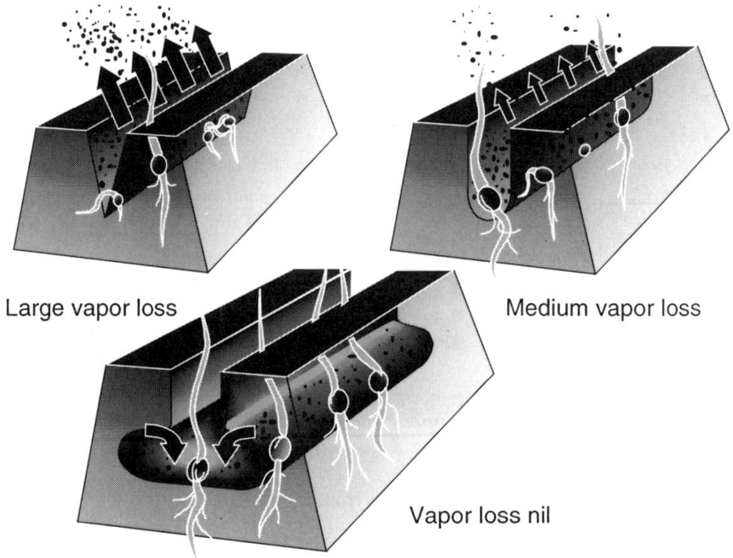

Figure 8.2. Effects on in-groove vapor moisture loss of opener designs: (a large vapor loss, double- or triple-disc opener; medium vapor loss, hoe opener; and low or nil vapor loss, inverted-T opener). (Taken from Choudhary, M. A. and Baker, C. J., *N. Z. J. Agric. Res.,* 23, 489–496, 1980.)

Other studies have also suggested that seedling establishment is related to the accuracy of seed placement within the seed groove and the consistency of depth of that groove. Choudhary et al.[18] and Campbell and Baker[24] developed an X-ray technique to determine in-groove positions of the sown seed. Sowing depth and its control for pasture legume seeds has been shown to be important to achieving high emergence counts.[25] In New Zealand seeding depth has special significance for pasture and crop establishment in soils with uneven surfaces caused by treading from cattle hooves, rutting from machinery wheels, and heavy levels of residue remaining from previous harvests.[18] Clover seeds are especially sensitive to accurate control of depth.[24]

8.4.3. Design of Drill Openers

Slot shape and the nature of soil and residue cover over the slot are the most important variables determining seedling emergence that can be controlled by opener design.[16] Until recently, most drill openers created essentially two shapes of soil slot with some minor variations within each. These are the "V" and "U" shapes.[14,21]

V-shaped slots are created by double- or triple-disc configurations. Their most advantageous features are an ability to physically handle surface residue without blockage and to incur low maintenance. These attributes are countered, however, by:

1. An inability to optimize the seed-zone microenvironment, particularly water vapor retention and oxygen movement in dry and wet soils,[23] respectively;
2. A tendency to smear and compact slot walls, which induces embryonic root stress at the near-vertical slot-wall interfaces;[19]
3. A tendency to tuck residue into the slot where it affects germination by releasing phenolic acids upon decomposition;[26]
4. A wedging action in the soil which creates little loose covering material;[19] and
5. A marked dependence on slow speeds to avoid "seed-flick" by the discs.

There is little opportunity for simultaneous but separate placement of fertilizer and seeds in V-shaped slots because the shape tends to concentrate them together at the base of the slot. Most designers have reluctantly utilized separate and additional fertilizer openers to avoid contact, but this has introduced additional demands on drill design and tractor power.

U-shaped slots are created by a wider range of openers than V-shaped slots; therefore, generalizations are more difficult. For example, hoes, flat-angled discs, dished discs, and power-till openers all create various configurations of U-shaped slots. These openers produce seed-zone microenvironments that are somewhat more optimal than V-shaped slots, but yet increase tillage, moisture loss, and randomized seed-soil contact. They all produce loose soil that can be utilized later for slot cover. However, little microcontrol is exercised over surface residue in the vicinity of the slot. The residues are indeed often swept aside as a prerequisite to avoiding opener blockage resulting in loss of moisture control, although sweeping aside also avoids residue tucking.

Separated fertilizer placement in the vertical plane of U-shaped slots has been achieved with modifications to hoe configurations.[11] The success of this action depends very much on soil plasticity and machine speed, and the soil below the seed is often quite fractured and of low density.[27] Disc and power-till openers have problems similar to those of V-shaped openers in this respect, although there is less tendency to collect seed and fertilizer together at the base of the wider U-shaped slots.

While U-shaped slots may experience base smearing and compaction when made with either hoe or power-till options this is usually not accompanied by sidewall compaction, and dished disc and angled flat disc options are mostly free of smearing tendencies. Because the slot wall interfaces of U-shaped slots include broader bases than V-shaped slots, root penetration is less restricted. Both U- and V-shaped slots benefit from pressing seeds into the base of the slot as this action partially eliminates the need for embryonic roots to negotiate any slot-wall interface, somewhat similar to a tilled soil in which slot interfaces are rare.

Hoe-type openers have difficulty passing through heavy residue without blocking unless individual openers are spaced widely apart, which creates undesirable spatial demands on drill designs. They are, however, usually constructed of inexpensive and less stressed components as compared to

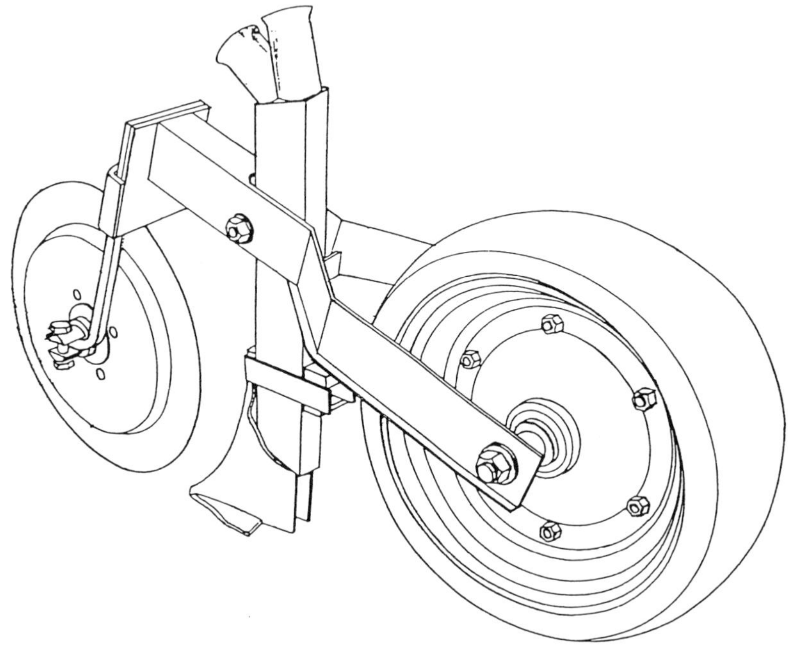

Figure 8.3. Schematic diagram of the original "Baker Boot" drill opener.

angled discs (flat or dished) and power-till alternatives. As with V-shaped slots, the measure of economic viability depends on the value put on the risk of stand loss, although such a risk is generally lower with U- than V-shaped slots.

The inverted-T slot was first discovered by Baker at Massey University when exploring geometrical options that contrasted with V or U shapes for comparative purposes (Figure 8.3). The wide top, narrow base features of V-shaped slots were reversed to produce a narrow top, wide base inverted-T shape. The first simple winged opener to achieve this ("Baker Boot") was commercialized by Aitchison Industries Ltd. of New Zealand. The "Baker Boot" configuration has consistently produced significantly better seedling emergence results than either V- or U-shaped slots,[17] from a biologist's point of view.

The inverted-T-shaped slot has especially increased the tolerance of seeds sown into soils that are suboptimal (too dry or too wet). The dry soil performance from the slot has been shown to be a function of diffusion resistance of the soil and residue slot cover to movement of water vapor. The increased germination and emergence of inverted-T-shaped slots can be attributed to the slot's ability to maintain residue-covered soil over the slot, thus controlling the seed zone microenvironment. This has been referred to as micromanagement of soil and residue.[14] Wet soil performance has also been linked to residue retention over the slot while simultaneously avoiding tucking it into the seed zone. In this case the residue greatly influences mobility and numbers of earthworms, which provided more oxygen diffusion and infiltration into the seed zone than other opener types.[26]

Biological performance and engineering results from the "Baker Boot" opener were combined to form the basis of the internationally patented Cross Slot™ opener design, similar to that shown in Figure 8.4. The inverted T opener, which is more sophisticated than the "Baker Boot," consists of a single 560 mm diameter flat disc (smooth or notched) running straight ahead, which cuts the residue and a vertical soil slot. Two winged side blades, one on each side, cuts a horizontal soil slot at the seeding depth and partially lifts the soil to allow seed and fertilizer, respectively, to fall between the blade and either side of the disc. Two depth-packer wheels immediately follow the blades to reset the raised soil and surface residue, close the slot, and firm the soil onto the seed. All soil and residues are replaced to almost their original position, even to the extent in some soils of preserving the integrity of layering between adjacent zones within the covering medium. The depth-packer wheels also maintain independent depth control of each opener. Fertilizer is displaced to one side and its depth of placement may be at or below the seed level. While the horizontal separation of seed and fertilizer is only about 10 to 20 mm, it is intersected by the vertical disc cut. Numerous field and laboratory studies have shown such separation to be effective from both crop stand and yield viewpoints. Placing fertilizer with the same opener that sows the seed greatly simplifies the complete drill design, and parallel banding with the seed row significantly increases fertilizer effectiveness and efficiency.[28,29]

Combined spring and winter wheat plant establishment and yields are summarized in Figures 8.5 and 8.6 for seven separate experiments at two locations in eastern Washington State.[10]. Comparisons were between double-disc (paired row, banded fertilizer) and Cross Slot™ openers (simultaneous separate placement of fertilizer by each opener). The Cross Slot™ opener consistently produced higher mean yields than the double disc. An unweighted average of all seven experiments, which included 206 individual plot samples, showed that the Cross Slot™ produced 4080 kg/ha as compared to 3624 kg/ha for the double disc, a 12.6% increase (Figure 8.6).

8.4.4. Commercial Production of No-Tillage Drills

There are three distinct markets for no-tillage machines in temperate New Zealand and Australia:

1. **Low-Cost Pasture Renovation Machines:** Because pasture renovation does not produce a cash crop for which the returns are easily measureable against the inputs, machines for sowing pasture seeds without cultivation generally have been characterized by low capital cost.

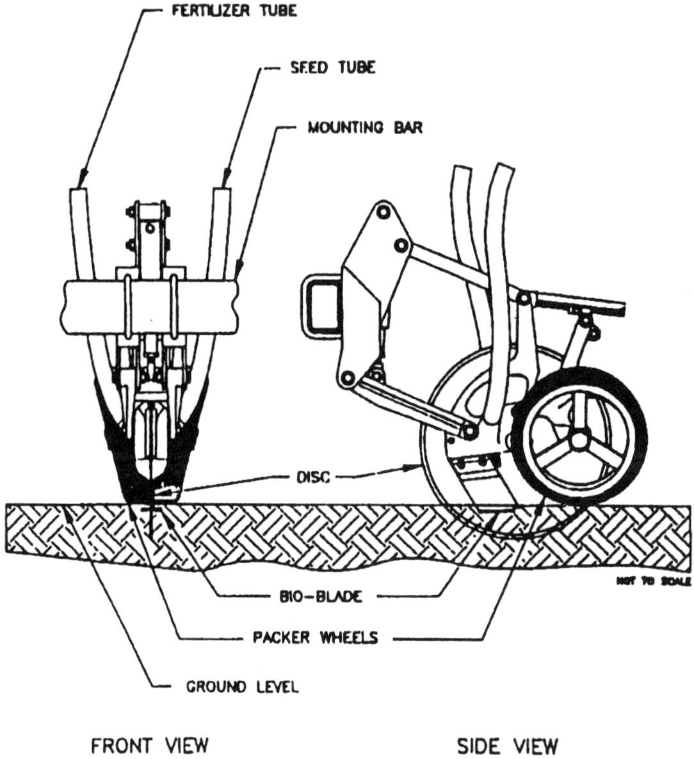

Figure 8.4. Schematic diagram of the Cross Slot™ drill (Taken from Baker, C. J. et al., *N.Z. J. Exp. Agric.*, 7, 1750, 1979.)

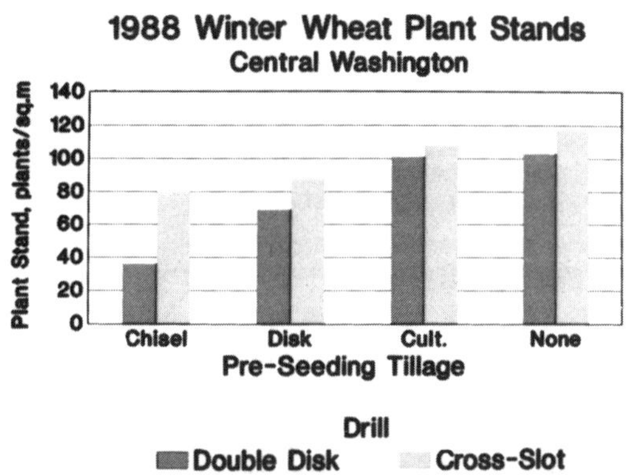

Figure 8.5. Winter wheat plant establishment with two conservation tillage opener designs in central Washington State, 1988. (Taken from Baker, C. J. and Saxton, K. E., *Am. Soc. Agric. Eng.*, 88, 1568, 1988.)

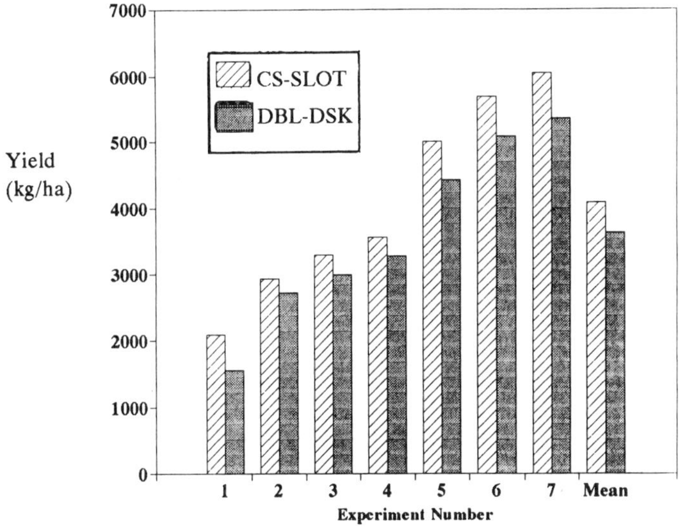

Figure 8.6. Wheat yields using conservation tillage drills in central and eastern Washington State. DBL-DSK = Double-disc drill, CS-SLOT = Cross Slot™ drill. (Taken from Saxton, K. E. and Baker, C. J., *Proc. Great Plains Conservation Tillage Symp.,* August 21–23, 1991, Bismark, ND, 1991.)

2. **Cheap No-Tillage Cropping Drills:** No-tillage, when first introduced to New Zealand for arable cropping in the 1960s, was promoted as a low-cost option. Understandably, the agricultural chemical companies, which bore the brunt of promotion of the technique, were not keen to associate it with expensive machinery options lest this compete with the sales opportunities for their agricultural chemicals.
3. **Sophisticated No-Tillage Drills:** Beginning in the 1960s, farmers in New Zealand have developed a willingness to pay for performance from herbicides by almost universally substituting glyphosate for the less expensive and narrower spectrum product, paraquat, in their weed control programs. This message has also registered with drill manufacturers, some of whom now realize that farmers will do likewise when sophisticated drills become available.

Clearly, the cost structures involved with sophisticated drills favor large annual usage of the machines. This contrasts with New Zealand's relatively small intensive farm units, even for arable cropping. Table 8.1. shows that even disregarding the increased crop yields possible with many sophisticated drills, at about 600 ha of use per year, little difference exists between the operating costs of drills: U.S. $8,500, $17,000 and $34,000, respectively. In fact, Saxton and Baker[15] reported 3 years of experiments in the United States that showed a 13% yield advantage for a more sophisticated drill, which tips the balance strongly in its favor at a low annual usage. The "horses for courses" approach

Table 8.1. Comparative Operating Costs of Conservation Tillage Machines in New Zealand (U.S. $/ha)[a]

Area Drilled	Simple Low Cost Drills (typical new cost U.S. $8500)		Conventional Direct Drills (typical new cost U.S. $17,000)		Heavy-Duty High-Technology Direct Drills (typical new cost U.S. $34,000)	
(ha/year)	24% Tax[b]	33% Tax[c]	24% Tax	33% Tax	24% Tax	33% Tax
50	46	41	71	62	121	106
100	30	26	41	36	63	56
200	21	19	26	23	35	30
400	17	15	19	17	20	18
600	16	14	17	14	15	14
800	15	13	15	13	13	12
1000	15	13	14	13	12	10

Source: Baker, C. J. et al, *Proc. 17th International Grassland Congress,* New Zealand and Queensland, Australia, (in press).

[a] Excludes tractor costs.
[b] This tax is the lower limit of New Zealand marginal tax for individuals.
[c] This tax is both the upper limit of New Zealand marginal tax for individuals and the New Zealand company tax rate.

to drill design has been the basis of promoting low-cost drills. This approach assumes that no single machine can have the universal capability of handling a wide range, let alone all, of the soil and residue conditions encountered in no-tillage. It is unrealistically based on hundreds of years of tillage experience in which farmers have owned several low-cost machines suited to specific conditions and have "mixed and matched" their machine choices to suit the conditions. In no-tillage, however, it is simply not economical for farmers to own several heavy-duty drills in order to cope with varying soil conditions day by day and field by field. Thus, successful no-tillage drills must have built into them a much wider range of capabilities and sophistication than their equivalent tillage tools.

The marketing principle that exporters should first establish a domestic market before exporting simply does not work for large, sophisticated, no-tillage drills in a small economy such as that of New Zealand. On the contrary, in order for the small domestic market to accept sophisticated machines, originating from New Zealand but sold in sufficient numbers elsewhere for the economies of scale to make them affordable in New Zealand, premarketing in volume offshore has had to take place. On the other hand, in the manufacture of new and sometimes experimental equipment, the short-line advantages of low-volume production and the variable soils and climates available within short distances, make New Zealand an ideal development zone for new farm technologies of this type.

8.5. CROP PERFORMANCE UNDER CONSERVATION TILLAGE

8.5.1. Crop Growth and Fertilizer Placement

Intensive continuous cropping by tillage in New Zealand, although practiced only on a small scale by international standards, has usually required applications of chemical fertilizers with successive crops. Cropping fertilizers can be applied by surface broadcasting, mixing with the seed at drilling, or separated placement alongside the seed at drilling.

With tilled soils the fertilizers applied close to the seeds have been most injurious when moisture was in the medium to low range of availability.[30] This is because of low dissolution, resulting in enhanced osmotic or chemical effects on germinating seeds. The dangers of extrapolating results from tilled soils to untilled soils was demonstrated by Baker and Afzal,[28] who found that placement of concentrated bands of fertilizer below or beside the seed in a no-tillage soil resulted in less seed damage and higher uptake and use-efficiency by plants than similarly placed fertilizer in previously tilled soils. Positive responses to seedling emergence were observed when granular nitrogenous fertilizers were separated from the seed horizontally by 10 to 20 mm as compared to vertical separation by the same distance. These effects were more pronounced as the soil became drier. Germination of a range of seeds was sensitive to fertilizer damage when the seed and fertilizer were mixed in the seed slot.[29,30]

Crops grown by conservation tillage methods often appear smaller, sometimes even stunted, in growth in the early stages of growth, but this visible difference usually diminishes at later stages. Fertilizer management must change under conservation tillage and take into account the increased potential for short-term immobilization of surface nitrogen and a decrease in plant-available nitrate-nitrogen as compared with conventional tillage. In untilled topsoil, fertility remains concentrated at the surface and is not mixed as in tilled soils, suggesting new requirements for at least the distribution, if not the total amount, of added fertilizers.[31]

In pasture renovation nitrogen is usually applied 2 to 4 weeks after sowing. This gives a timely boost to tillering in autumn and growth in spring and early summer, and avoids accelerating the growth of the competing resident species at the expense of the early development of the introduced seedlings.

8.5.2. Yield Sustainability

Before the success or failure of a system can be established, results from long-term trials are necessary to reach final conclusions. Stand establishment of crops is markedly influenced by the efficacy of seed germination and seedling emergence. For conservation tillage, especially no-tillage, to succeed, biological risks of stand establishment and growth must be less or no worse than those of conventional tillage. This is more so for crop establishment than pastures as the effects on crop yields and evenness of maturity are more obvious and sensitive to consistent emergence, row spacing, and plant density.

There have been four notable long-term no-tillage experiments in New Zealand, two in the Manawatu and two in Canterbury. Unfortunately, subsequent data on the effects of drill openers on crop stands and yields suggest that the results from some of these studies may have reflected more the inadequate qualities of drill design at sowing than any true soil, climate, seedbed- preparation, or location effects. For example, a 6-year double-cropping experiment on an imperfectly drained Tokomaru silt loam soil in Manawatu suggested that even when drilled with inverted-T slots yields of maize (*Zea mays* L.) declined under no tillage with time as compared with tillage, regardless of how much fertilizer was applied by surface broadcasting at the time of sowing and/or side-dressed after emergence.[32] However, when fertilizer was banded as described earlier using the Cross Slot™ opener, yield comparisons between the tilled and untilled soils were identical. A later experiment confirmed that after 6 years, the untilled soil had a greater response to placement of fertilizer as compared to surface broadcasting than the same soil in a tilled state.[33] It had been shown earlier that in untilled soils, preferential flow pathways (mainly from root and earthworm channels) tended to leach soluble nutrients past the root zones of crops.[34] By contrast, tilling of the soil destroys preferential channels and instead encourages even infiltration of surface-applied nutrients through the

Table 8.2. Mean Forage Yield from a 10-Year Maize/Oats Double Crop Rotation Under Three Tillage Systems (1978–1988) on a Silt Loam

	Tillage Systems (yields in t Dm/ha)			
Crop	Conventional Tillage	Minimum Tillage	Zero Tillage	LSD (0.05)
Maize (*Zea mays* L.)	15.2	14.7	12.8	0.6
Oats (*Avena sativa* L.)	7.4	7.2	7.0	0.6

Source: Adapted from Hughes, K. A. et al., *Soil Tillage Res.*, 22, 145–157, 1992.

artificially homogenized soil. This is thought to explain the clear preference of no-tillage (by summer crops at least) to the banding of starter fertilizer at drilling.[31]

In other long-term (10 years) trials on heavy soils in Canterbury[35] and Manawatu[36] similar trends in summer crop yields occurred, favoring tilled soils over untilled soils (Table 8.2) in the absence of banded fertilizer. In these instances, fertilizer was not banded close to the seed at drilling. Indeed, Hughes[36] reported that the double-disc openers used for no-tillage also resulted in stand reductions and poor root growth, which he felt were the main causes of poor summer crop yields. Interestingly, none of the winter crops in any of these long-term experiments showed consistent trends for or against no-tillage,[35,37] or in two cases minimum tillage.[33,36] Further, Janson[35] reported similar yields between tillage and no-tillage, even with summer crops, without banded fertilizer grown on a light, well-aerated soil, possibly because less well-defined preferential flow pathways existed in the more friable conditions (Table 8.3). These studies also indicated that organic matter, total nitrogen, and soil physical characteristics were maintained better under no-tillage than tillage. The size and number of earthworms also increased more under no-tilled cropping than those under conventionally sown crops (Table 8.4).[35] Slot compaction and low earthworm populations and activity had been associated with the action of triple-disc openers.[19,23] Measurements from a farm property in Manawatu, which had experienced 15 years of continuous double-cropping by no-tillage, showed that earthworm populations had been maintained at approximately 50% of the level found in an adjacent 30-year continuously stocked and undisturbed pasture.

8.5.3. Weed and Pest Control

The primary reasons given for tillage are to control weeds, to manage surface residues, and to improve soil physical conditions. The most important prerequisite for the success of conservation tillage is good weed control, particularly during the initial 4 weeks after planting.

In New Zealand both annual and perennial weeds are prominent and grow actively year-round. It has been necessary to develop special treatments for the control of species such as dock (*Rumex obtusifolius* L.), dandelion (*Taraxacum*

Table 8.3. Crop Yields from a 6-Year Crop Rotation under Two Tillage Systems on a Light, Stony Lismore Soil (1978–1984)

Year (Crop)	Zero Tillage (t/ha)	Conventional Tillage (t/ha)
1 (Linseed)	3.4	3.1
2 (Wheat)	5.47	5.25
3 (Clover)	0.30	0.30
4 (Wheat)	5.31	5.42
5 (Peas)	3.81	4.04
6 (Barley)	6.28	6.19

Source: Adapted from Janson, C. G., *Proc. Monsanto Conservation Tillage Seminar,* Christchurch, NZ: Monsanto Agricultural Products Co., 1984, pp. 39–54.

Table 8.4. Earthworms and Grass Grub (*Costelytra zealandica*) Population after 5 Years under Zero Tillage or Conventional Cultivation on a Light Soil in Canterbury, New Zealand

	Zero Tillage	Conventional Cultivation
Earthworms[a]		
(no./m^2)	1198	536
(g/m^2)	306	128
Grass grub		
(no./m^2)	331	19

Source: Adapted from Janson, C. G., *Proc. Monsanto Conservation Tillage Seminar,* Christchurch, NZ: Monsanto Agricultural Products Co., 1984, pp. 39–54.

[a] Earthworms include both *Allolobophora caliginosa* and *Lumbricus rubellus* species.

officinale Wiggers), storksbills (*Erodium cicutarium* L.), and yarrow (*Achillea millefolium* L.), which are effectively controlled by tillage. Sometimes in an area that is badly infested with perennial weeds, it is better to first control the weed problem with a one-off tillage regime before attempting to initiate reduced tillage methods of new crop establishment. Conversely, some perennial grass weeds such as couch (*Agropyron repens* L. Beauv.) and Californian thistle (*Cirsium arvense* L. Scop.) are more effectively controlled by no-tillage, especially where a double-spray herbicide program can be used.

Postemergence weed problems differ according to the type of crops planted. For example, where grain crops are grown, normal postemergence selective herbicide treatments still are required in conservation tillage. On the other hand, in forage crop establishment, some weeds make useful contributions to available feed and further control measures are not required. Forward planning in New Zealand is essential for good weed control. For example, because a winter forage crop is often grown ahead of the spring planting of a summer cereal crop, the choice of winter forage species and how it is managed can greatly influence the cost and effectiveness of the chemical weed and pest control required to establish the cereal crop.

Early successes in pasture renovation were achieved using dalapon or amitrole, but problems were encountered because of the length of time chemical residues remained in the soils.[39] The introduction of the short soil life herbicide, paraquat, in the early 1960s stimulated fresh interest. Emphasis then shifted to the control of white clover (*Trifolium repens* L.), using selective herbicides such as dicamba, in addition to the desiccation of grasses with paraquat.[40] As drill technology developed, conservation tillage, especially no-tillage, became an accepted technique. Agronomic recommendations have stressed the importance of delaying regrowth of the existing sward for as long as possible. For example, split applications of small doses of paraquat plus dicamba produced better results than single applications of this narrow-spectrum herbicide. Later, broader spectrum translocated herbicides such as glyphosate have almost totally replaced paraquat as a conservation tillage weed control adjunct. Where only annual weeds are present, effective control has been achieved with split applications of low rates of glyphosate followed by paraquat.

Tillage generally helps destroy or reduce soil fauna that otherwise may be directly or indirectly harmful or advantageous to crop growth. No-tillage in particular encourages the survival of useful pathogens that control soil invertebrates, predators, and parasites of pests, and increases populations of earthworms,[35] mites and springtails which decompose organic matter and thus release nutrients,[40] and nematodes.[41] Conservation tillage also encourages harmful pests e.g., slugs (*Deroceras reticulatum*),[42] through retention of crop residues and through transferring root and seedling pests, e.g., Argentine stem weevil (*Hyperodes bonariensis*), greasy cutworms (*Agrotis ypsilon*), and cereal aphid (*Rhopalosiphum pady*).

Grasses and clovers have been found to be particularly susceptible to pest damage when established by no-tillage.[43] In New Zealand, Argentine stem weevil, cutworm, and aphid can also cause problems in spring drilling of pastures and cereals sown by no-tillage where the previous crop has been a pasture or other species susceptible to the pest. Insect damage to maize seedlings sown by no-tillage has also been reported.[44]

Killing of existing vegetation with herbicides in the spring forces invertebrate pests to feed on the sown species.[40] Therefore, it is necessary to either apply appropriate granular insecticides in the slot with the seed or plan and change crop management to ensure that host crop species do not precede susceptible crops. For example, the Argentine stem weevil does not overwinter on brassica forage crops, so susceptible cereals can be safely planted following a winter brassica crop. Similarly, peas (*Pisum sativum* L.) are not susceptible to Argentine stem weevil or aphid and therefore can be safely grown after a winter pasture or forage cereal, e.g., oats (*Avena sativa* L.).

Conservation tillage seed slots, especially those that have a humid microenvironment, inadvertently provide an ideal habitat for moisture-seeking mollusks.[45] Metaldehyde, in the form of bait, is often used to successfully lure slugs out of resident vegetation and seed slots. It is, therefore, best applied on the ground surface after drilling. Phorate and methiocarb are both contact poisons used to kill insects and slugs, respectively. Because they are contact poisons they are most advantageously applied with the seed in the slot. However, phorate and methiocarb (but not metaldehyde) are also injurious to earthworms, which tends to conflict with some of the objectives of conservation tillage.

8.6. AVAILABILITY OF KNOWLEDGE AND EXPERTISE

Conservation tillage is gaining acceptance among New Zealand farmers as an efficient method of crop establishment. While erosion control and moisture conservation are the dominant advantages in Australia, North America, and Africa, for New Zealand, as in Europe, management flexibility and time conservation are seen as the most important advantages. In any case, if predictability and reliability of crop yield are to be achieved, farmers must have access to appropriate advice and be encouraged to avoid shortcuts.

A traditional cultivation system is more forgiving and repairable if incorrect decisions are made while preparing a seedbed. Mistakes can be made that are not easily visible from neighboring farms. With no-tillage, however, failure to complete either of the single critical operations — spraying and drilling — correctly is not usually obvious until it is too late and is then not easily repaired or hidden.

A recent survey of farmers who have used no-tillage in New Zealand identified support expertise as one of the key inputs into a successful no-tillage program.[47] Crop failures often relate to inappropriate seeding equipment, unsuitable soils (especially poor drainage), unsatisfactory herbicide application, failure to control pests, variations in soil moisture in the field, poor weather conditions at drilling, and incorrect application methods and rates of fertilizer. Therefore, to become acceptable as a recommended technique, conservation tillage must be capable of being applied by a wide range of farmers under a wide range of farming conditions with knowledgeable support expertise. Unfortunately, in New Zealand the success of the technique will remain constrained until professionally skilled consultants are available in sufficient numbers to give individual advice in the field. Ironically, New Zealand is one of the predominant international sources of scientific information on the topic, but the demand for information in the field is growing slowly. It is likely that as the benefits of the more sophisticated technologies become apparent (Figure 8.7), an increase in demand for skilled consultants will follow. Until now many farmers have had to learn the importance of correct management techniques using conservation tillage as a result of experience, especially failures. There

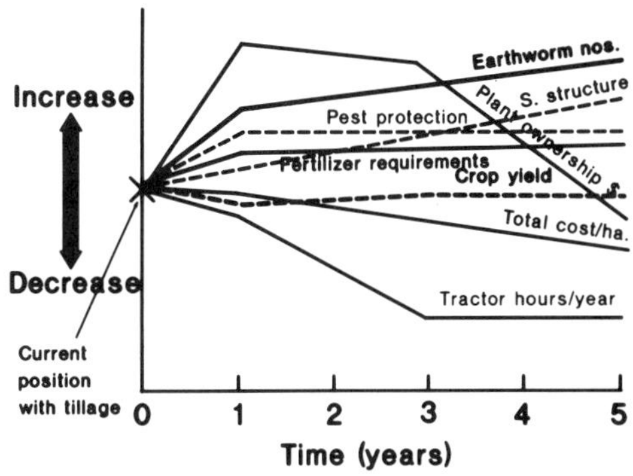

Figure 8.7. Likely trends from adoption of no-tillage in temperate New Zealand.

has been an unfortunate tendency for many to implement no-tillage systems on the basis of their experience with conventional tillage. The often poor outcome has, in turn, reflected badly and unfairly on conservation techniques in general. Often a farmer will accept failures with conventional tillage as "bad luck," whereas failures with a new technique such as no-tillage are attributed to failure of the new system as a whole rather than to the component cause of the failure. Again, this reflects lack of knowledge of the specific functions and weaknesses of the individual components of the system.

Training of skilled consultants who can apply no-tillage techniques and management in the field is urgently required. These consultants will be required to work with farmers to implement no-tillage practices, particularly during the transition phase; to carry out site-specific research, particularly in the areas of fertility and pest and disease control; and to provide the forum for group discussion between the innovators and early adopters, who will inevitably have the greatest impact on the successful embracing of no-tillage by farmers in general.

8.7. GENERAL SUMMARY AND CONCLUSIONS

In most temperate countries of the world where conservation tillage is needed urgently, the pressure has been enormous on technologists to invent the tools necessary to facilitate the practice. Herbicidal technology has advanced rapidly and has mostly fulfilled the requirements for reliable weed control, although the search continues for a wider spectrum of more environmentally friendly chemicals and indeed, nonchemical, means of weed control. Advances in electronic selective field-recognition of weeds also promises much by reducing overall use and wasting of herbicides.

With seed drilling technology, advancement has been generally slower. Time constraints have forced most design engineers to take short cuts and wrongly (as it turns out) assume that the requirements of seeds and plants under no-tillage, for example, could be extrapolated from what was known under tillage, with adjustments mainly confined to robustness. New Zealand's limitations are its small domestic market; the unpredictability of its temperate climate, which makes the tolerance of seeding equipment to rapidly changing soil conditions important; prolific year-round weed growth; a small, but advanced pool of knowledge and expertise; heavy crop residues; the need for fertilizer placement at seeding; pest control; and the low demand for conservation tillage. Countering these disadvantages are the advantages of a modern, if small, manufacturing capability, a wide range of miniclimates within convenient traveling distances, a high technical literacy among farmers, and the luxury of time for research and development that the low demand for conservation tillage provides.

Research and development have painstakingly redefined the biological requirements of seeds, seedlings, and productive plants in untilled soils. In so doing, unique and innovative seed drill technologies have arisen. These technologies now match current chemical technologies and will form the basis of the next and first major generation of true conservation tillage drills. The new generation machines deliberately target and utilize the unique potential of untilled soils and surface crop residues. They have already raised the potential biological reliability of no-tillage to a higher level than that achievable under conventional tillage.

8.8. REFERENCES

1. Ross, C. A., and Wilson, A. D. "Conservations Tillage-Soils and Conservation," *N.Z. Agric. Sci.* 17(3):283–287 (1985).
2. Newsome, P. F. J. "The Vegetative Cover of New Zealand," *Water Soil,* Misc. Publ. No. 112 (1987).
3. Ritchie, W. R. "Pasture Renovation by Overdrilling," in *Proc. N.Z. Grassl. Assoc.* 47:(1986), pp 159–164.
4. Ross, C. W., and K. A. Hughes. "Maize/Oats Forage Rotation under Three Cultivation Systems, 1975–83. II. Soil Properties," *N.Z. J. Agric. Res.* 28:209–219 (1985).
5. Chan, K. Y., and J. A. Mead. "Surface Physical Properties of a Sandy Loam Soil under Different Tillage Practices," *Aust. J. Soil Res.* 26:549–559 (1988).
6. Choudhary, M. A., and C. J. Baker. "Physical Effects of Direct Drilling Equipment on Undisturbed Soils. II. Seed Groove Formation by a Triple Disc Coulter and Seedling Performance," *N.Z. J. Agric. Res.* 34:189–195 (1981).
7. Lindwall, C. W., and D. C. Erbach. "Residue Cover, Soil Temperature Regimes and Corn Growth," *Am. Soc. Agric. Eng.* 83-1024 (1983).
8. Lynch, J. M. "Production and Phytotoxicity of Acetic Acid in Anaerobic Soils Containing Plant Residues," *Soil Biol. Biochem.* 10:131–135 (1978).
9. Chaudhry, A. D., and C. J. Baker. "Barley Seedling Establishment by Direct Drilling in a Wet Soil. I. Effects of Openers under Simulated Rainfall and High Water Table Conditions," *Soil Tillage Res.* 111:43–61 (1988).

10. Baker, C. J., and K. E. Saxton. "The 'Cross Slot' Conservation Tillage Grain Drill Opener," *Am. Soc. Agric. Eng.* 88: 1568 (1988).
11. Hyde, G. M., D. E. Wilkins, K. E. Saxton, J. Hammel, G. Swanson, R. Hermanson, E. A. Dowding, J. B. Simpson and C. L. Peterson. "Reduced Tillage Seeding Equipment Development," in *STEEP — Conservation Concepts and Accomplishments,* L. F. Elliot (Ed.) (Olympia: Washington State University Press, 1987).
12. Choudhary, M. A. "Modelling of Field Seedling Establishment as a Function of Seeding Techniques and Soil Moisture Interactions," in *Proc. 11th International Soil Tillage Research Organization Conf.,* Vol. 1 (Edinburgh, Scotland, 1988), pp. 355–360.
13. Baker, C. J., E. M. Badger, and J. H. McDonald. "Develoment with Seed Drill Coulters for Direct Drilling. I. Trash Handling Properties of Coulters," *N.Z. J. Exp. Agric.* 7:1750–1784 (1979).
14. Baker, C. J., and M. A. Choudhary. "Seed Placement and Micro-Management of Residue in Dryland No-Till," in *Challenges in Dryland Agriculture — a Global Perspective,* Unger, P.W., T. V. Sneed, W. R. Jordan, and R. Jensen (Eds.) Amarillo/Bushland, TX, 1988), pp. 544–546.
15. Saxton, K. E., and C. J. Baker. "The Cross Slot Opener for Conservation Tillage," in *Proc. Great Plains Conservation Tillage Symp.* (Bismark, N.D. USA, August 21 to 23, 1991), pp. 65–72.
16. Baker, C. J. "Experiments Relating to Techniques for Direct Drilling of Seeds into Untilled Dead Turf," *J. Agric. Eng. Res.* 32:133–145 (1976).
17. Choudhary, M. A., and C. J. Baker. "Effects of Drill Coulter Design and Soil Moisture Status on Emergence of Wheat Seedlings," *Soil Tillage Res.* 2:131–142 (1982).
18. Choudhary, M. A., Pei Yu. Guo and C. J. Baker. "Seed Placement Effects on Seedling Establishment in Direct Drilled Fields," *Soil Tillage Res.* 6:79–93 (1985).
19. Baker, C. J., and T. V. Mai. "Physical Effects of Direct Drilling Equipment on Undisturbed Soils V. Groove Compaction and Seedling Root Development," *N.Z. J. Agric. Res.* 25:51–60 (1982).
20. Baker, C. J. "Technical Potentialities of Overdrilling for Hill Pasture Improvement and Renovation," in *3rd Anim. Sci. Congr. Asian Australasian Association of Animal Production Societies (AAAP),* Part 1, (Seoul, South Korea, 1985). pp. 211–218.
21. Choudhary, M. A., and C. J. Baker. "Physical Effects of Direct Drilling Equipment on Undisturbed Soils: I. Wheat Seedling Emergence from a Dry Soil under Controlled Climates," *N.Z. J. Agric. Res.* 23:489–496 (1980).
22. Choudhary, M. A., and C. J. Baker. "Physical Effects of Direct Drilling on Undisturbed Soils. III. Wheat Seedling Performance and In-Groove Micro-Environment in a Dry Soil," *N.Z. J. Agric. Res.* 24:183–187 (1981).
23. Chaudhry, A. D., C. J. Baker, and J. A. Springett. "Barley Seedling Establishment by Direct Drilling in a Wet Soil: II. Effects of Earthworms, Residue and Openers," *Soil Tillage Res.* 9:123–133 (1986).
24. Campbell, A. J., and C. J. Baker. "An X-Ray Technique for Determining Three Dimensional Seed Placement in Direct Drilled Soils," *Trans. Am. Assoc. Agric. Engr.* 32(2):379–384 (1985).

25. Campbell, B. D. "Winged Coulter Depth Effects on Overdrilled Red Clover Seedling Emergence," *N.Z. J. Exp. Agric.* 28:7–17 (1985).
26. Baker, C. J., A. D. Chaudhry and J. A. Springett. "Barley Seedling Establishment by Direct Drilling in a Wet Soil. III. Comparison of Six Seed Sowing Techniques," *Soil Tillage Res.* 11:167–181 (1988).
27. Wilkins, D. E., G. A. Muilenburg, R. R. Allmaras and C. E. Johnson. "Grain Drill Opener Effects on Wheat Emergence," *Trans. Am. Soc. Agric. Eng.* 26(3):651–660 (1983).
28. Baker, C. J., and M. A. Afzal. "Dry Fertilizer Placement in Conservation Tillage: Seed Damage in Direct Drilling (No-Tillage)," *Soil Tillage Res.* 7:241–250 (1986).
29. Choudhary, M. A., C. J. Baker and W. Steifel. "Dry Fertilizer Placement in Direct Drilling," *Soil Tillage Res.* 12:213–221 (1988).
30. Carter, O. "The Effects of Fertilizers on Germination and Establishment of Pasture and Fodder Crops," *J. Wool Tech. Sheep Breeding,* July: 69–75 (1969).
31. Malhi, S. S., and M. Nyborg. "Evaluation of Methods of Placement for Fall-Applied Urea under Zero-Tillage," *Soil Tillage Res.* 15:383–389 (1990).
32. Sims, R. E. H., and C. J. Baker. "Comparison of Reduced Time and Energy Seedbed Preparation Systems: Year 1," *N.Z. J. Exp. Agric.* 9:299–305 (1981).
33. Baker, C. J., and C. M. Afzal. "Some Thoughts on Fertilizer Placement in Direct Drilling," in *Proc. Monsanto Conservation Tillage Seminar,* (Christchurch, NZ: Monsanto Agricultural Products Co., 1981), pp. 343–354.
34. Kanchanasut, P., D. R. Scotter and R. W. Tillman. "Preferential Solute Movement through Larger Soil Voids. II. Experiments with Saturated Soil," *Aust. J. Soil Res.* 16:269–276 (1978).
35. Janson, C. G. "Conservation Tillage Studies under Intensive Arable Cropping," in *Proc. Monsanto Conservation Tillage Seminar,* (Christchurch, NZ: Monsanto Agricultural Products Co., 1984), pp. 39–54.
36. Hughes, K. A. "Maize/Oats Forage Rotation under 3 Cultivation Systems, 1978–83. I. Agronomy and Yield," *N.Z. J. Agric. Res.* 28:201–207 (1985).
37. Hughes, K. A., D. J. Horne, C. W. Ross and J. F. Julian. "A 10-Year Maize/Oats Rotation under Three Tillage Systems. II. Plant Population, Root Distribution and Forage Yields," *Soil Tillage Res.* 22:145–157 (1992).
38. Robinson, G. S., and M. W. Cross. "Improvement of Some New Zealand Grassland by Oversowing and Overdrilling," in *Proc. 8th International Grassland Congress,* (1960), pp. 402–405.
39. Ivens, G. W. "The Latest Developments in Weed and Pest Aspects of Low Energy Cropping," in *Proc. International Conference on Energy Conservation in Crop Production,* Massey University, NZ (1977), pp. 231–233.
40. Pottinger, R. P. "The Role of Soil Fauna and Pest Control in Conservation Tillage Systems," in *Proc. Conservation Tillage Technical Seminar* (Christchurch, NZ: Monsanto Agricultural Products Co., 1979).
41. Yeates, G. W., and K. A. Hughes. "Effects of Three Tillage Regimes on Plant and Soil Nematodes in an Oats/Maize Rotation," *Pedobioligia* 34:379–387 (1990).
42. Edwards, C. A. "Effect of Direct Drilling on the Soil Fauna," *Outlook Agric.* 46:243–244 (1975).
43. Charlton, J. R. L. "Initial Trials for Improvement of North Island Hill Country by Conservation Tillage," in *Proc. Conservation Tillage Seminar* (Christchurch, NZ: Monsanto Agricultural Products Co., 1981), pp. 27–34.

44. Carpenter, A., W. M. Kain, C. J. Baker and R. E. H. Sims. "The Effects of Tillage Techniques on Insect Pests of Seedling Maize," in *Proc. 31st New Zealand Weed and Pest Control Conf.* (1978), pp. 89–91.
45. Baker, C. J. "Techniques of Overdrilling for the Introduction of Improved Species into Temperate Grasslands," in *Proc. 14th International Grassland Congress,* KY (1981), pp. 542–544.
46. Baker, C. J., M. A. Choudhary and K. E. Saxton. "Inverted-T Drill Openers for Pasture Establishment by Conservation Tillage," in *Proc. 17th International Grassland Congress,* New Zealand and Queensland, Australia, (in press).
47. Ritchie, W. R. Personal communication.

CHAPTER 9

Role of Conservation Tillage in Sustainable Agriculture in the Southern United States

Donald D. Tyler
University of Tennessee; Jackson, Tennessee

Michael G. Wagger
North Carolina State University; Raleigh, North Carolina

Daniel V. McCracken and William L. Hargrove
University of Georgia; Griffin, Georgia

TABLE OF CONTENTS

0-87371-571-3/94/$0.00+$.50

9.1. INTRODUCTION

The southeastern United States historically has had severe soil erosion and subsequent surface water quality problems. Many soils have root-restrictive fragipans or are underlain by shallow limestone, sand, or heavy clay. Potential losses in productivity on these soils and the water quality problems from soil erosion would indicate that sustainability was not achieved in the past. The advent of conservation tillage and effective residue management offered a good potential solution to many of the constraints to sustainable cropping systems. The following discussions involve major resource areas in the southeastern United States and their unique problems and subsequent need for conservation tillage for economically and environmentally sound row crop production. The areas to be intensively discussed in this chapter are the coastal plain, the Piedmont, and the interior low plateau. The western coastal plain areas are covered by loess and referred to as the Southern loess belt. Much of the information is also applicable to regions 3, 4, and 5 of Figure 9.1.

9.2. SOIL AND CLIMATIC FACTORS

9.2.1. Loess and Limestone Areas

The southern loess belt consists of loess-derived soils in Kentucky, Tennessee, Mississippi, and Louisiana.[1] These soils are typically >2% slope ranging up to as high as 30% or greater in the deeper loess areas. Historically they have had very high soil erosion rates when cropped.[2] The soils in this region are mainly alfisols.[3] The productivity of a large hectarage of these soils is affected by accelerated erosion because of the presence of root-restrictive fragipans. Approximately 55% of the soils on these highly erodible uplands have fragipans. In some areas, such as Fayette County in western Tennessee, fragipans are present in about 70% of the uplands.[4] Inherent productivity losses can occur on these soils as the fragipan depth decreases as severe erosion proceeds over time.[5-8] Because these soils are loessially derived and the underlying coastal plain sediments are inherently unproductive, soil erosion results in permanent depletion of the soil resource.[9]

Other areas of the southeastern United States with high erosion and potential losses in productivity are central Kentucky, central Tennessee, and northwestern Alabama.[2] These soils are generally loess over sandstone or limestone. In some cases, they are formed from residuum of limestone or in a thin layer of

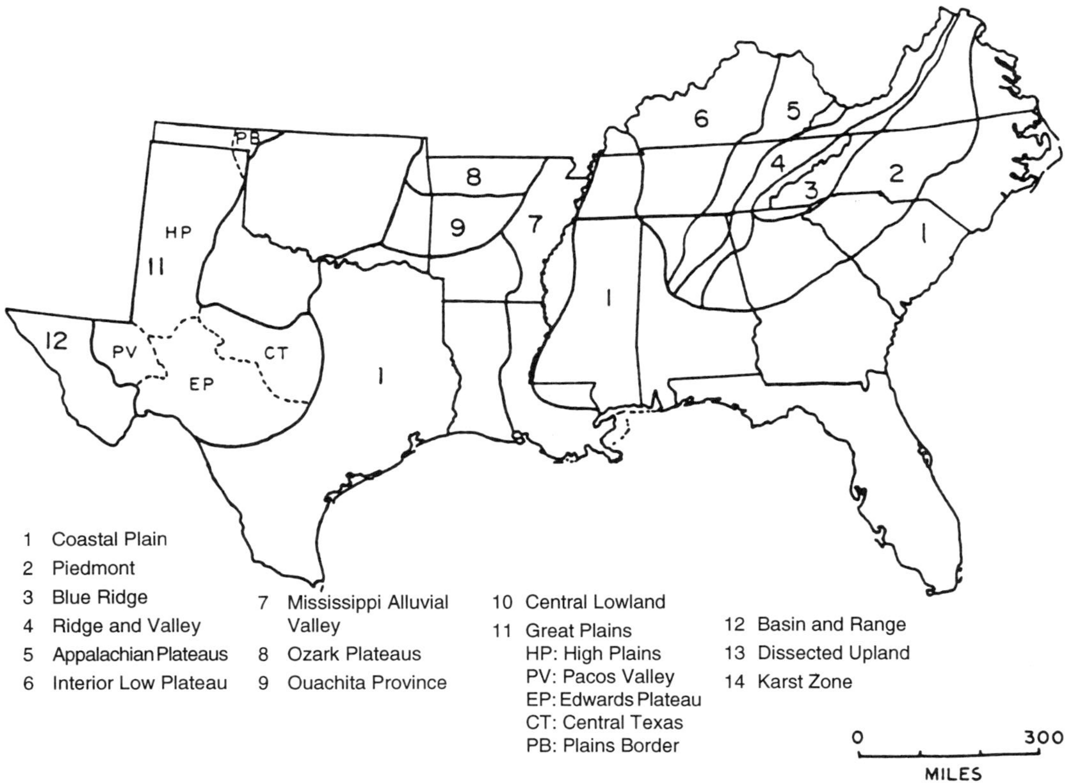

Figure 9.1. Physiographic division of the southeastern states. (Adapted from Buol, S.W., *Soils of the Southern States and Puerto Rico,* North Carolina State University, 1973. With permission.)

loess over rock.[10] The losses in productivity on these soils as a result of erosion have been documented by Frye et al.[11,12] Erosion rates vary in these areas relative to rainfall amount and intensity. This was quantified by Wischmeier and Smith.[13] They quantified the erosive potential of a storm by the product of the total kinetic energy and the 30-min intensity of the storm. For much of the southern United States, the highest values occur from April through July.[14] Total annual precipitation ranges from 115 to 130 cm/year. Rainfall is fairly evenly distributed, but extended periods without rain resulting in drought stress in mid- to late summer are common.

9.2.2. Atlantic Coastal Plain

From a physiographic perspective, the relatively flatter lands of the coastal plains do not possess as severe an erosion hazard as the broadly sloping lands of the Piedmont and Appalachian plateau. Moreover, the Atlantic coastal plain is generally warmer, more humid, more diverse in soil types, and more diverse in cropping systems than sloping areas of the southeastern United States. Total rainfall is generally adequate, but its distribution is often erratic, leading to periods of drought during most growing seasons.

Agricultural soils in well-drained areas of the Atlantic coastal plain are typically udults characterized by coarse-textured A and E horizons and the finer-textured, acidic B horizon.[15] These soils often have dense layers or pans in the E horizon, apparently formed genetically or from the weight of farm equipment.[15,16] The coarse-textured Ap horizons hold low amounts of plant-available water and the pans restrict root growth into subsoil moisture, thereby limiting nutrient uptake as well.

9.2.3. Southern Piedmont

In a broad band between the mountains and the Atlantic coastal plain, the southern Piedmont extends from south-central Virginia through the Carolinas and Georgia into east-central Alabama.[17,18] Surface features of the region have been shaped by geologic erosion, and the resulting physiography is characterized by gently rolling interfluves dissected by the steeper and deeper valleys of modern streams.[17] Large areas of flat land on uplands and along streams are rare.

Soils of the southern Piedmont are dominated by well-drained ultisols that developed in the residuum of diverse igneous, metamorphic, and sedimentary rocks.[19] The Cecil (clayey, kaolinitic, thermic typic kanhapludults) and related udults are common soils throughout the region. These soils are highly weathered, and relatively low in natural fertility and organic matter content. In the absence of liming, they are moderately to strongly acid throughout the pedon. They possess clayey subsurface horizons, and, if erosion has not been severe, their relatively thin surface horizons are coarse textured.

From the time of European settlement in the 1700s and early 1800s through the first quarter of the 20th century, land clearing and intensive tillage for cotton (*Gossypium* sp.), tobacco (*Nicotiana* sp.), and corn (*Zea mays* L.) resulted in accelerated soil loss throughout the southern Piedmont.[18] On the more severely eroded sites, the plow layer consists of material derived mainly from below the original A horizon, and shows a striking increase in clay content.[20] With loss of topsoil and truncation of soil profiles by erosion came depletion of soil productivity under the practices in common use. Worn-out and gullied fields were abandoned. The consequences of erosion contributed to the dramatic net withdrawal since 1920 of land from row crop production.[18]

The climate of the southern Piedmont is warm and humid.[21] In winter, the ground seldom freezes for more than a few days at a time, if at all. High annual rainfall amounts (1100 to 1400 mm/year) are distributed evenly on average throughout the year, except for a somewhat drier late summer-early autumn period. Crop residues decompose rapidly in the warm, humid climate, making year-round soil protection a challenge and compounding the threat of soil loss on the region's highly erodible soils. Much of the more erosive rain falls during the spring and early summer, times when the land can be left loose and bare by tillage for warm-season crops.[22] Despite the reasonably even distribution of rainfall throughout the year, short periods of drought frequently limit agricultural production in the southern Piedmont.[23] Groundwater supplies are typically low yielding, and this confines irrigation mainly to small acreages of high-value crops that can be supplied by farm ponds.[24] The likelihood of yield-limiting drought, the sloping character of much of the region's agricultural land, and the damage to the soil resource caused by past mismanagement and erosion make soil and water conservation key elements in the development of a sustainable agriculture in the southern Piedmont.

9.3. ADOPTION OF CONSERVATION TILLAGE

9.3.1. Loess and Limestone Areas

The lack of a need for deep tillage has been a distinct advantage in the loess- and limestone-derived soil in the southern United States. Successful conservation tillage systems have been developed for soybeans (*Glycine max* L. Merr.), corn, grain sorghum (*Sorghum bicolor* L. Moench), and cotton.

The use of conservation tillage, especially in conjunction with residue and cover crop management, offers one of the best ways to maintain good surface water quality.[25] Conservation tillage and residue management can effectively conserve soil water and reduce soil erosion.[14,26] One common system used in the south is wheat (*Triticum* sp.) double-cropped with soybean. In this system, wheat is planted in mid- to late autumn. It is harvested for grain the following June, with soybeans planted in the wheat residue with a no-tillage planter. In the case of the medium-textured soils the slit is made using only a rippled coulter. Increases in hectarage of single-cropped soybeans planted in previous crop residue has also occurred. No-tillage hectarage of corn and grain sorghum is also very high. When other systems, defined as conservation tillage (which maintain 30% residue), such as ridge-till and mulch-till cover, are included percentages for many systems exceed 70%.

The most dramatic increases in conservation tillage use in Tennessee is occurring with cotton. In 1988, conservation tillage in cotton production was used on only about 1% of the hectarage.[27] Of this 1%, about 0.8% was no-tillage using only a coulter. In 1991, this figure had increased to 7%, with 6% being no-tillage. No-tillage cotton hectarage in 1992 was 17%.[28]

The continued adoption of conservation tillage is in part due to the 1985 Food Security Act. This legislation connects farm subsidy payments to whether erosion is being controlled. The Soil Conservation Service and United States Department of Agriculture (USDA) Extension Service are responsible for writing farm plans for highly erodible land which must be fully implemented by 1995. This part of the law is called conservation compliance.[29] The use of conservation tillage in farm plans has been extensive and many crops such as cotton have been targeted for increased no-tillage research.[30]

Cropping systems that conserve soil water and afford minimum soil disturbance and maximum soil cover, especially during the period of April through July, are essential to produce sustainable agricultural systems.

9.3.2. Atlantic Coastal Plain

In-row subsoiling is common in conservation tillage systems on the coastal plain. With drought stress often an important factor limiting the yield of warm-season annual crops, there is a need to utilize natural precipitation as efficiently as possible. Strategies for avoiding water stress in plants include increasing plant use of stored water in the soil, increasing the capture of water from summer rainfall, and conserving soil water that might otherwise be lost via evaporation. In-row subsoiling is practiced to alleviate compacted zones and increase root exploitation of subsoil water. Relative yield increases due to subsoiling have been most notable for corn, ranging from 14 to 197%.[31,32] Differences in relative yield increases between and within experiments can be largely attributed to differences in rainfall and soil characteristics. The advantage of subsoiling is substantially greater in dry growing seasons than in years with adequate, well-distributed rainfall.[33-35]

Conservation tillage practices have shown variable results on coastal plain soils. Corn and soybean yields during a 5-year period were unaffected by conventional tillage or no-tillage on a Eunola sandy loam in North Carolina.[36] Even though two out of five growing seasons were relatively dry, Wagger and Denton concluded that on a soil with no surface crusting or associated infiltration limitations, the water conservation that may occur with no-tillage is of minimal consequence after canopy closure. In related work on an Aycock fine sandy loam, a soil generally considered susceptible to crusting and runoff, no-tillage corn and double-cropped soybean yields were higher than conventional tillage in 1 out of 3 years and were similar the other 2 years.[37] NeSmith et al.[38] reported higher soil water contents under no-tillage compared to a moldboard plow/disc system for double-cropped soybean grown on a Greenville sandy clay loam. Soil bulk density in excess of 1.7 Mg/m^3 under no-tillage, however, reduced water extraction and plant growth. The authors suggested that tillage pan amelioration may be needed for sandy ultisols with poorly developed structure.

Winter annual cover crops have been used in southern agriculture for many years to reduce soil erosion. In addition, the nitrogen provided by leguminous cover crops to summer crops such as corn, grain sorghum, and cotton can be substantial.[39-41] Maintaining a living, vegetative cover just prior to planting in the spring, however, can have detrimental consequences on soils with low water-holding capacity or shallow rooting depth due to root-restrictive layers. Campbell et al.,[42] working with a predominant coastal plain soil (Norfolk), found significantly greater soil-water depletion 15 days after corn planting for a rye (*Secale cereale* L.) cover chemically or mechanically (disced) killed at planting compared to a conventional tillage system with no cover crop. The practice of in-row subsoiling to loosen highly compacted soil layers and cover management (removed vs intact) was examined by Ewing et al.[43] Compared with fallow treatments, crimson clover (*Trifolium incarnatum* L.) depleted soil water in the surface 15 cm, just prior to corn planting, by 28 and 55% in the two site-years (Table 9.1). In 1 out of 2 years, the corn grain yield reduction in the presence of crimson clover was entirely overcome by subsoiling (Table 9.2). On soil such as that described above, minimizing the effects of soil water depletion can be achieved by early herbicide desiccation, animal grazing, or mechanical harvesting.

9.3.3. Southern Piedmont

The problem of soil water depletion is much less in the southern Piedmont. Numerous reviews cover aspects of the development and adaption of conservation tillage to the soil and climatic constraints of the southern United States.[44-49] Tillage systems that leave considerable quantities of previous and cover crop residue on the soil surface appear necessary for restoring and/or maintaining the productivity of erodible southern Piedmont soils.[46,50] Langdale et al.[46] concluded that fertilization and liming alone are incapable of restoring production on eroded southern Piedmont soils. Conservation tillage, which leaves at least 30% of the soil surface covered with crop residues after planting, augments the effect of improved fertility by enhancing the crop environment in many ways that further increase yield.[51] No-tillage, which maximizes soil coverage by crop residues, is well adapted to most soils of the southern

Table 9.1. Soil Water Content in the Surface 15 cm at Corn Planting as Affected by Crimson Clover Management

Treatment	1985 (m^3/m^3)	1986 (m^3/m^3)
No cover	0.16 a[a]	0.18 a
Top growth intact	0.11 c	0.08 b
Top growth removed[b]	0.14 b	0.13 b

Source: Adapted from Ewing, R. P., et al., *Soil Sci. Soc. Am. J.,* 55:1083, 1991. With permission.

[a] Means followed by the same letter in each column are not significantly different ($p = 0.01$).
[b] Top growth removed 1 week before corn planting.

Table 9.2. Corn Grain Yield as Affected by Subsoiling and Crimson Clover Management.

Treatment	1985 (Mg/ha)	1986 (Mg/ha)
No cover/nonsubsoiled	5.5	2.8
No cover/subsoiled	6.2	4.7
Crimson clover intact/nonsubsoiled	4.3	1.2
Crimson clover intact/subsoiled	6.5	3.1
Crimson clover removed/nonsubsoiled	4.8	2.1
Crimson clover removed/subsoiled	6.4	4.2
Orthogonal contrasts		
No cover vs clover	*	**
Subsoiled vs nonsubsoiled	**	**
Top growth intact vs removed	NS	**

Source: Adapted from Ewing, R. P., et al., *Soil Sci. Soc. Am. J.,* 55: 1084, 1991. With permission.

Note: *,** indicates F test significant at 0.05 and 0.001 probability level, respectively.

Piedmont, and is the conservation tillage method most extensively adopted by farmers in the region (Table 9.3).

On southern Piedmont upland soils, no-tillage improves the supply of soil moisture to crops and has been associated with increases in grain yield.[52] In a 3-year tillage trial with a wheat-double-crop soybean-corn rotation on a Pacolet sandy clay loam (clayey, kaolinitic, thermic kanhapludults), the yield of warm-season crops was consistently increased by no-tillage.[52] Average corn grain yield was 32% greater with no-tillage (7.93 Mg/ha) than with chisel plow/disc tillage (6.01 Mg/ha). Similarly, the average double-crop soybean yield was 43% greater with no-tillage (2.58 Mg/ha for no-tillage vs 1.67 Mg/ha for chisel plow/disc tillage). In general, yield differences due to tillage reflected differences in soil water content measured during the growing season. Greater soil water contents with no-tillage were attributed to reduced soil crusting, less runoff, greater infiltration, and lower evaporative moisture losses with no-tillage management.

Table 9.3. Southern Piedmont Areal Tillage Estimates by State, 1992

Southern Piedmont (by state)	Conventional Tillage[a]	Conservation Tillage (ha) No-Till[b]	Ridge-Tillage[c]	Mulch-Tillage[d]
Alabama	5,548	1,223	0	2,219
Georgia	44,289	16,851	644	20,380
South Carolina	54,385	14,013	20	13,123
North Carolina	276,433	137,446	121	43,257
Virginia	82,839	29,384	0	13,164
Total	463,493	198,916	785	92,143

Source: Adapted from CTIC.[28]

[a] Tillage types that leave less than 30% residue cover after planting.
[b] Planting or drilling in a narrow seedbed or slot. Soil is left undisturbed from harvest to planting; >30% residue cover after planting.
[c] Planting is done in a seedbed on ridges; >30% residue cover after planting.
[d] Use of chisels, field cultivators, discs, sweeps, or blades to disturb total soil surface while leaving >30% of the soil surface covered by crop residues.

Recent studies in the southern Piedmont suggest that the vegetative cover and lack of soil disturbance afforded by no-tillage retard surface crusting, serving to maintain infiltration and reduce runoff.[53-55] In the absence of vegetative cover, Cecil and Pacolet soils formed surface crusts readily under raindrop impact, which rapidly reduced infiltration rates and produced runoff.[54,56] In contrast, vegetative cover maintained high infiltration rates until infiltration was limited by subsurface layers that restricted internal drainage.[55,56] On typic Kanhapludults, West et al.[53] measured runoff from rainfall simulation applied to no-tillage crimson clover-grain sorghum, disc-tillage monocrop grain sorghum and disc-tillage monocrop soybean soon after killing the crimson clover in the spring. Rainfall simulations were made to plots with residue and with residue removed. Far less runoff was measured from no-tillage crimson clover/grain sorghum with or without residue than from the two disc-tillage cropping systems, both before and after they were tilled. While the tillage comparison was confounded with differences in cropping system, the results of West et al.[53] suggest that the greater aggregate stability of the no-tillage soil played a role in resisting the formation of a surface seal and permitting greater infiltration. In an unreplicated study on an unterraced 2.71-ha watershed, Langdale et al.[57,58] measured greater annual runoff amounts from conventional-tillage winter fallow-monocrop soybean (17.3% of rainfall total) than from no-tillage winter barley-grain sorghum (7.0% of rainfall total).

Unrestricted traffic patterns may excessively compact the surface of no-tillage soils, increasing runoff and reducing (or eliminating) the moisture-supplying advantage of no-tillage on affected southern Piedmont soils.[52] In-row chiseling may offset this effect while maintaining crop residues on the soil surface.[59,60]

While the utility of surface residue coverage in the southern Piedmont is apparent, especially in the spring and early summer, it is hard to generalize about the need for in-row chiseling. Despite significantly lower average grain yield with moldboard plow-disc tillage (7.45 Mg/ha), no consistent differences among corn yields were found with no-tillage, in-row chiseling, or in-row subsoiling (3-year average: 8.51, 8.47, and 8.85 Mg/ha, respectively).[59] However, in this test irrigation was applied when the soil moisture tension in the surface 15 cm exceeded 0.03 MPa, and this may have masked potential yield benefits from in-row tillage. Langdale et al.[60] examined the response to tillage of soybean and grain sorghum rotational sequences in double-crop systems with winter wheat. During the first 4-year rotation cycle of this 8-year study, soybean yields overall were not affected by tillage, but average sorghum yields were higher with in-row chiseling than with no-tillage or disc-tillage (4.89 vs 4.58 and 4.39 Mg/ha). However, during the second 4-year cycle, grain sorghum yields were greatest for no-tillage (5.14, 4.74, and 4.40 Mg/ha). Langdale et al.[60] suggested that after several years the cumulative activity of crop roots, soil animals, and microorganisms under no-tillage management may have raised sorghum yields by gradually improving soil macroporosity and increasing water capture. In a floodplain soil in the Georgia Piedmont, higher annelid and soil arthropod densities were measured under long-term no-tillage than under long-term conventional tillage.[61,62]

The warm, humid climate of the southern Piedmont favors rapid crop residue decomposition. Current strategies for maintaining effective soil erosion control usually focus on a combination of conservation tillage and year-round crop production. Adapted grain production systems which maximize the time that crops and residues cover the land include no-tillage double-cropped soybean or sorghum after a small grain, usually wheat.[63] Similarly, winter annual cover crops provide surface coverage during winter periods that otherwise are fallow. When legume cover crops are used, they typically enrich the soil with nitrogen, and can provide the subsequent crop with much or all of its nitrogen requirement.[40]

9.4. ENVIRONMENTAL ISSUES

9.4.1. Surface Water Quality

Over 60% of the total soybean hectarage harvested in 1991 was planted after wheat in double-cropped systems using no-tillage.[64] For loess and limestone areas, a list of commonly used conventional tillage soybean systems and no-tillage systems using a rippled coulter that have been studied relative to soil erosion by Shelton et al. are listed in Table 9.4.[14] Measurements of runoff and erosion from these systems are presented in Table 9.5. Soil losses were reduced for the two no-tillage systems on all three dates. On June 11, 1981, an intense

Table 9.4. Description of Soybean Tillage Systems Used on Loess- and Limestone-Derived Soils

Symbol	Crop(s)	Row Spacing (m)	Tillage
CTSC	Soybeans	1	Conventional (chisel, disc, plant, cultivate)
NTDC	Soybeans, wheat (double cropped)	0.5	No-till in wheat stubble
DHSC	Soybeans	0.2	Disc harrow, drill
CTDC	Soybeans, wheat (double cropped)	1	Conventional (plow, disc, plant, cultivate)
NTSS	Soybeans	0.5	No-till in soybean stubble

Source: Adapted from Shelton et al., *J. Soil Water Conserv.*, 38:426, 1983. With permission.

Table 9.5. Rainfall, Runoff, and Soil Loss by Soybean Tillage System for Three Natural Storms

	Rainfall		Runoff		Soil Loss
Storm and System[a]	(mm)	(mm/hr)[b]	(mm)	(mm/sec)[c]	(t/ha)
June 23, 1980					
CTSC	34	65	25	0.013	2.87
NTDC	34	65	32	0.024	0.16
DHSC	34	65	22	0.016	0.49
CTDC	34	65	17	0.016	0.70
NTSS	34	65	26	0.021	0.16
July 3, 1980					
CTSC	33	76	8	0.011	0.29
NTDC	33	76	5	0.005	0.02
DHSC	33	76	10	0.012	0.27
CTDC	33	76	19	0.015	0.22
NTSS	33	76	8	0.011	0.04
June 11, 1981					
CTSC	64	127	51	0.032	25.59
NTDC	64	127	46	0.036	0.14
DHSC	64	127	51	0.035	33.67
CTDC	64	127	38	0.020	0.11
NTSS	64	127	56	0.038	0.37

Source: Adapted from Shelton et al., *J. Soil Water Conserv.*, 38:426, 1983. With permission.

[a] See Table 9.4 for description of tillage system.
[b] Maximum 5 min intensity.
[c] Peak rate.

thunderstorm produced very large erosion losses on the two conventionally-tilled systems. The no-tillage, double-cropped soybeans and wheat (NTDC) and conventional tillage double-cropped soybeans and wheat (CTDC) systems both had unharvested wheat present, and the no-tillage, single-cropped soybeans (NTSS) had a good soil cover of previous crop residue.

The ability to maintain high yields with conservation tillage on loess- and limestone-derived soils, with only a small disturbed planting area using a rippled coulter, allows for the maintenance of most of the crop residue present, which effectively prevents soil erosion. Subsoiling rarely results in yield increases on these medium-textured silt loam soils in contrast to prior discussions on the effects of subsoiling in the Piedmont and the coastal plain.[65-68] As it does elsewhere, no-tillage effectively reduces soil erosion losses in the southern Piedmont.[45,47,53,57,58,69] Control of soil loss with no-tillage in the southern Piedmont is related primarily to the lack of soil disturbance and the maintenance of crop residue cover during the spring and summer periods when intense thunderstorms are most likely.[20,22,53,57] As previously discussed, sediment is usually reduced with conservation tillage. Runoff may or may not be reduced depending on antecedent moisture conditions, surface sealing, and other factors that may affect surface infiltration. Runoff comparison of no-till and conventional till soybeans in Kentucky indicated a runoff reduction of 50% with no-tillage.[70]

9.4.2. Groundwater Quality

Potential environmental problems may arise from conservation tillage as a result of increased chemical use and increased percolation to groundwater. Unfortunately, information on groundwater quality based on actual field measurements are limited and somewhat contradictory. Isensee et al.[71] reported that atrazine, and to a lesser extent alachlor and cyanazine, leached into shallow groundwater under no-tillage corn. Macropores were suggested as contributing to early, rapid transport of atrazine in this Maryland study. In an Illinois study on coarse-textured sand, higher pesticide concentrations were found in groundwater for chisel plow compared to no-tillage treatments.[72]

The sandy coastal plain soils are intensively used for crop production and receive considerable amounts of fertilizer nitrogen and sometimes irrigation during a growing season. As a result, these soils present a large potential for NO_3 leaching and subsequent contamination of groundwater. This situation may be further exacerbated when summer drought or excess fertilization results in high levels of residual fertilizer nitrogen in soil. For example, the probable leaching depth of NO_3 under a corn crop grown on a Norfolk soil was estimated at 0 cm for April through July, but increased from 8 to 188 cm during the months of August through March.[73]

In the above example, the period of high leaching potential coincides with the growing of a winter annual cover crop. Consequently, properly managed cover crops may fill an environmental niche by effectively recovering and thus conserving nitrogen within the soil-plant system. Ryegrass (*Lolium perenne* L.) reduced NO_3 leaching by 50 to 70% following wheat and corn.[74] Muller and Sundman[75] reported a 70% reduction in soil NO_3 concentration when rye was grown compared to a fallow treatment. Recent investigations by Shipley et al.[76] on the Maryland Atlantic coastal plain evaluated the ability of winter cover crops to recover residual corn fertilizer nitrogen. By mid-April of the following spring, fertilizer nitrogen recovery values in above-ground biomass were 45% for cereal rye, 27% for annual ryegrass (*Lolium multiflorum* Lam.), 10% for hairy vetch (*Vicia villosa* Roth), and 8% for crimson clover. Given the greater capacity of cereal rye and annual ryegrass to conserve fertilizer nitrogen, the authors recommended the integration of grass cover crops into cropping systems in humid climates.

Finally, the combination of weather, management, and soil factors affecting NO_3 leaching represents a dynamic system. Analysis of these interrelationships could be enhanced through the use of a model. Hubbard et al.[77] illustrated this concept by comparing model simulations of different nitrogen management systems on NO_3 movement into shallow groundwater. The CREAMS (Chemicals Runoff and Erosion from Management Systems) model was used to predict NO_3 loads from a Georgia coastal plain sand for two nitrogen fertilizer rates (168 and 336 kg/ha/year) and application methods (sidedressed and fertigation), with and without a winter cover of annual ryegrass.[78] Total cumulative leaching increased with nitrogen rate and decreased with both winter cover and fertigation. The mean total NO_3 load also decreased more with a winter cover than with fertigation.

Conservation tillage reduces runoff and soil erosion, lessening the threat of stream and lake siltation and the undesirable effects of sediment-borne contaminants on surface water quality. However, reduced runoff and lessened evaporation with no-tillage promote greater internal drainage and increase the threat to groundwater quality posed by the leaching of agricultural chemicals. On instrumented, tile-drained plots located on nearly level (0 to 2% slope) Cecil sandy loam soil, the effects of tillage (moldboard plow-disc vs no-tillage) and winter cover cropping (fallow vs rye) on NO_3 leaching from land devoted to corn production were investigated. First-year results indicate that a rye cover crop significantly limited NO_3 leaching loss (Table 9.6) by reducing both the volume and the NO_3-N concentration of water that leached through the root zone (Figure 9.2). Tillage treatments were imposed at the end of the first winter season. Leaching losses of NO_3 were greater for no-tillage than for conventional tillage during the corn growing season (Table 9.6) because of higher NO_3 concentration through the middle of August and greater drainage volumes with no-tillage (Figure 9.3). While quantities of NO_3-N in drainage were small, NO_3-N concentrations generally remained above 10 mg NO_3-N/L during the summer season. These preliminary results suggest that no-tillage in the southern Piedmont may necessitate a higher level of management if stringent control of NO_3 leaching is required.

The effect of conservation tillage in the loess and limestone areas is important because of its effect on surface water quality. The soils discussed in this section are predominantly silty or clayey in texture and usually moderately to strongly structured. This, in conjunction with the enhanced macroporosity that may develop in the absence of tillage, has resulted in implications that greater leaching of nitrates and pesticides could possibly occur under continuous

Table 9.6. Tillage and Winter Cover Cropping Effects on Loss of NO_3-N in Tile Drainage from Land Devoted to Corn Production at Watkinsville, GA

Cover Crop or Fallow Period (18 Oct 1991 to 23 Apr 1992)			Corn Growing Season (24 Apr 1992 to 30 Oct 1992)		
Tillage	Winter Cover Crop	Loss of NO_3-N in Tile Drainage	Tillage	Previous Winter Cover Crop	Loss of NO_3-N in Tile Drainage
CT	Rye	3.6(2.0) a[b]	CT	Rye	23(2.7) a
			NT	Rye	35(2.8) b
CT	Fallow	11.8(3.2) b[c]	CT	Fallow	27(7.9) a
			NT	Fallow	34(8.9) b

[a] CT = conventional tillage (moldboard plow and disc), NT = no-tillage.
[b] Where letter postscripts differ within a column, means are significantly different ($p < 0.01$).
[c] Values in parentheses are sample standard deviations.

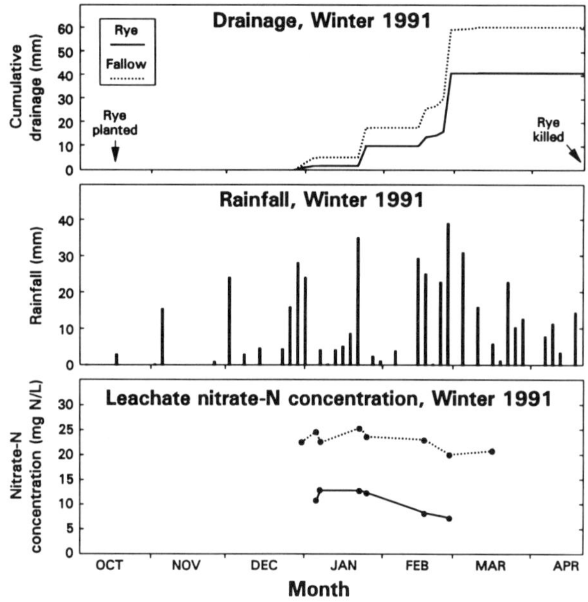

Figure 9.2. Preliminary results from tile drains under various cropping systems in the winter months of 1991 and 1992.

no-tillage cropping.[79-81] Studies to determine potential differences in leaching of pesticides and nitrates are underway in Tennessee and Kentucky.[82,83] The roles of organic matter and leaching are also being studied. Soil organic matter increases in no-tillage cropping, especially in the top 5 to 10 cm.[68,84] Increases in organic matter may result in greater storage of residual nitrogen and possibly greater adsorption of pesticides, subsequently reducing the potential leaching.[85]

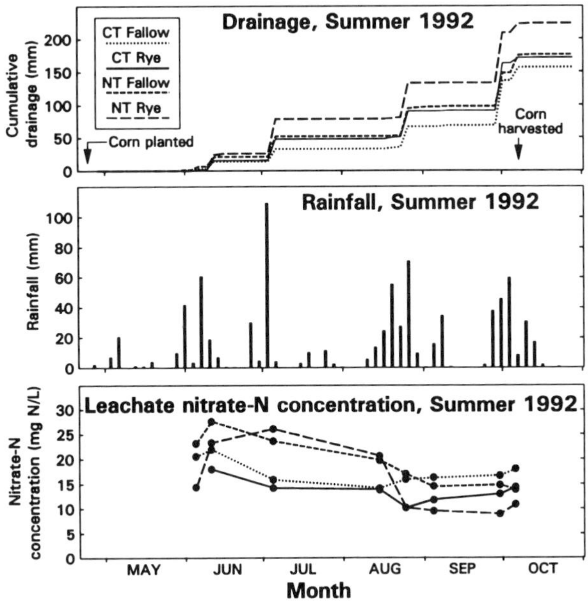

Figure 9.3. Preliminary results from tile drains under various cropping systems in the summer months of 1992.

Research on the role of no-tillage and residue management, especially with winter cover crops relative to weed control, is reviewed by Worsham.[86] The use of crop residues and cover crops in appropriate crop rotation systems may allow a reduction in the number and amount of herbicides used, especially preemergence herbicides. The lack of tillage results in fewer weed seeds brought to the surface. The subsequent accumulation of plant residues and the potential allelopathic substances they release may also have beneficial effects on weed control. Postemergence herbicides will probably be required. These are used at generally low rates, are foliar applied, and usually have low to very low groundwater contamination potential.[87]

9.5. GENERAL SUMMARY AND CONCLUSIONS

The use of conservation tillage is dramatically increasing in the southern United States. Residue management benefits to the soil system are well verified. Its role in reducing soil erosion is also well documented. Specifics of its use relative to the type of equipment needed is very much related to soil type. Typically, the loess, limestone, and in many cases the Piedmont soils do not require subsoiling. This results in an easy transition from tillage to no-tillage. In most cases, if subsoiling is needed on the sandier coastal plain soils, it can be accomplished and still leave a >30% residue cover.

The use of cover crops for residue enhancement and in double-crop systems offers many advantages in conservation tillage management. The expense of cover crop establishment and potential water depletion on low water-holding capacity soils offer some disadvantages. Advantages include improved soil organic matter, improved water infiltration, and usually an improvement in overall soil tilth. If the winter cover crop is an adapted nitrogen-fixing legume, additional nitrogen can be recycled into the system for cotton or grain crops. The increases in carbon and nitrogen and the recycling of other nutrients are essential in establishing sustainability.

The important role of conservation tillage, especially no-tillage, in developing sustainable agricultural systems in the southeastern United States is apparent. The surface water quality problems, potential losses in soil productivity, and present federal environmental regulations all point to conservation tillage and residue management for sustainability. Variable results under specific soil conditions on crop yields and water quality impacts do occur; however, most conservation tillage problems have been identified and solved. The future of no-tillage was discussed by Thomas.[88] The important practices such as crop rotation, along with appropriate inexpensive herbicides, in making conservation tillage sustainable is discussed. Proper incorporation of sound principles for crop production in conjunction with conservation tillage, at present, offer the only solution for continued row crop production on large acreages in the southeast.

9.6. REFERENCES

1. Buol, S. W. (Ed.) *Soils of the Southern States and Puerto Rico.* Southern Coop. Ser. Bull. No. 174 (Raleigh: North Carolina State University, 1973).
2. Langdale, G. W., H. P. Denton, A. W. White, J. W. Gilliam and W. W. Frye. "Effects of Soil Erosion on Crop Productivity of Southern Soils," in *Soil Erosion and Crop Productivity,* R. J. Follet and B. A. Stewart, (Eds.) (Madison, WI: American Society of Agronomy, 1985), pp. 251–271.
3. Springer, M. E., and J. A. Elder. "Soils of Tennessee," Agric. Exp. Stn. Bull. 596 (Knoxville: University of Tennessee, 1980).
4. Flowers, R. L., J. A. Phillips, W. C. Mangrum, R. K. Moore and L. A. Dungan. "Soil Survey of Fayette County, Tennessee" (Washington, DC: USDA Soil Conservation Service, 1964).
5. Denton, H. P. "The Effects of Erosion and Slope Characteristics on Soybean Yields on Memphis, Grenada, Lexington, and Loring Soils," M.S. thesis, (Knoxville, University of Tennessee, 1978).
6. Rhoton, F. E. "Soybean Yield Response to Various Depths of Erosion on a Fragipan Soil," *Soil Sci. Soc. Am. J.* 54:1073–1079 (1990).
7. Rhoton, F. E. and D. D. Tyler. "Erosion-Induced Changes in the Properties of a Fragipan Soil," *Soil Sci. Soc. Am. J.* 54:223–228 (1990).
8. Tyler, D. D., J. G. Graveel, and J. R. Jones. "Southern Loess Belt," in *Soil Erosion and Productivity,* J. W. Gilliam and G. D. Bubenzer (Eds.) (Madison, WI: Wisconsin Agriculture Experimental Station, 1992), pp. 36–43.

9. Schumn, S. A., and M. D. Harvey. "Natural Erosion in the USA," in *Determinants of Soil Loss Tolerance* ASA Spec. Publ. No. 45, B. L. Schmidt, R. R. Allmaras, J. V. Mannering and R. J. Papendick (Eds.) (Madison, WI: American Society of Agronomy, 1979), pp. 15–22.
10. Blevins, R. L., D. V. Midkiff and W. W. Frye. "Interior Low Plateaus," in *Soil Erosion and Productivity,* J. W. Gilliam and D. D. Bubenzer (Eds.) (Madison, WI: Wisconsin Agriculture Experimental Station, 1992), pp. 44–52.
11. Frye, W. W., S. A. Ebelhar, L. W. Murdock and R. L. Blevins. "Soil Erosion Effects on Properties and Productivity of Two Kentucky Soils," *Soil Sci. Soc. Am. J.* 46:1051–1055 (1982).
12. Frye, W. W., L. W. Murdock and R. L. Blevins. "Corn Yield-Fragipan Depth Relations on a Zanesville Soil," *Soil Sci. Soc. Am. J.* 47:1043–1045 (1983).
13. Wischmeier, W. H., and D. D. Smith. "Predicting Rainfall Erosion Losses — A Guide to Conservation Planning," Agric. Handb. No. 537 (Washington, DC: U.S. Dept. of Agriculture, 1978).
14. Shelton, C. H., F. D. Tompkins and D. D. Tyler. "Soil Erosion from Five Soybean Tillage Systems," *J. Soil Water Conserv.* 38:425–428 (1983).
15. Campbell, R. B., D. C. Reicosky and C. W. Doty. "Physical Properties and Tillage of Paledults in the Southeastern Coastal Plain," *J. Soil Water Conserv.* 29:220–224 (1974).
16. Van Diepen, J. C. "Corn Response to Different Tillage Practices in Selected North Carolina Soils," M.S. thesis (Raleigh: North Carolina State University, 1980).
17. Daniels, R. B., B. L. Allen, H. H. Bailey and F. H. Beinroth. "Physiolography," in *Souls of the Southern States and Puerto Rico,* Southern Coop. Ser. Bull. No. 174, S. W. Buol (Ed.) (Raleigh: North Carolina State University, 1973), pp. 3–16.
18. Trimble, S. W. "Man-Induced Soil Erosion on the Southern Piedmont, 1700–1970," (Ankeny, IA: Soil Conservation Society of America, 1974).
19. Perkins, H. F., H. J. Byrd and F. T. Ritchie, Jr. "Ultisols — Light-Colored Soils of the Warm Temperate Forest Lands," in *Soils of the Southern States and Puerto Rico,* Southern Coop. Ser. Bull. No. 174, S. W. Buol (Ed.) (Raleigh: North Carolina State University, 1973), pp. 73–86.
20. White, A. W., Jr., R. R. Bruce, A. W. Thomas and G. W. Langdale, H. F. Perkins. "Characterizing Productivity of Eroded Soils in the Southern Piedmont," in *Erosion and Soil Productivity,* D. K. McCool (Ed.) (St. Joseph, MI: American Society of Agricultural Engineers, 1985), pp. 83–95.
21. Austin, M. E. "Land Resource Regions and Major Land Resource Areas of the United States (exclusive of Alaska and Hawaii)," USDA Agric. Handb. 296 (Washington, DC: U.S. Dept. of Agriculture, 1965).
22. Wischmeier, W. H., and D. D. Smith. "Predicting Rainfall Erosion Losses: A Guide to Conservation Planning," USDA Agric. Handb. 537 (Washington, DC: U.S. Dept. of Agriculture, 1978).
23. van Bavel, C.H.M., and J. R. Carreker. "Agricultural Drought in Georgia," Tech. Bull. N.S. 15 (Athens: Georgia Agricultural Experiment Station, 1957).
24. Bruce, R. R., J. L. Chesness, T. C. Keisling, J. E. Pallas, Jr., D. A. Smittle, J. R. Stansell and A. W. Thomas. "Irrigation of Crops in the Southeastern United States," USDA-SEA ARM-S-9 (Washington, DC: U.S. Dept. of Agriculture, 1980).

25. Langdale, G. W., R. L. Blevins, D. L. Karlen, D. K. McCool, M. A. Nearing, E. L. Skidmore, A. W. Thomas, D. D. Tyler and J. R. Williams. "Cover Crop Effects on Soil Erosion by Wind and Water," in *Cover Crops for Clean Water*, W. L. Hargrove (Ed.) (Akeny, IA: Soil and Water Conservation Society, 1991), pp. 15–22.
26. Blevins, R. L., D. Cook, S. H. Phillips and R. E. Phillips. "Influence of No-Tillage on Soil Moisture," *Agron. J.* 63:593–596 (1971).
27. "1988 National Survey of Conservation Tillage Practices" (West Lafayette, IN: Conservation Technology Information Center, 1988).
28. "1992 National Survey of Conservation Tillage Practices" (West Lafayette, IN: Conservation Technology Information Center, 1992).
29. Cook, K. "Commentary: The 1985 Farm Bill: A Turning Point for Soil Conservation," *J. Soil Water Conserv.* 40:218–220 (1985).
30. Heard, L. P., and P. W. Dillard. "How Goes Conservation Compliance in Mississippi," *J. Soil Water Cons.* 44:415–416 (1989).
31. Box, J. E., Jr., and G. W. Langdale. "The Effects of In-Row Subsoil Tillage and Soil Water on Corn Yields in the Southeastern Coastal Plain of the United States," *Soil Tillage Res.* 4:67–68 (1984).
32. Cassel, D. K., and E. C. Edwards. "Effects of Subsoiling and Irrigation on Corn Production," *Soil Sci. Soc. Am. J.* 49:996–1001 (1985).
33. Trouse, A. C., Jr. "Observation on Under-the-Row Subsoiling after Conventional Tillage," *Soil Tillage Res.* 3:67–81 (1983).
34. Anderson, S. H., and D. K. Cassel. "Effect of Soil Variability on Response to Tillage of an Atlantic Coastal Plain Ultisol," *Soil Sci. Soc. Am. J.* 48:1411–1416 (1984).
35. Vepraskas, M. J., G. S. Miner and G. F. Peedin. "Relationships of Soil Properties and Rainfall to Effects of Subsoiling on Tobacco Yield," *Agron. J.* 79:141–146 (1987).
36. Wagger, M. G., and H. P. Denton. "Crop and Tillage Rotations: Grain Yield, Residue Cover, and Soil Water," *Soil Sci. Soc. Am. J.* 56:1233–1237 (1992).
37. Wagger, M. G., and H. P. Denton. "Tillage Effects on Grain Yields in a Wheat, Double-Crop Soybean, Corn Rotation," *Agron. J.* 81:493–498 (1989).
38. Nesmith, D. S., W. L. Hargrove, D. E. Radcliffe, E. W. Tollner and H. H. Arioglu. "Tillage and Residue Management Effects on Properties of an Ultisol and Double-Cropped Soybean Production," *Agron. J.* 79:570–576 (1987).
39. Touchton, J. T., D. H. Rickerl, R. H. Walker and C. E. Snipes. "Winter Legumes as a Nitrogen Source for No-Tillage Cotton," *Soil Tillage Res.* 4:391–401 (1984).
40. Hargrove, W. L. "Winter Legumes as a Nitrogen Source for No-Till Grain Sorghum," *Agron. J.* 78:70–74 (1986).
41. Wagger, M. G. "Cover Crops Management and Nitrogen Rate in Relation to Growth and Yield of No-Till Corn," *Agron. J.* 81:533–538 (1989).
42. Campbell, R. B., D. L. Karlen and R. E. Sojka. "Conservation Tillage for Maize Production in the U.S. Southeastern Coastal Plain," *Soil Tillage Res.* 4:511–529 (1984).
43. Ewing, R. P., M. G. Wagger and H. P. Denton. "Tillage and Cover Crop Management Effects on Soil Water and Corn Yield," *Soil Sci. Soc. Am. J.* 55:1081–1085 (1991).

44. Bruce, R. R., S. R. Wilkinson and G. W. Langdale. "Legume Effects on Soil Erosion and Productivity," in *The Role of Legumes in Conservation Tillage Systems, Proceedings of a National Conference,* J. F. Power (Ed.) (Ankeny, IA: Soil Conservation Society of America, 1987), pp. 127–138.
45. Hargrove, W. L., and W. W. Frye. "The Need for Legume Cover Crops in Conservation Tillage Production," in *The Role of Legumes in Conservation Tillage Systems, Proceedings of a National Conference,* J. F. Power (Ed.) (Ankeny, IA: Soil Conservation Society of America, 1987), pp. 1–5.
46. Langdale, G. W., H. P. Denton, A. W. White, Jr., J. W. Gilliam and W. W. Frye. "Effects of Soil Erosion on Crop Productivity of Southern Soils," in *Soil Erosion and Crop Productivity,* R. F. Follett and B. A. Stewart (Eds.) (Madison, WI: American Society of Agronomy, 1985).
47. Langdale, G. W., R. L. Blevins, D. L. Karlen, D. K. McCool, M. A. Nearing, E. L. Skidmore, A. W. Thomas, D. D. Tyler and J. R. Williams. "Cover Crop Effects on Soil Erosion by Wind and Water," in *Cover Crops for Clean Water, Proceedings of an International Conference,* W. L. Hargrove (Ed.) (Ankeny, IA: Soil and Water Conservation Society, 1991), pp. 15–22.
48. Reicosky, D. C., D. K. Cassel, R. L. Blevins, W. R. Gill and G. C. Naderman. "Conservation Tillage in the Southeast," *J. Soil Water Conserv.* 32:13–19 (1977).
49. Sojka, R. E., G. W. Langdale and D. L. Karlen. "Vegetative Techniques for Reducing Water Erosion of Cropland in the Southeastern United States," *Adv. Agron.* 37:155–181 (1984).
50. Langdale, G. W., J. E. Box, R. A. Leonard, A. P. Barnett and W. G. Fleming. "Corn Yield Reductions on Eroded Southern Piedmont Soils," *J. Soil Water Conserv.* 34:226–228 (1979).
51. "1983 National Survey Conservation Tillage Practices" (Washington, DC: Conservation Tillage Information Center, 1983).
52. Denton, H. P., and M. G. Wagger. "Interaction of Tillage and Soil Type on Available Water in a Corn-Wheat-Soybean Rotation," *Soil Tillage Res.* 23:27–39 (1992).
53. West, L. T., W. P. Miller, G. W. Langdale, R. R. Bruce, J. M. Laflen and A. W. Thomas. "Cropping System Effects on Interrill Soil Loss in the Georgia Piedmont," *Soil Sci. Soc. Am. J.* 55:460–466 (1991).
54. Miller, W.P., C.C. Truman and G.W. Langdale. "Influence of Previous Erosion on Crusting Behaviour of Cecil Soils," *J. Soil Water Conserv.* 43:338–341 (1988).
55. Radcliffe, D. E., E. W. Tollner, W. L. Hargrove, R. L. Clark and M. H. Golabi. "Effect of Tillage Practices on Infiltration and Soil Strength of a Typic Hapludult Soil After Ten Years," *Soil Sci. Soc. Am. J.* 52:798–804 (1988).
56. Radcliffe, D. E., L. T. West, G. O. Ware and R. R. Bruce. "Infiltration in Adjacent Cecil and Pacolet Soils," *Soil Sci. Soc. Am. J.* 54:1739–1743 (1990).
57. Langdale, G. W., A. P. Barnett and J. E. Box, Jr. "Conservation Tillage Systems and Their Control of Water Erosion in the Southern Piedmont," in *Proc. 1st Annu. Southeastern No-Till Systems Conference,* J. T. Touchton and D. G. Cummins (Eds.) (Athens, GA: University of Georgia, 1978), pp. 20–29.
58. Langdale, G. W., A. P. Barnett, R. A. Leonard and W. G. Fleming. "Reduction of Soil Erosion by the No-Till System in the Southern Piedmont," *Trans. Am. Soc. Agric. Eng.* 22:82–86, 92 (1979).

59. Hargrove, W. L. "Influence of Tillage on Nutrient Uptake and Yield of Corn," *Agron. J.* 77:763–768 (1985).
60. Langdale, G. W., R. L. Wilson, Jr. and R. R. Bruce. "Cropping Frequencies to Sustain Long-Term Conservation Tillage Systems," *Soil Sci. Soc. Am. J.* 54:193–198 (1990).
61. Parmelee, R. W., M. H. Beare, W. Cheng, P. F. Hendrix, S. J. Rider, D. A. Crossley, Jr. and D. C. Coleman. "Earthworms and Enchytraeids in Conventional and No-Tillage Agroecosystems: A Biocide Approach to Assess Their Role in Organic Matter Breakdown," *Biol. Fertil. Soils* 10:1–10 (1990).
62. Hendrix, P. F., R. W. Parmelee, D. A. Crossley, Jr., D. C. Coleman, E. P. Odum and P. M. Groffman. "Detritus Food Webs in Conventional and No-Tillage Agroecosystems," *BioScience* 36:374–380 (1986).
63. Langdale, G. W., W. L. Hargrove and J. Giddens. "Residue Management in Double-Crop Conservation Tillage Systems," *Agron. J.* 76:689–694 (1984).
64. "1991 National Survey of Conservation Tillage Practices" (West Lafayette, IN: Conservation Technology Information Center, 1991).
65. Mullins, J. A., T. C. McCutchen, W. L. Parks, J. J. Ball and S. Parks. "A 5-Year Comparison of Seedbed Preparation Systems for Cotton on Memphis and Collins Silt Loams," *Tenn. Farm Home Sci.* 92:14–17 (1974).
66. Tyler, D. D., and T. C. McCutchen. "The Effect of Three Tillage Methods on Soybeans Grown on Silt Loam Soils with Fragipans," *Tenn. Farm Home Sci.* 114:23–26 (1980).
67. Tyler, D. D., and J. R. Overton. "No-Tillage Advantages for Soybean Seed Quality during Drought Stress," *Agron. J.* 74:344–347 (1982).
68. Tyler, D. D., J. R. Overton and A. Y. Chambers. "Tillage Effects on Soil Properties, Diseases, Cyst Nematodes, and Soybean Yields," *J. Soil Water Conserv.* 38:374–376 (1983).
69. Mills, W. C., A. W. Thomas and G. W. Langdale. "Conservation Tillage and Season Effects on Soil Erosion Risk," *J. Soil Water Conserv.* 46:457–460 (1991).
70. Rasnake, M. "Soybean Tillage and Cover Crop Effects on Water Runoff and Soil Erosion," in *Cover Crops for Clean Water,* W. L. Hargrove (Ed.) (Ankeny, IA: Soil and Water Conservation Society, 1991), pp. 55–56.
71. Isensee, A. R., C. S. Helling, T. J. Gish, P. C. Kearney, C. B. Coffman and W. Zhuang. "Groundwater Residues of Atrazine, Alachlor, and Cyanazine under No-Tillage Practices," *Chemosphere* 17:165–174 (1988).
72. Bicki, T. J., and A. Felsot. "Influence of Tillage System and Water Management Practices on Leaching Alachlor, Cyanazine, and Nitrates in Sandy Soil: Preliminary Results," in *Proc. Agric. Impacts on Groundwater Conf.* (Dublin, OH: National Water Well Association, 1988), pp. 115–127.
73. Smith, S. J., and D. K. Cassel. "Estimating Nitrate Leaching in Soil Materials," in *Managing Nitrogen for Groundwater Quality and Farm Profitability,* R. F. Follett, D. R. Keeney and R. M. Cruse (Eds.) (Madison, WI: Soil Science Society of America, 1991), pp. 165–188.
74. Martinez, J., and G. Giraud. "A Lysimeter Study of the Effects of a Ryegrass Catch Crop, during a Winter Wheat/Maize Rotation, on Nitrate Leaching and on the Following Crop," *J. Soil Sci.* 41:5–16 (1990).

75. Muller, M. M., and V. Sundman. "The Fate of Nitrogen-15 Released from Different Plant Materials during Decomposition under Field Conditions," *Plant Soil.* 105:133–139 (1988).
76. Shipley, P. R., J. J. Meisinger and A. M. Decker. "Conserving Residual Corn Fertilizer Nitrogen with Winter Cover Crops," *Agron. J.* 84:869–876 (1992).
77. Hubbard, R. K., G. J. Gascho, J. E. Hook and W. G. Knisel. "Nitrate Movement into Shallow Ground Water through a Coastal Plain Sand," *Trans. ASAE* 29:1564–1571 (1986).
78. Knisel, W. G. (Ed.). "CREAMS: A Field-Scale Model for Chemicals, Runoff, and Erosion from Agricultural Management Systems," Conserv. Res. Rep. 26 (Washington, DC: USDA 1980).
79. Thomas, G. W., R. L. Blevins, R. E. Phillips and M. A. McMahon. "Effect of Killed Sod Mulch on Nitrate Movement and Corn Yield," *Agron. J.* 63:736–739 (1973).
80. Thomas, G. W., K. L. Wells and L. Murdock. "Fertilization and Liming," in *No-Tillage Research: Research Reports and Reviews*, R. E. Phillips (Ed.) (Lexington: University of Kentucky, 1981), pp. 43–54.
81. Tyler, D. D., and G. W. Thomas. "Lysimeter Measurement of Nitrate and Chloride Losses from Conventional and No-Till Corn," *J. Environ. Qual.* 6:63–66 (1977).
82. Wilson, G. V., D. D. Tyler, J. Logan and K. Turnage. "Tillage and Cover Crop Effects on Nitrate Leaching," in *Cover Crops for Clean Water*, W. L. Hargrove (Ed.) (Ankeny, IA: Soil and Water Conservation Society, 1991), pp. 71–73.
83. Tyler, D. D., W. V. Wilson, J. Logan, G. W. Thomas, R. L. Blevins, W. E. Caldwell and M. Dravillis. "Tillage and Cover Crop Effects on Nitrate Leaching," in *Proc. 1992 Southern Conservation Tillage Conf.*, M. D. Mullen and B. N. Duck (Eds.) (Milan and Jackson, TN, 1992), pp. 1–5.
84. Blevins, R. L., G. W. Thomas and P. L. Cornelius. "Influence of No-Tillage and Nitrogen Fertilization on Certain Soil Properties after Five Years of Continuous Corn," *Agron. J.* 69:383–386 (1977).
85. Bouwer, H. "Agricultural Chemicals and Groundwater Quality," *J. Soil Water Conserv.* 45:184–189 (1990).
86. Worsham, A. D. "Role of Cover Crops in Weed Management and Water Quality," in *Cover Crops for Clean Water,* W. L. Hargrove (Ed.) (Ankeny IA: Soil and Water Conservation Society, 1991), pp. 141–145.
87. Weber, J. B. "Potential Problems for North Carolina Groundwater from Herbicides: A Ranking Index," *Proc. Weed Sci. Soc., N.C.* 8:30–46 (1990).
88. Thomas, G. W. "The Future of No-Tillage," in *The Rising Hope of our Land, Proc. of Southern Region No-Till Conf.* (Griffin: University of GA, 1985), pp. 242–247.

CHAPTER 10

Conservation Tillage in the Southeastern Australian Wheat-Sheep Belt

Graham R. Steed and Anthony Ellington
Rutherglen Research Institute; Rutherglen, Victoria, Australia

James E. Pratley
Charles Sturt University; Wagga Wagga, New South Wales, Australia

TABLE OF CONTENTS

0-87371-571-3/94/$0.00+$.50

10.1. INTRODUCTION

Farming systems in Australia have evolved from the time of European settlement. The production of grain crops, alone or together with livestock production, has been part of this evolution. Australian soils are old and very fragile. Most tillage has taken place where land has been cleared of trees and on soils that are prone to degradation. On these soils crop production systems based on cultivation are not sustainable. In southeastern Australia the problem of soil degradation resulting from inappropriate tillage practices was dramatically illustrated in the 1982 to 1983 drought. Severe wind erosion blew valuable topsoil off farms and preceded extreme water erosion when the drought broke.

This chapter follows the development of conservation tillage systems for southeastern Australia. Unlike the traditional, cultivation-based tillage systems, conservation tillage systems permit more sustainable and economic crop production without causing soil degradation. The process of developing these systems started in the 1960's and continues strongly today. Vigorous research and extension programs have resulted in the development of necessary herbicides, introduction of grain legume crops, improved levels of soil fertility and crop establishment techniques which do not use cultivation. More recently, the emphasis has been on retaining crop stubble for soil and water conservation. Factors that have influenced the adoption of these techniques are also discussed, as are the advantages of conservation tillage systems, the constraints to adoption of these systems, and their environmental impact.

10.2. AGROECOSYSTEMS IN THE SOUTHEAST WHEAT-SHEEP BELT

10.2.1. Soil Types and Climate

The farmed land described in this chapter is on the inland lower slopes of the Great Dividing Range and on the Riverine Plain from the Goulburn River, south of the Murray River, to north of the Murrumbidgee River (Figure 10.1). The farming systems include wheat (*Triticum aestivum* L.) and a range of other dryland grain crops in rotation with legume-based pastures grazed by sheep, and by beef cattle in some areas. The systems also include irrigated farming, which comprises broad-area crops, dairying, fruit, and horticulture.

The soils are described in general terms by Butler et al.[1] and by Walker et al.,[2] and in detail by Stace et al.,[3] while Oades et al.[4] give detail on land management practices. The soils of the Riverine Plain are on alluvial material, but the plains and some slopes are covered with a wind-deposited mantle (parna) on which the surface soils are formed. The soils range from mildly leached to strongly leached and highly differentiated (duplex) in texture. The sodic red-brown earths occur in drier areas, and with increasing rainfall soils grade through noncalcic brown soils, solodized solonetz, solodics, and soloths into red podzolic and yellow podzolic types. In the north, near Wagga Wagga, red earths also occur. All except the red earths are duplex, i.e., they have strong texture differentiation down the profile, with an abrupt change to high clay contents at depths of 30 to 70 cm. Most of the soils have low cation exchange capacity (CEC) in the surface. The soils tend to be hard setting and are sodic, with low organic matter and calcium contents. They are alkaline or neutral in drier areas and acid, at least in the surface, in wetter areas.

The climate is often described as Mediterranean, but this is misleading; the summers are hotter and have more rain than typical Mediterranean types. The variability of rainfall between and within seasons is high. Annual rainfall ranges from about 400 mm in the west to 650 mm at the eastern and southern fringes of the cropping area. About two thirds of the rain falls between April and October. The growing season is from the end of April to early November. Summer rains often come from subtropical systems and may be of high intensity, creating a soil erosion hazard, but rarely wetting the soil sufficiently for plant growth.

Evaporation may be less variable than rainfall, ranging from about 1500 to 2000 mm/year (class A pan). Evaporation ranges from 1 to 2 mm/day in winter to up to 13 mm/day in summer. Mean daily air temperatures range from about 7°C in winter to 23°C in summer. Thus, soil temperatures are low enough to restrict growth for part of the time in winter, and transpiration rates then are low. Temperature inversions are common,[5] as are winter frosts.

10.2.2. Cropping Systems

When the Europeans arrived, the vegetation was open savannah with perennial grasses, maintained by the burning practices of aboriginals. The long-term effect of such burning was the acceleration of erosion.[6] These old soils were naturally fragile in structure and did not react kindly to the introduction of European methods of farming.[7] Overgrazing and trampling by hard-hoofed animals, followed by tree clearing, plowing for crops, and long fallowing all resulted in loss of structure and fertility of soil. Current farming systems are

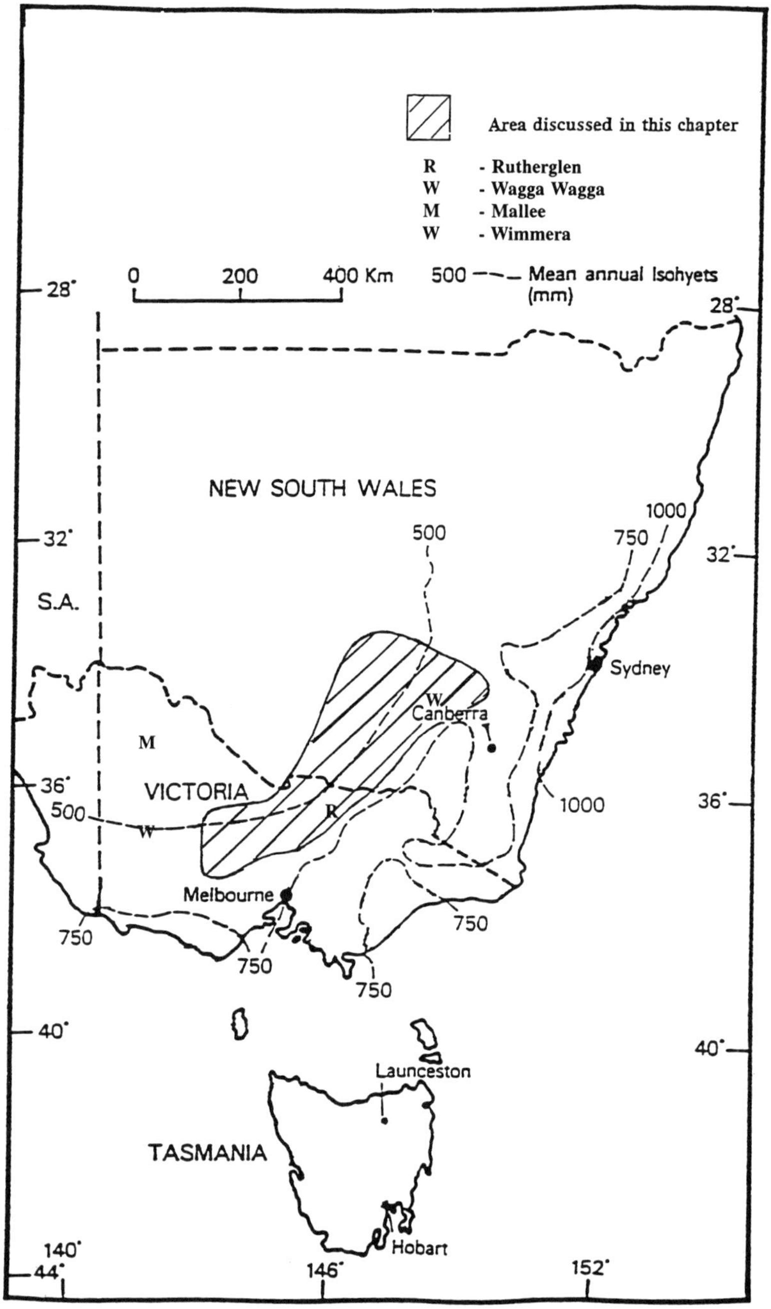

Figure 10.1. Southeastern Australia; shaded area is the subject of this chapter.

mostly based on pasture-crop rotations.[8] The pastures generally consist of introduced annual legumes (usually subterranean clover [*Trifolium subterranean* L.]) and grasses that regenerate each year. They are usually grazed by sheep, but sometimes by beef cattle; on irrigation areas the pastures are perennial and are commonly grazed by dairy cattle.[9] At Rutherglen, the introduction of clover-based pasture has trebled the stock-carrying capacity of the land and nearly trebled the following wheat yields.[10] Nitrogen and soil structure benefits associated with legume-cereal crop rotations have been described by Ellington et al.[11]. For cropping systems, use of pasture and grain legumes in rotation enabled dispensation of the long fallows of the 1930s through the 1950s. Also, tillage operations were reduced from 10 to 12 (plowing, scarifying etc.) to a minimum of one to two operations. Kohn et al.[12] showed long fallows to be uneconomic where growing season rainfall exceeded 380 mm, but the practice persists, particularly where rainfall is less than 500 mm/year.

Thus, since the 1950s and 1960s the pasture-crop rotation system of three to four cereal crops followed by five to eight years of pasture has been widely adopted. This length of pasture phase has helped to overcome the widespread soil erosion in the region in the 1930s.[48] In this pasture-crop rotation stubble land was usually cleared by burning, then scarified and harrowed two to five times to prepare a seedbed and to kill weeds. Long-term experiments failed to show any benefit to crop yields from retaining cereal stubble[13] because of problems of weed control and crop emergence. For crop production after the pasture phase the tillage system was the same as for stubble land, except that moldboard or disc plows were used in the preliminary tillage operation.

10.3. DEVELOPMENT OF CONSERVATION TILLAGE SYSTEMS

Between 1961 and 1975 three significant changes to crop production took place in southeastern Australia. The first was the development and increased availability of herbicides for pre- and postsowing weed control. The second, due to both imposition of wheat quotas in 1969 and the need for higher nitrogen inputs, was the recognition of the usefulness of grain legume and other noncereal crops in rotation with cereals. Crops such as field peas (*Pisum sativum* L.), lupins (*Lupinus angustifolius* L.), and faba beans (*Vicia faba* L.) were evaluated. These studies demonstrated the benefits of grain legumes for improving soil nitrogen and reducing the incidence of cereal diseases in subsequent wheat crops.[14,15] The net effect was a 35% boost to grain yields in cereal crops following grain legumes. The third change, and the focus of this chapter, was a move toward reduced tillage crop establishment techniques.

10.3.1. Reduced Tillage

Much of the impetus for development of reduced tillage cropping systems came from rapidly rising fuel costs resulting from the worldwide energy crisis of the 1970s. The concept that cultivation damaged soil structure was important, but the concern over fuel costs was a major factor.[16] Traditional cultivation practices were more expensive and more time consuming, thus the practice of reducing the number of cultivations from four to six to one to two (called minimum tillage cropping) was sold to farmers on the basis of reduced fuel costs without adverse effect on weed control. Substitution of herbicides for cultivation was shown to be a realistic option.

10.3.2. Direct Drilling

Early success with minimum tillage techniques prompted the question of whether cultivation was needed at all. Two companies ICI (initially) and Monsanto (later) developed knock-down herbicides which could be used to control weeds before sowing. These companies worked with research organizations to develop the direct drill (DD) system, which required no presowing cultivation. Much of this early work concentrated on the basic agronomy required for successful DD. Depth, time, and rates of sowing and crop varieties were evaluated as was the need to modify traditional sowing equipment. In Victoria, removal of the cultivating tines was recommended, as was the fitting of narrow points (35 mm) for ease of soil penetration. In New South Wales, however, full soil disturbance at sowing was maintained by use of cultivating tines and traditional points (100 mm).

The DD cropping system allowed land to be used for grazing up until sowing time, which is economically important for mixed farms in this area. DD was also much more time efficient than traditional tillage because of herbicide effectiveness and the ability to sow within hours of spraying. Farmers were able to sow more crop and also sow more of their crop at the optimum sowing time. Without DD, a late autumn break delayed sowing until all cultivations had been performed. Often this meant that the cropping program ran into wet winter problems, causing the fields to be waterlogged and crops sown either very late or not at all. With DD, this risk was minimized as sowing could take place on the first rains.

Early research compared conventional cultivation (CC) with DD as crop establishment techniques on both continuous wheat and wheat-lupin rotations. For the first 1 to 3 years of these experiments, crop yields tended to be unaffected by tillage system.[17-19] Hence, the extension message concentrated on the flexibility and time savings of the DD system. After these experiments had run for 3 to 4 years, the common trend was for DD treatments to outyield CC treatments due largely to the improved physical condition of soils expressed through water-stable aggregates, trafficability, infiltration of water, reduced losses of nitrogen and carbon (Table 10.1), and reduced soil-borne fungal diseases of wheat.

In terms of encouraging adoption of DD it was unfortunate that many soils were not immediately suitable for DD. Overcultivation caused structural damage

Table 10.1. Soil Total Nitrogen (kg/ha, 0–20 cm) Changes Over 4 Years under Continuous Wheat or Lupin-Wheat Rotations, with Cultivation or Direct Drilling

		Cultivated	Direct Drilled
1979	Mean	2452	
1983	Continuous wheat	2072	2315
	Wheat-lupin	2109	2385
	Lupin-wheat	2069	2378
	Mean	2087	2359
Mean nitrogen loss (1979–1983)		365	93

Source: After Ellington, A., Australian Institute of Agricultural Science, Occ. Publ. No. 28 (1987). With permission.

which contributed to the common crop failures due to poor emergence in the early years of DD. Attention thus became focused on soil structure and the easily observable cultivation-induced damage to many of the cropping soils in southeastern Australia. Soil physical degradation was expressed as surface soil crusting, excessive water runoff, soil erosion, and periodic waterlogging. The estimated losses of crop yield ran into the millions of dollars. An extensive survey of crops in 1980[20] showed the widespread nature of the crop "yellowing" phenomenon, in which large areas of yellow, stunted crop were surrounded by areas of vigorous, healthy crop. The survey showed that the areas of poor growth were characterized by increased soil acidity, poor root development, nitrogen deficiency, subsurface hardpans, and waterlogging. These factors reduced crop yield by 25% over the whole region.

Concurrent with this research work, a widespread extension campaign in New South Wales and Victoria caused a major shift to conservation tillage cropping systems. Farmers were encouraged to identify the soil factors limiting crop production, to correct them, and then to implement more sustainable cropping systems to prevent recurrence of problems and to improve long-term productivity. Corrective measures included:

- Deep ripping for hardpans;
- Applying lime to soils where pH was below 5.3 (water); and
- Returning to the pasture phase if soil surface structure was degraded or if weed problems were excessive.

More sustainable systems were then put in place, such as:

- DD for reduced soil surface degradation;
- Use of legume-cereal crop rotations to maintain soil nitrogen fertility and reduce disease; and
- Identifying paddocks most suited to cropping (soil structure, fertility, better drainage) and avoiding paddocks of lower productive potential.

With the widespread adoption of new cropping systems came the need to refine techniques to accommodate the unique characteristics of the region such as soil type and rainfall. Thus, different conservation tillage systems were developed for the Wimmera and Mallee regions of Victoria, for northern Victoria, and for southern and central New South Wales. Although coverage of tillage systems in the Wimmera, Mallee, and irrigation regions of southeastern Australia is beyond the scope of this chapter, some information is relevant. Average annual rainfall in the Wimmera is 400 to 450 mm and in the Mallee, 325 mm. In these regions the two important soil types are self-mulching gray clays and Mallee sand. Neither soil type is as prone to physical degradation caused by cultivation as are the red brown earths, red earths, and other soils in northern Victoria and southern New South Wales. The major advantages of conservation tillage systems for these soils are soil moisture conservation and protection against wind erosion.

In the irrigation areas research was dictated by a knowledge of limiting factors in the soil. There was much transfer of ideas with dryland researchers. Management of red-brown earths under irrigation was reviewed by Cockroft and Martin,[9] who identified drainage, subsoil loosening, gypsum, and organic matter as necessary to stabilize these soils. After stabilization, these soils required management to encourage biological activity and therefore maximized productivity. The minimum tillage systems that resulted, including the Tatura raised bed system, have been reviewed by Tisdall and Huett.[21]

10.3.3. Stubble Retention

As both the scientific and farming communities gradually became more expert with DD and other conservation tillage techniques, the conventional practice of stubble burning was questioned. Using crop stubble rather than destroying a potentially valuable resource by burning was the obvious next step in the evolution of conservation tillage systems. In southern Australia the traditional practice of burning crop stubble occurred as soon as weather conditions allowed it to be done safely, usually in early autumn. Its main purpose was to allow sowing equipment to establish the next crop although disease prevention and weed control were also considered as benefits. Pressure from farmers and industry representatives in 1980 resulted in new research activity into stubble retention practices for the region. Improved organic matter levels, protection of the soil surface, moisture conservation and "just not needing to burn stubble" were all among the reasons for setting this priority.

The first investigations at Rutherglen, Victoria showed that successful wheat and lupin production was possible by sowing the crop into the stubble of the previous crop as long as the sowing equipment was not the limiting factor. The problems of straw blockage, incorrect sowing depth, and poor seed-soil contact occurred because of unsuitable equipment. The availability of appropriate sowing equipment was the most limiting factor facing farmers who wanted to adopt this form of cropping. In southern Australia both the traditional combines (sowing equipment with cultivating tines in front of sowing tines) and those modified for DD (sowing tines only) consisted of two or more gangs of tines arranged so that the row spacing was 15 to 18 cm. In a wheat stubble the effect was more like a hay rake than a seeder.

The practice of stubble retention had been adopted in the summer rainfall areas of northeastern Australia largely to prevent soil erosion from the high-intensity summer storms. The transfer of stubble retention to the southern winter rainfall areas was complicated, however, by two main factors. First, the crop yields are higher in the south, thus producing a much greater stubble mass through which sowing implements must pass. Second, the dry summers in the south provide little opportunity for stubble breakdown, with the result that most of the stubble remains at sowing time. New technology was thus required for successful adoption of stubble retention in the south as well as careful management of stubble at harvest. Many modifications of existing equipment were tried with variable success. Disc seeders were the most successful for crop establishment, particularly in cereal stubble. Less difficulty was experienced in lupin and canola (*Brassica napus* L. and *Brassica campestris* L.) stubble. The choice of seeder is often region dependent. Equipment suited to the red brown earth soils, for example, was unsatisfactory in more sandy soils.

It became apparent that successful stubble retention farming depended on the presowing management of the stubble.[22] Options practiced include:

1. Incorporating stubble into the top 10 cm of soil using a disc plow, usually in late summer or early autumn;
2. Creating a stubble mulch by chopping or slashing standing stubble in smaller lengths (5 to 10 cm);
3. DD into the standing stubble of the previous crop; on mixed farms this option is often preceded by grazing with sheep; and
4. Leaving the straw standing, but using a blade plow to control weeds before sowing.

Stubble incorporation is favored by some farmers who believe that straw breakdown is quicker with this treatment. In addition, stubble incorporation results in a cultivated seedbed which is more "acceptable" to many and causes few problems with straw blockages at sowing time. Stubble incorporation, however, does not conserve soil water, and soil nitrogen tie-up, or immobilization, is likely. A further important drawback of stubble incorporation is the physical damage to soil structure caused by the disc plow. In wet seasons stubble incorporation often renders a paddock untrafficable. For these reasons stubble incorporation is not a recommended cropping technique.

The major advantages of stubble mulching are soil water conservation through improved infiltration and reduced evaporation and less difficulty at sowing as compared with standing stubble. This extra soil water in autumn can often facilitate sowing lupins and winter wheats at the optimum sowing time, whereas stubble burning followed by cultivation can delay sowing if soil water is marginal. The use of equipment such as prickle chains to bash the stubble before sowing is a common variation of the stubble mulching technique. The major disadvantages of stubble mulching are the extra time and equipment needed to conduct the operation.

DD into standing stubble is the simplest and cheapest option. Standing stubble is very effective for soil water conservation and provides the best defenses against wind and water erosion. Practical difficulties again relate to straw blockages during sowing, but this can be largely overcome by suitable treatment of the stubble at harvest. Livestock can be used for a mulching effect to help control weeds and to clean up unharvested grain. Strategic livestock management can be very effective in reducing the amount of stubble to be handled. After harvest, immediate introduction of sheep to the stubble ensures maximum utilization before digestibility deteriorates. Cattle can be used later to digest lower quality feed.[23] All grazing should be undertaken during dry conditions to minimize the effects of soil compaction.

Blade plows leave the stubble on the soil surface, but cut off weed roots just below the soil surface. Their usefulness is confined to soils where the structure is not susceptible to degradation by cultivation or to the formation of plow pans (e.g., Mallee sands).

The harvest operation is a vital step in successful stubble retention cropping. Without some form of straw spreader attached to the back of the header, chaff and straw are concentrated into windrows, which severely hamper the sowing operation. Chaff and straw need to be spread evenly over the largest possible area.

10.3.4. The Role of Grain Legumes

In southeastern Australia all cropping systems now rely increasingly on grain legumes for inputs of nitrogen and for a cereal disease break. A secondary benefit of lupins is that animal nutrition is improved by grazing the lupin stubble for spilt grain. Grain legumes are also agronomically suited to conservation tillage systems, including stubble retention, rarely suffering from poor emergence in heavy stubble or from reduced yields in wider row spacings.

Two potential problems are particularly associated with grazing grain legume stubble. The first is that the stubble are generally light, so overgrazing can leave the soil very exposed and susceptible to erosion. The second is that the disease "lupinosis" can be contracted by sheep grazing lupin stubble. Recently released varieties of lupins are less likely to cause lupinosis, and stock management techniques to avoid the disease are commonly practiced.

10.4. IMPROVEMENT OF SOIL CONDITIONS UNDER CONSERVATION TILLAGE

One of the main reasons for developing conservation tillage systems was the need to arrest the various forms of soil degradation caused by the traditional tillage systems. After more than 3 decades of experience with conservation tillage systems (including stubble retention, DD and legume-cereal rotations) some conclusions can be drawn about the impact of these systems on soil conditions and thus on crop productivity (present and future) in southeastern Australia.

Overall soil health and potential to produce can be defined in terms of the physical, biological, and chemical state of that soil. Conservation tillage systems can improve all three characteristics relative to traditional systems based on cultivation and stubble burning. A combination of DD and stubble retention improves the soil surface physical characteristics of the fragile red brown earths and red earths.[24] Improvements in infiltration rate, depth of wetting, and time to runoff have been measured, while runoff, sediment loss, and bulk density declined (Table 10.2).

More earthworms are found in DD soils and where stubble is retained rather than burned (Table 10.3).[25] Microbiological activity, microbial biomass, and soil organic matter tend to increase under minimum tillage and stubble retention systems, but more research is needed to clarify these effects. DD can impede the decline in soil organic matter, in comparison to conventional tillage, in the years initially following pasture.[26] In continuous cropping systems, however, DD may have only marginal influence on organic matter increase and soil physical stability at the soil surface.[27] In warm climates, the overriding influence of soil temperature and moisture, which regulates plant production, crop residue levels, and decomposition rates would tend to minimize tillage-induced increases in soil organic matter.[28]

The use of pasture legumes in rotation with cereal crops contributes to the maintenance of soil nitrogen. Use of both crop and pasture legumes further increases sustainability of production through an effect on cereal disease control.[11,15] Stubble retention systems can initially immobilize soil nitrogen, causing significant yield loss; however, the combination of legume-cereal in rotation together with stubble retention can maintain or improve soil nitrogen fertility.[49]

10.5. CONSTRAINTS TO ADOPTION OF CONSERVATION TILLAGE

While defining the advantages of conservation tillage systems has been very important in the development of those systems, defining and overcoming the constraints has also been an important part of this process.

Table 10.2. Improvements in Soil Physical Properties under a Combination of Direct Drilling and Stubble Retention

		Rainfall Simulation Experiments[a]			
Tillage System[b]	Bulk Density (Mg/m^3)	Sorptivity ($m \times 10^{-5}\ s^{-1/2}$)	Time to Steady State Runoff (min)	Sediment in Runoff (g/mm)	Wetting Depth (mm)
CCB	1.41	22.8	25.2	3.29	58.3
DDB	1.33	24.3	37.7	3.71	84.3
DDR	1.23	49.9	64.0	1.53	170.5

Source: Adapted from Carter, M. R. and Steed, G. R., *Aust. J. Soil Res.,* 30, 505, 1992. With permission.

[a] Rainfall intensity of 46.0 ± 0.8 mm/hr.
[b] CCB: conventional tillage, stubble burned; DDB: direct drilled, stubble burned; DDR: direct drilled, stubble retained.

Table 10.3. Influence of Stubble Management Practice on Earthworm Number and Number of Pores >2 mm.

Tillage System[a]	Total No. of Earthworms (m^2)	No. of Pores >2 m^2
CCB	117 ± 31	96 ± 19
DDB	123 ± 36	81 ± 33
DDR	275 ± 69	123 ± 11

Source: Adapted from Haines, P. J. and Uren, N. C., *Aust. J. Soil Res.,* 30, 505, 1992. With permission.

[a] CCB: conventional tillage, stubble burned; DDB: direct drilled, stubble burned; DDR: direct drilled, stubble retained.

10.5.1. Soil Constraints

The main constraint to the adoption of conservation tillage has been the wide occurrence of hard-setting soils.[29,30] Hardsetting is caused by high contents of fine sand and silt, low organic matter levels, low meso-faunal activity, sodicity, and extremes of wetting and drying. Where hard-setting is severe, penetration of soils by sowing equipment is difficult. Soils are usually dry in autumn, low infiltration rates of rain are common, and the water-holding capacity of soil is much reduced.[31] Runoff rates are higher than with more friable soil, leading to increased erosion and waterlogging in low lying areas. Although it was thought that cultivation improved water penetration, we now know that the reverse is true.[24] Where surface crusts are present initial infiltration can be improved by a cultivation, but generally water penetration is improved where soils are not disturbed. Farmers also cultivated to ensure that sowing can proceed and to kill weeds. However, subsequent trafficability on cultivated land can be poor in wet conditions.

Periodic subsurface waterlogging is common in the areas receiving more than 500 mm rain per year because of impermeable subsoils and/or dispersive subsoils.[9,32] Strongly bleached subsoils occur, and these can readily acidify because of their low cation exchange capacities. They have low bearing capacity when waterlogged, thus limiting the timeliness of operations.

10.5.2. Climate Constraints

Farmers traditionally have perceived the need to ensure that in autumn there is adequate soil moisture for germination while soil temperatures are warm enough for rapid plant growth. Long fallow was always thought to be the way to conserve water. Additional cultivations in autumn were to ensure that rains penetrated the soil adequately and to kill weeds. Soil bared by overgrazing or cultivation may erode with high-intensity rain, and eroded soils were usually recultivated to level the seedbed. Recent evidence[24,33-36] shows that both long fallow and autumn cultivation only benefit soil water storage through weed control and are inefficient methods for improving moisture storage. With time, cultivation contributes to soil degradation.

Frost or low temperatures in winter can sometimes be a problem with stubble retention systems of cropping. The insulating effect of a stubble layer can result in reduced soil warming by the sun in winter, and greater severity of frost above the stubble layer at night, resulting in some reduction of plant growth.[37] The effects are greater in dry seasons.

A normal constraint on crop production is the "frost-drought" syndrome. Crop sowing in autumn must be timed to ensure that flowering in spring occurs as soon as possible after the last frost in order to maximize grain filling time before the summer drought sets in. Frost at flowering can cause total crop loss, so farmers in southeastern Australia tend to accept the yield penalties of delayed sowing to reduce the frost danger. The interval for optimum sowing time is short, as the opening rains usually come in late April, with optimum sowing time for most crops being early May. To some extent this time pressure encourages the use of DD or minimum tillage systems, as sowing can be completed faster and closer to the optimal time of sowing.

10.5.3. Technology Transfer

The change to conservation tillage cropping systems has been slow but steady. Research and extension programs that have demonstrated the benefits of conservation tillage have been a vital element of the adoption process, as have the inputs from chemical companies and farmers themselves. It is important to realize that the change from cultivation-based systems to those of minimum tillage or DD has often been in conjunction with the use of other soil management techniques such as increased use of grain legumes, lime application, and deep ripping. Farmers were attracted to this package of techniques as their awareness of the need to conserve soil increased.

Coventry and Brooke[38] reviewed the development of minimum tillage systems and noted that farmer interest in conservation tillage systems was evident from the first research program at Rutherglen, Victoria (1961), but at that stage weed control problems prevented any adoption. Farmer interest and support continued until eventually a research planning meeting at Rutherglen (1980) involving 20 farmers provided the impetus for the current stubble retention research program.

Apart from the soil and climate constraints to the adoption process, many farmers were naturally reluctant to change tillage systems because of the need to learn completely new methods of crop production. The weed control techniques required in conservation tillage programs require specialized knowledge, machinery, and expertise. With time the extensive use of on-farm demonstration areas increased farmer experience (learning from neighbors), and extension techniques based on groups of farmers has helped to affect a gradual change in cropping techniques. By 1992, most farmers in southeastern Australia cultivated much less than they did 20 to 30 years before. DD is routinely practiced by about 30% of farmers and the use of stubble retention has progressed past the innovators to become relatively common. Farmers recognize that their long-term viability depends largely on their ability to conserve the soil and improve the condition of that soil. For this reason adoption of conservation tillage systems will continue to increase.

10.6. ENVIRONMENTAL ISSUES IN CONSERVATION TILLAGE ADOPTION

10.6.1. Agricultural Chemicals

The techniques involved in conservation tillage involve the use of knock-down herbicides for preplanting weed control in preference to the use of tillage. The reduction in the amount of cultivation is important in maintaining the structure of the surface soil and reducing the risk of erosion. However, the introduction of chemicals does raise environmental concerns in the community.

The knock-down herbicides used for this purpose are the bipyridyls paraquat and diquat and the organophosphate glyphosate. The bipyridyls act more or less by desiccating the plant tissues on contact, whereas glyphosate is a systemic nonselective chemical. In both cases the chemicals are inactivated on contact with soil and hence there is no significant residual activity. Long-term studies by ICI show no build-up of bipyridyl residues which break down with time to harmless products such as CO_2 and water. Many reports, including Haines and Uren,[25] link conservation tillage with improved earthworm activity;

however, in the laboratory Eberbach and Douglas[39] measured some inhibition of plant emergence after very heavy applications (ten times normal) of some herbicides. Wardrop[40] reviewed some other environmental effects of herbicides on the environment. The bipyridyls are classified under the Australian poisons legislation as schedule 7 chemicals (high to very high hazard), while glyphosate is listed as a schedule 5 (low to medium hazard). In both cases, application of approved safety procedures are more than adequate to protect the users and the environment.

In conservation tillage systems farmers now expect to control any weed problem by use of appropriate herbicides, and much extension activity is required to sell the message of weed management involving a range of techniques of which chemicals are only one.

The importance of using a diverse range of weed management techniques is highlighted by the development of herbicide resistance in Australia.[41] Farmers are faced with having to choose options other than herbicides, particularly for weed control in the pasture phase, in order to overcome this problem. Resistance to herbicides is prevalent in annual ryegrass (*Lolium rigidum* Gaud.), but also occurs in wild oats (*Avena fatua* L.), silver grass (*Vulpia bromoides* L., S.F. Gray), capeweed (*Arctotheca calendula* L. Levyns), and barley grass (*Hordeum leporinum* Link), and can be expected in other species given the selection pressures imposed on weed populations by herbicide applications. One weed management strategy involves rotating herbicides through the different chemical groups. It also involves the minimization of seed set by weeds considered to have resistance and possibly a reversion to tillage to stimulate weed germination, which can then be controlled before sowing with a knockdown herbicide.

Farmer confidence in the use of herbicides has prompted them to look for chemical solutions to other problems such as crop diseases and pests. Insecticides are not used extensively in this region except for red-legged earthmite (*Halotydeus destructor* (Tucker)) control in the establishment phase of some crops and pastures, aphid control in canola crops, and for *Heliothis* spp. in some years. Use of fungicides to control cereal diseases is increasing. The incidence of these diseases can be expected to increase as productivity of crops also rises, but sound rotation practices are still important for crop disease control.[42] Control of alternate hosts in the pasture, in the fallow, or in the previous crop has a substantial effect on the impact of the take-all disease (*Gaeumannomyces graminis* var. *tritici* Walker) in wheat. A single cultivation appears to reduce the effect of Rhizoctonia (*Rhizoctonia solani* Kuhn), while DD substantially reduces the incidence of eyespot (*Pseudocercosporella herpotrichoides* (fron) Deighton) lodging.[43]

Another potential problem is the carryover of chemical from one season to another in the crop stubble. In years where little rain falls between harvest and sowing, significant quantities of herbicide can leach into the seedbed sufficient to impair establishment of susceptible species. A similar effect may occur where allelochemicals are present in the crop stubble or associated residues of weed species.[44] The effect is greatest in the trails of stubble concentrated by harvesting machinery. These chemical problems are reduced in importance in years in which significant rain occurs during the summer to autumn period.

In summary, conservation tillage systems increase both the dependence on and the usage of agricultural chemicals. While they remain very useful tools of productivity, they are not a panacea. Chemicals cannot and will not replace the need for good agronomic practices, but will remain an essential part of the farming systems of this region into the foreseeable future.

10.6.2. Soil Erosion

There can be no doubt that soil erosion is substantially reduced by the use of conservation tillage practices. Susceptibility to erosion is increased by cultivation, as soils are more easily washed, and by poor structural stability resulting from excessive tillage.[45] In many soils in this environment, the surface can seal over, resulting in reduced infiltration and thus greater runoff.

The introduction of reduced tillage systems has resulted in soils being less vulnerable to erosion as they are protected by surface vegetative matter for longer perods, and soil surface structures have improved allowing a greater proportion of the rainfall to infiltrate. This reduction in surface runoff has resulted in less replenishment of surface water stores for livestock and may create drinking water shortages in some years.

In the drier areas of lighter soils, wind erosion can be a significant problem. It has long been the practice to retain standing stubble on these soils to prevent saltation. This residue retention in conjunction with minimum tillage practice has reduced the vulnerability of soils to wind erosion. This greater stability is associated with improved soil organic matter levels and higher productivity.

10.6.3. Groundwater Implications

It is important to note that in the southeastern dryland cropping zone of Australia, yield is limited largely by the availability of water. Potential crop yields can be estimated on the basis of water use efficiency, a common rule of thumb for cereals being 10 to 15 kg of grain per millimeter of rainfall for the April to October period.[46]

As previously indicated, one of the consequences of conservation tillage is improved soil structure and improved infiltration of rainfall into the soil.[24,47] If plants cannot use this extra water it can reach the groundwater, which may rise, bringing excess salts to the root zone. Hence, it becomes imperative that water entering the system must also leave the system via transpiration. This involves the planting of deep-rooting species such as trees, perennial pastures including lucerne and grasses, and most importantly increased productivity during the cropping phase.[44]

The development of conservation tillage however, does provide positive contributions to these environmental considerations. First, the reduction in surface runoff and hence, surface erosion, reduces the eutrophication by phosphate and nitrogen that are usually associated with such surface soils. Second, nitrogen in uncultivated soils is more likely to be stored in organic form which reduces the extent of leaching and consequent acidification and eutrophication. Third, the improved soil structure and higher organic matter component provides an improved water storage capability in the root zone.

10.6.4. Fertilizer Use

In this environment, soils are inherently infertile and any raised level of productivity involves the input of phosphatic fertilizers. Nitrogen is supplied by pasture and crop legumes, or occasionally applied as fertilizer in the later years of the cropping phase, particularly where grain legume crops have not been included in the rotation.

DD bands fertilizer at the same depth each year. This may result in a phosphorus-rich zone in the soil in which plant uptake of that phosphorus is enhanced. Conversely, in DD soils initial root growth may be slower than where the soil has been cultivated, causing inhibition of phosphorus uptake.

Nitrogen fertilizers are soluble and hence prone to leaching. In practice most nitrogen fertilizer is applied prior to or at sowing and, depending on the amount of rainfall at sowing time, can be leached out of reach of newly establishing crop plants. The efficiency of utilization of nitrogen fertilizers is variable and dependent on many factors such as fertilizer type, rainfall, soil pH, and method of application. The unused fraction can be lost to the atmosphere in gaseous form (e.g., from urea) or be a source of eutrophication to watercourses. Much more work needs to be done exploring split applications and foliar applications to increase the efficiency of utilization.

10.7. GENERAL SUMMARY AND CONCLUSIONS

Conservation tillage systems are now widely accepted by grain producers in southeastern Australia. As with any changed farming practice there has been a range of adoption rates amongst individuals; however, in the early 1990s it is very rare to find grain production based solely on the old system of stubble burning and numerous cultivations. Similarly, the practice of summer fallowing prior to autumn cropping has nearly disappeared as the association between cultivation and soil degradation is more widely recognized.

DD, or a greatly reduced cultivation regime, is more widely practiced than stubble retention. Reduced grain prices since 1989 have not encouraged farmers to purchase or modify the sowing equipment needed for successful stubble retention, but evidence from farmer discussion groups shows that stubble retention is very high on their list of priorities. This gradual move toward conservation tillage has been associated with and helped by a groundswell of interest among farmers and the general community in all conservation issues. Sustainable agriculture has been the focus of the past 10 years, and fortunately the push toward this goal shows no signs of abating.

10.8. ACKNOWLEDGMENTS

The Grains Research and Development Corporation (previously Wheat Research Committee of Victoria) and the Land and Water Resources Research and Development Corporation (previously National Soil Conservation Programme) provided funds for conservation cropping research at Rutherglen Research Institute.

10.9. REFERENCES

1. Butler, B. E., G. Blackburn and G. D. Hubble. "Murray-Darling Plains (VII)," in *Soils: An Australian Viewpoint* (Melbourne: CSIRO/London:Academic Press, 1983), pp. 231–239.
2. Walker, P. H., K. D. Nicolls and F. R. Gibbons. "South-Eastern Region and Tasmania (VIII)," in *Soils: An Australian Viewpoint* (Melbourne:CSIRO/London:Academic Press, 1983), pp. 241–250.
3. Stace, H. C. T., G. D. Hubble, R. Brewer, K. H. Northcote, J. R. Sleeman, M. J. Mulcahy and E. G. Hallsworth. *A Handbook of Australian Soils* (Glenside, South Australia: Rellim Technical Publications, CSIRO, and International Society for Soil Science, 1968), pp. 152–244, 311–382.
4. Oades, J. M., D. G. Lewis and K. Norrish (Eds). *Red-Brown Earths of Australia* (Adelaide: University of Adelaide and CSIRO, 1981), pp. 168.
5. Moriarty, W. W. *Survey of Low Level Winds and Air Dispersion Characteristics of the Albury-Wodonga Area.* Bull. 50. (Canberra: Dept. of Science and Environment, Bureau of Meteorology, AGPS, 1979), pp. 227.
6. Hughes, P. J., and M. E. Sullivan. "Aboriginal Landscape," in *Australian Soils: The Human Impact,* J. S. Russel, and R. F. Isbell (Eds.) (St. Lucia: University of Queensland Press, 1986), pp. 117–133.
7. Jenkin, J. J. "Western Civilization," in *Australian Soils: The Human Impact,* (St. Lucia: University of Queensland Press, 1986), pp. 134–156.
8. Smith, A. N. "Management under Higher Rainfall (Victoria and New South Wales)," in *Red-Brown Earths of Australia.* J. M. Oades, D. G. Lewis and K. Norrish (Eds.) (Adelaide: University of Adelaide and CSIRO, South Australia, 1981), pp. 117–132.
9. Cockroft, B., and F. M. Martin. "Irrigation," in *Red-Brown Earths of Australia* J. M. Oades, D. G. Lewis and K. Norrish (Eds.) (Adelaide: University of Adelaide and CSIRO, South Australia, 1981), pp. 133–147.

10. Morrow, J. A., and R. H. Hayman. "Clover-Ley Farming for Mixed Farming Areas of Moderate Rainfall," *J. Agric. Vic.* 38(5):205–210 (1940).
11. Ellington, A., T. G. Reeves, K. A. Boundy and H. D. Brooke. "Increasing Yields and Soil Fertility with Pasture/Wheat/Grain Legume Rotations and Direct Drilling," in *Proc. National Direct Drilling Conference 1979* (Melbourne: ICI Australia Operations Pty Ltd, 1979), pp. 128–142.
12. Kohn, G. D., R. R. Storrier and E. G. Cuthbertson. "Fallowing and Wheat Production in Southern New South Wales," *Aust. J. Exp. Agric. Anim. Husb.* 6:233–241 (1966).
13. Sims, H. J. "Cultivation and Fallowing Practices," in *Soil Factors in Crop Production,* J. S. Russel and E. L. Graecen (Eds.) (St. Lucia: Queensland University Press, 1977), pp. 243–261.
14. Kollmorgen, J. F., J. B. Griffiths and D. N. Walsgott. "The Effect of Various Crops on the Survival and Carryover of the Wheat Take-All Fungus *Gaeumannomyces graminis* var *tritici*," *Plant Pathol.* 32:73–77 (1983).
15. Reeves, T. G., A. Ellington and H. D. Brooke. "Effect of Lupin-Wheat Rotations on Soil Fertility, Crop Disease and Crop Yields," *Aust. J. Exp. Agric. Anim. Husb.* 24:595–600 (1984).
16. Ellington, A. "Energy Requirements and Operating Costs of Reduced Versus Conventional Tillage," *Agric. Eng. Aust.* 8:87–99 (1979).
17. Reeves, T. G., and A. Ellington. "Direct Drilling Experiments with Wheat," *Aust. J. Exp. Agric. Anim. Husb.* 14:237–240 (1974).
18. Rowell, D. L., G. J. Osborne, P. G. Matthews, W. C. Stonebridge and A. A. McNeill. "The Effects of a Long Term Trial of Minimum and Reduced Cultivation on Wheat Yields," *Aust. J. Exp. Agric. Anim. Husb.* 17:802–811 (1977).
19. Stonebridge, W. C., I. E. Fletcher and D. B. Lifroy. "Sprayseed: The Western Australian Direct Sowing System," *Outlook Agric.* 7:155 (1973).
20. Ellington, A., T. G. Reeves and K. I. Peverill. "Chlorosis and Stunted Growth of Wheat Crops in North-East Victoria," in *Proc. Soil Management Conference.* (Dookie, Victoria: Australian Society for Soil Science, 1981), pp. 91–109.
21. Tisdall, J. M., and D. O. Huett. "Tillage in Horticulture," in *Tillage, New Directions in Australian Agriculture,* P. S. Cornish and J. E. Pratley (Eds.) (Melbourne: Inkata Press, 1987), pp. 72–93.
22. Pratley, J. E., and P. S. Cornish. "Conservation Farming — A Crop Establishment Alternative or a Whole-Farm System," *Proc. 3rd Australian Agronomy Conf.,* (1985), pp. 95–111.
23. Mulholland, J. G., J. B. Coombe, M. Freer and W. R. McManns. "Supplementation of Sheep Grazing Wheat Stubble with Urea, Molasses and Minerals: Quality of Diet, Intake of Supplements and Animal Response," *Aust. J. Exp. Agric. Anim. Husb.* 19:23–31 (1979).
24. Carter, M. R., and G. R. Steed. "The Effects of Direct Drilling and Stubble Retention on Hydraulic Properties at the Surface of Duplex Soils in North East Victoria," *Aust. J. Soil Res.* 30:505–516 (1992).
25. Haines, P. J., and N. C. Uren. "Effects of Conservation Tillage Farming on Soil Microbial Biomass, Organic Matter and Earthworm Populations in North-Eastern Victoria," *Aust. J. Exp. Agric.* 30:365–371 (1990).

26. Ellington, A., and T. G. Reeves. "Regulation of Soil Nitrogen Release for Wheat, by Direct Drilling following Clover Pasture," *Soil Tillage Res.* 17:125–142 (1990).
27. Carter, M. R., and P. M. Mele. "Changes in Microbial Biomass and Structural Stability at the Surface of a Duplex Soil under Direct Drilling and Stubble Retention in North-Eastern Victoria," *Aust J. Soil Res.* 30:493–503 (1992).
28. Mele, P. M., and M. R. Carter. "Effect of Climatic Factors on the Use of Microbial Biomass as an Indicator of Changes in Soil Organic Matter," in *Soil Organic Matter Dynamics and Sustainability of Tropical Agriculture,* K. Mulongoy and R. Merckx (Eds.) (Chichester: John Wiley and Sons), pp. 57–63.
29. Northcote, K. H. "Morphology, Distribution and Classification," in *Red-Brown Earths of Australia,* J. M. Oades, D. G. Lewis and K. Norrish (Eds.) (Adelaide: University of Adelaide, and CSIRO, South Australia, 1981), pp. 11–28.
30. Hubble, G. D., R. F. Isbell and K. H. Northcote. "Features of Australian Soils," in *Soils: An Australian Viewpoint,* (Melbourne/Academic Press:CSIRO, London: 1983), pp. 17–47.
31. Graecen, E. L. "Physical Properties and Water Relations," in *Red-Brown Earths of Australia,* J. M. Oades, D. G. Lewis and K. Norrish (Eds.) (Adelaide: University of Adelaide, and CSIRO, South Australia, 1981), pp. 83–96.
32. Northcote, K. H. "Soil Resources" (Map), in *Atlas of Australian Resources, 2rd Series.* (Canberra: Divivision of Natural Mapping, 1978).
33. Cornish, P. S., and J. E. Pratley. "Tillage Practices in Sustainable Farming Systems," in *Dryland Farming a Systems Approach: An Analysis of Dryland Agriculture in Australia,* V. Squires and P. Tow (Eds.) (Sydney: Sydney University Press, 1991).
34. Griffith, J. B. "Water Use of Wheat in the Victorian Mallee," in *Proc. 4th Australian Agronomy Conf.* (1987), p. 296.
35. Incerti, M. "Fallowing for Moisture Conservation in the Victorian Mallee," in *Proc. 5th Australian Agronomy Conf.* (1989), p. 583.
36. Steed, G. R., and G. A. Robertson. "Alternative Conservation Cropping Systems in Other Areas of Australia," in *Proc. Aust. Conf. Agricultural Engineering for Conservation Cropping, QDPI,* (1988) pp. 85–94.
37. Steed, G. R. "Stubble Retention Research in NE Victoria," in *Recent Advances in Crop Residue Management, Occ. Publ. 2,* J. E. Prately and P. S. Cornish (Eds.) (Southern Conservation Farming Group, 1986), pp. 27–28.
38. Coventry, D. R., and H. D. Brooke. "Development of a Minimum Tillage System: Rutherglen Experience," in *Proc. 5th Australian Agronomy Conf.* (1989), pp. 245–251.
39. Eberbach, P. L., and L. A. Douglas. "Effect of Herbicide Residues in a Sandy Loam on the Growth, Nodulation and Nitrogenase Activity of *Trifolium subterraneum,*" *Plant Soil* 131:67–76 (1991).
40. Wardrop, A. J. "Environmental Effects of Herbicides Used in Conservation Cropping Systems: A Review," Tech. Rep. Ser. No. 129 (Victoria: Dept. of Agricuture and Rural Affairs 1986).
41. Powles, S. B., and J. A. M. Holtum. "Herbicide Resistant Weeds in Australia," *Proc. 9th Aust. Weeds Conf. Adelaide* (1989), pp. 185–193.

42. deBoer, R. F., G. R. Steed, J. F. Kollmorgen and B. J. Macauley. "Effects of Rotation, Stubble Retention and Cultivation on Take-All and Eyespot of Wheat in Northeastern Victoria, Australia," *Soil Tillage Res.* 25:263–280 (1993).
43. Rovira, A. D. "Tillage and Soil-Borne Root Diseases of Winter Cereals," in *Tillage — New Directions in Australian Agriculture,* P. S. Cornish and J. E. Pratley (Eds.) (Melbourne: Inkata Press, 1987), pp. 335–354.
44. Pratley, J. E. "Soil, Water and Weed Management — The Key to Farms' Productivity in Southern Australia," *Plant Protect. Q.* 2:21–30 (1987).
45. Hamblin, A. "The Effect of Tillage on Soil Physical Conditions," in *Tillage — New Directions in Australian Agriculture,* P. S. Cornish and J. E. Pratley (Eds.) (Melbourne: Inkata Press, 1987), pp. 128–170.
46. Cornish, P. S., and G. M. Murray. "Low Rainfall Rarely Limits Wheat Yields in Southern New South Wales," *Aust. J. Exp. Agric.* 29:77–83 (1989).
47. Carter, M. R., P. M. Mele and G. R. Steed. "The Effects of Direct Drilling and Stubble Retention on Water and Bromide Movement and Earthworm Species in a Duplex Soil in Southeastern Australia," *Soil Sci.* (in press).
48. Bath, G. J. Personal communication, 1979.
49. Steed, G. R. Unpublished data.

Mainly Sub-Humid to Semi-Arid Continental Climates

CHAPTER 11

Conservation Tillage Systems in the Northernmost Central United States

Raymond R. Allmaras and Steve M. Copeland
U.S. Department of Agriculture, University of Minnesota; St. Paul, Minnesota

J. F. Power
U. S. Department of Agriculture, University of Nebraska; Lincoln, Nebraska

Donald L. Tanaka
U.S. Department of Agriculture, Northern Great Plains Laboratory; Mandan, North Dakota

TABLE OF CONTENTS

11.1. INTRODUCTION

The geographic area envisioned in the northernmost central United States (NmCUS) is that land area between 41° to 48°N latitude and 90° to 105°W longitude. All of North Dakota (ND), South Dakota (SD), Nebraska (NE), and Iowa (IA), and the eastern part of Montana (MT) and nearly all of the arable area in Minnesota (MN) are included. There are excellent transitions of climate, soils, and adapted crops in the NmCUS because there are no mountain ranges or large water bodies to dominate local gradients of climate. The moisture environment ranges from moist subhumid in the eastern part to semiarid in the western part.[1] The thermal environment ranges from cold in the northernmost part (monthly averages of –18.9° and 18.9°C in January and July, respectively) to moderate in the southernmost parts (monthly averages of –2.2° and 25.6°C in January and July, respectively). Soils in the intense croplands of the NmCUS are mostly deep Mollisols with only small areas of Alfisols in the southeastern portion. Entisols occur more frequently in western parts of the NmCUS, where cropland is often a small percentage of the land in farms. The NmCUS includes the northwestern and western portions of the Corn Belt and the northern and northeastern portions of the Great Plains. Corn (*Zea mays* L.) and soybean (*Glycine max* L.) are major crops in the Corn Belt, while wheat (*Triticum aestivum* L.) is the major crop of the Great Plains. Adoption of conservation tillage in its various forms is sensitive to climate, crop systems (adapted crops and crop rotations), soil properties, topography, and pest problems, as well as the needs for water conservation and control of wind and water erosion.[2,3] All of these factors are linked to conservation tillage adoption in the NmCUS.

11.2. DESCRIPTION OF THE AGROECOSYSTEM

11.2.1. Climate

Total annual precipitation, warm-season precipitation (April through September inclusive), and length of frost-free period (Figures 11.1 to 11.3) are

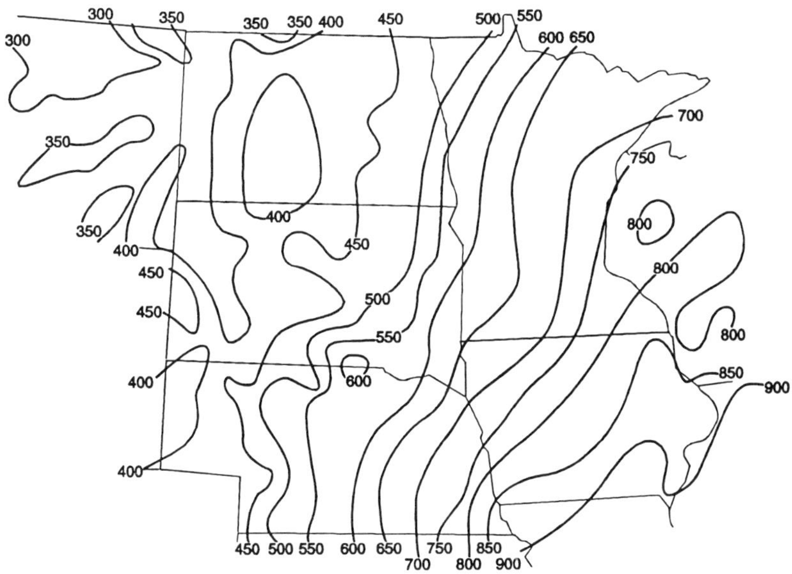

Figure 11.1. Total annual precipitation (mm) in the NmCUS. (Data from "Climatic Data of Iowa, Minnesota, Montana, Nebraska, North Dakota, and South Dakota," National Oceanic and Atmospheric Administration, Asheville, NC, 1992.)

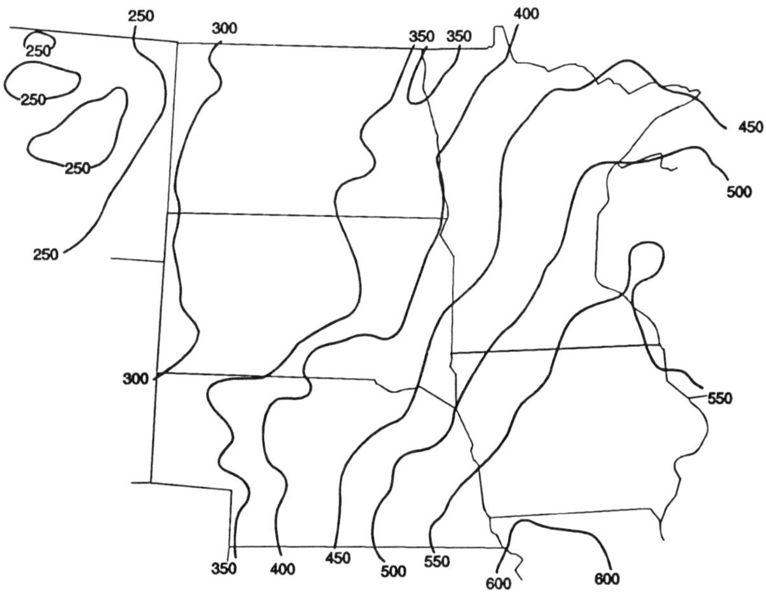

Figure 11.2. Warm season (1 April to 30 September) precipitation (mm) in the NmCUS. (Data from "Climatic Data of Iowa, Minnesota, Montana, Nebraska, North Dakota, and South Dakota," National Oceanic and Atmospheric Administration, Asheville, NC, 1992.)

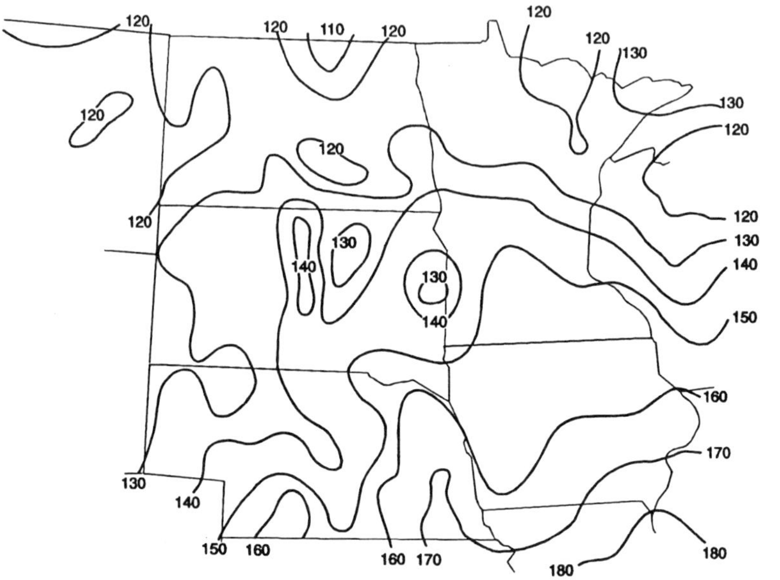

Figure 11.3. Mean number of frost-free days in the NmCUS. (Data from "Climatic Data of Iowa, Minnesota, Montana, Nebraska, North Dakota, and South Dakota," National Oceanic and Atmospheric Administration, Asheville, NC, 1992.)

three of the most critical elements of climate in the NmCUS. Isohyetals (Figures 11.1 and 11.2) showing a strong north-south direction are a manifestation of an increasing aridity progressing east-west, although there is some deviation from this north-south pattern, especially in IA. The precipitation range east-west is roughly 900 to 300 mm/year. Warm-season precipitation ranges from 60 to 75% of the annual total, but the absolute amount of cold-season precipitation ranges from about 200 to 80 mm east-west. This variation in cold-season precipitation significantly impacts supplementation of warm-season precipitation for a crop with a nominal root zone of 150 cm, which contains about 25 cm of available water in a silt loam. Not only is the cold-season precipitation less, but the evaporation potential (Figure 11.4) is also greater in the westernmost parts of the NmCUS, both in the warm and cold seasons. A fully recharged soil profile (of the root zone) at spring planting is nearly a certainty in the easternmost part of the NmCUS, but rarely is the soil profile recharged in the westernmost part of the NmCUS. Hence, the practice of summer fallow in some parts of the NmCUS. The frost-free period ranges from 170 days in the south to less than 110 days in the north (Figure 11.3). These contour lines have a much stronger east-west orientation than the isohyetals. The last killing frost ranges from 25 April to 30 May, while the first killing frost in the autumn ranges from 10 September to 15 October across the NmCUS. A killing frost is assumed when the air temperature at shelter height (1 to 2 m) falls below 0°C.

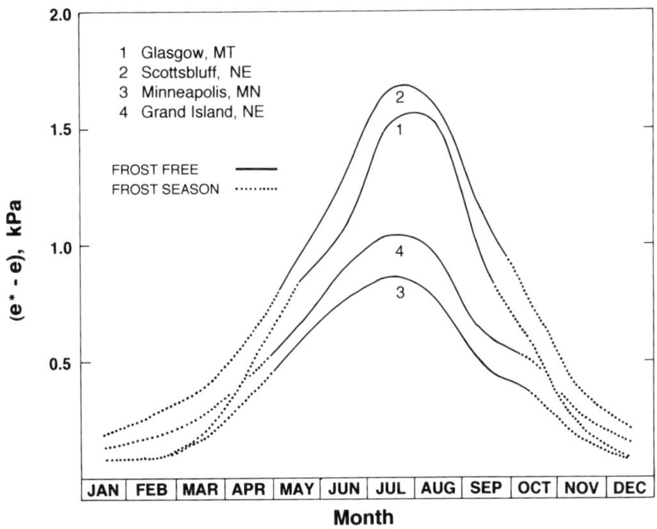

Figure 11.4. Saturation vapor deficit as a measure of evaporative potential (sensitive to radiation, temperature, advection, and cloudiness) at selected sites in the NmCUS. (Data from "Comparative Climatic Data Through 1977," National Oceanic and Atmospheric Administration, Asheville, NC, 1978.)

11.2.2. Soils and Landscape

Soils of the NmCUS vary from frigid (mean annual temperature at the 50 cm depth is <8°C and the difference between mean winter and summer soil temperature is >5°C) in the north to mesic (mean annual soil temperature ranges between 8° and 15°C at 50 cm depth and the range between winter and summer mean temperature is >5°C) in the central and southernmost parts of the NmCUS. The moisture regime of these soils is usually ustic or udic, with some aquic regimes. The control section (10 to 30 cm) in the ustic moisture regime is dry (<–1500 kPa water potential) in some or all parts for 90 or more cumulative days in the mesic and frigid thermal regime; in the udic regime the soil moisture control section is not dry for more than 45 consecutive days during the summer season in mesic and frigid thermal regimes. Aquic regimes usually have a capillary fringe at the soil surface for significant periods of the year.[6] Cosper[7] has shown that soil moisture and thermal regimes can be used to predict the success of various forms of conservation tillage in the Corn Belt. The soil moisture regime expressed as a drainage classification was shown to influence the yield of soybean-corn in the Corn Belt as related to various forms of conservation tillage.[3] Because of the different ecological adaptation of wheat, these relations may need modification for conservation tillage systems involving wheat.

Within the NmCUS, there are 25 major land resource areas (MLRAs), each of which is delineated based on a common combination of climate, topography, soils, land use, and cultural practices.[8] Cropland as a percent of farmland ranges between 50 and 75% in the northern and eastern two thirds of ND, northern and eastern half of SD, and in a small band of eastern IA and MN.[9] Less than 15% of the farmland is cropland in southwestern ND, the western half of SD, and the north central and northwestern parts of NE. More than 75% of NmCUS farmland consists of cropland as opposed to pastureland. These farmland types relate quite well (not shown) to the delineations of the MLRA. Later comparisons with geographic distributions of specific adapted crops show the influence of soil and topography as delineated in the MLRA.

Potential percolation (movement beyond the root zone under rainfed agriculture) is less than 2.5 cm in ND, SD, MT, and most of NE, except the Nebraska Sand Hills, and a band along the eastern one sixth of the state. The remaining part of the NmCUS has a percolation of 2.5 to 7.5 cm, except for southeastern MN (the Mississippi Valley Loess Hills) and the southeastern half of IA, where potential percolation is from 7.5 to 17.5 cm.[10] These estimates are by MLRAs. Average annual runoff is less than 2.5 cm west of the 550 mm isohyetal (Figure 11.1); it ranges from 7.5 to 17.5 cm in southeastern MN and northeastern IA (the eastern Iowa and Minnesota Till Prairies) and south central IA; otherwise, average annual runoff ranges from 2.5 to 7.5 cm.[10] Frost penetration ranges from 120 cm in the northernmost parts to 50 cm in the southernmost parts; the isolines are strikingly east-west.

11.2.3. Adapted Crops

The production or adoption of some major crops (corn, soybean, and wheat) in the NmCUS in 1987, shown in Figures 11.5 to 11.7, is taken as an indicator of adaptation. Production for each crop is shown as a percentage of harvested cropland. When a crop constitutes less than 10% of the harvested cropland, it is a miscellaneous supporting crop, and the limit of adaptation is delineated approximately. When a crop constitutes more than 50% of the harvested cropland, it should be considered a hub crop[11] because it has the greatest comparative advantage economically and ecologically. When a crop constitutes >10 and <30% of the harvested cropland, it can still be a hub crop, especially when it is a row crop.

Within the NmCUS, corn constitutes >50% of the cropland in two areas, in IA and NE (Figure 11.5), but it constitutes >30% of the cropland in the southern half of MN, all of IA, the southeastern half of NE, and the southeastern fifth of SD. Climatic properties of the NmCUS indicate that the northern and western limits of corn adaptation are driven by both rainfall and length of growing season (Figures 11.1 through 11.3).

Soybean production that constitutes >30% of the harvested cropland is limited to MN and IA, with only a small part of eastern SD and NE included (Figure 11.6). The northern limit of adaption is similar to that of corn, but the western limit follows rainfall isohyetals more closely than corn. A zone of

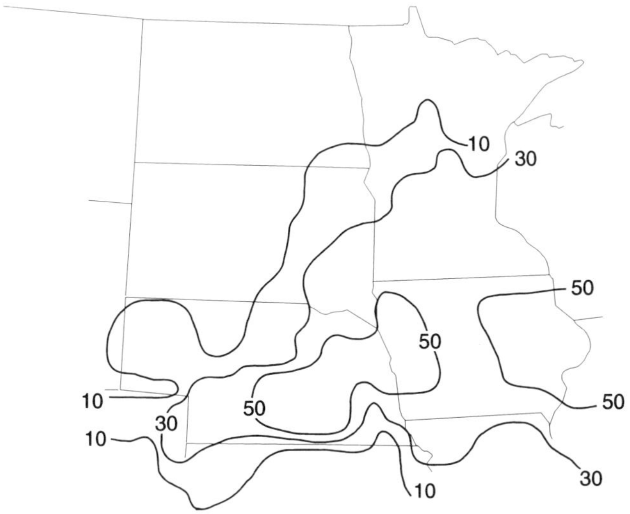

Figure 11.5. Corn as a percentage of the planted cropland in the NmCUS in 1987. (Data from Bureau of Census, "1987 Census of Agriculture, Vol. 2, Subject Series, Part 1, Atlas of the United States," U.S. Department of Commerce, Washington, D.C., 1987.)

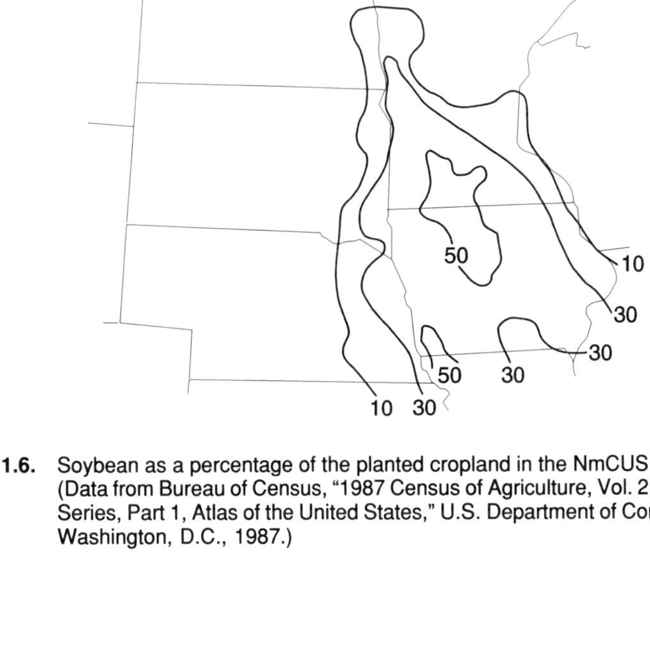

Figure 11.6. Soybean as a percentage of the planted cropland in the NmCUS in 1987. (Data from Bureau of Census, "1987 Census of Agriculture, Vol. 2, Subject Series, Part 1, Atlas of the United States," U.S. Department of Commerce, Washington, D.C., 1987.)

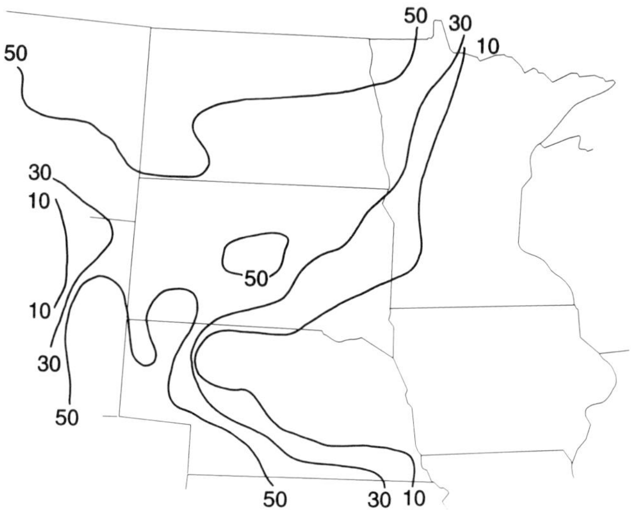

Figure 11.7. Wheat as a percentage of the planted cropland in the NmCUS in 1987. (Data from Bureau of Census, "1987 Census of Agriculture, Vol. 2, Subject Series, Part 1, Atlas of the United States," U.S. Department of Commerce, Washington, D.C., 1987.)

>50% of the cropland in IA and MN and a 10 to 30% zone in the Red River Valley of ND and MN suggest somewhat more tolerance than corn for short length of growing season. Soybean adaptation drops off markedly in eastern MN and IA, a region with Alfisols dominating. Corn is apparently less sensitive to this soil environment.

The >50% share of wheat in harvested cropland indicates where spring and winter wheat are adapted in the NmCUS (Figure 11.7). Except for an area in north central NE, the 30% adaptation of the two crops merge, again suggesting that the comparative advantage of both crops is competing along a climate zone controlled by precipitation and length of growing season combined. A rather striking but unexplained feature in the adaptation of all three crops (Figures 11.5 to 11.7) is the narrow width of the 10 to 30% production as a share of the harvested cropland. The band that delineates the 5 to 10% zone is indeed very narrow compared to the 10 to 30% band.[9]

Some minor crops are widely scattered and others are concentrated in small subareas. Some of these crops may facilitate conservation tillage by crop rotation, but others may suppress conservation tillage by intolerance to disease hazards associated with surface residue. Oat (*Avena sativa* L.) is scattered throughout MN, IA, ND, the eastern two thirds of SD, and the northeast corner of NE. Barley (*Hordeum vulgare* L.) is concentrated in ND, SD, and MT. Most production of barley in MN is in the Red River Valley. Very little barley is grown in NE and IA. Dry edible bean (*Phaseolus vulgaris* L.), potato (*Solanum tuberosum* L.), and sugar beet (*Beta vulgaris* L.) are all concentrated in the Red

River Valley. Dry edible bean is produced also in eastern ND outside the Red River Valley. Sunflower (*Helianthus annuus* L.) is grown intensively in the Red River Valley and the eastern half of ND and SD. Sorghum (*Sorghum bicolor* L.) constitutes up to 20% of the harvested cropland in south central NE.

Barley and oat can facilitate conservation tillage because they can break disease cycles in wheat or corn associated with surface residues.[12] Common root rot (*Cochliobolus sativus* or *Bipolaris sorokiniana*) is increasing in areas of wheat and barley production, but the disease is only mildly sensitive to surface residues of barley and wheat ranging from 4 to 50% cover.[13]

Sunflower is well suited to planting into small grain residues,[14] but dry edible bean requires a firm seedbed without surface residue because of its sensitivity to low temperature and weeds.[15] Small grains are easily planted into the sparse residue of dry edible bean and sunflower. Because surface residues of these two crops are a ready source of inoculum for white mold (*Sclerotinia* spp.) disease in dry edible bean, sunflower, and soybean,[15] the tendency is to rotate only one of these crops with small grains.

Crop adaptation can be linked to MLRA somewhat as related to soil and landscape. Soybean production drops off sharply where alfisols dominate in the northern Mississippi Valley Loess Hills; both intense argillic horizons and steep slopes can be the causative factors. Soybean production is also significantly less in the steep and highly erosive loess deposits of the Iowa and Missouri Deep Loess Hills. Of the total land area, less than 15% is cropland and more than 50% of the harvest on cropland is hay in the Rolling Soft Shale Plains, Rolling Pierre Shale Plains, Nebraska Sand Hills, and Pierre Shale Plains of western ND, SD, and NE. Many of the soils in these MLRAs are entisols. The percentage of farmland in cropland as discussed above also follows the MLRA delineations.

11.3. CROP ROTATIONS

Assessments of conservation tillage adoption have emphasized the critical role of crop rotations.[3] Gill and Daberkow[16] statistically sampled 1989 and 1988 crops preceding the 1990 plantings of corn, soybean, wheat, and potato. Corn followed corn on 32, 27, 17, and 68% of the planted corn in IA, MN, SD, and NE, respectively; in these same states corn followed soybean on 61, 54, 38, and 20% of planted corn. Corn after corn after corn occupied 73% of the planted corn under irrigation in NE, but corn after corn occupied only 22% of the dryland planted corn. In most states corn after corn was most often 3 years of successive corn. Interestingly, corn followed wheat in 14% of the planted corn in SD, but in the other states corn following wheat was negligible. Corn after wheat is hazardous because of stalk rot in corn and head scab in wheat caused by *Fusarium graminearum*. This disease is easily spread by surface residue (crown or basal stem parts) of either crop; small amounts of surface residue cover can risk the disease in a moist environment.[17,18] Soybean followed corn on 91, 77, and 76%, respectively, of the soybean planted in 1990 in IA, MN, and NE. On the average only 5% of the planted soybean followed soybean. These comparative soybean vs corn plantings illustrate the relatively greater hazard of soybean monocrop. Crop rotation is a control for three serious pests: brown stem rot (*Philophora gregata*), *Phytophthora* spp. root rot, and soybean cyst nematode (*Heterodera glycines*) in soybean; none are related to surface residue. Corn and soybean growth responses in rotation compared to monocrop of either crop are suggested as the removal of some negative factor present in the monocrop.[19]

Wheat after corn or barley occupied 21, 5, 10, and 29%, respectively, of the 1990 planted spring wheat in MN, MT, ND, and SD. Again, the rotation of wheat and corn is more predominant in SD. In MN and SD, 28 and 22%, respectively, of the wheat followed soybean, which explains the northern distribution of soybean as a percent of cropland in these states. It is a reflection of the reduced hazard of head scab in wheat where inoculum from the corn is not available. Wheat after fallow occupied 74 and 30% of the planted wheat in MT and ND, respectively. Three years of wheat monocrop occupied 11 to 15% of the wheat planted in 1990 in these four states. Even though this is not a major practice, it may drop rapidly if conservation tillage expands. Tan spot in wheat (*Pyrenophora/Dreschlera tritici-repentis*) is a risk in winter wheat even if there is only 5% cover of residue from diseased wheat.[20] Field studies in infested wheat straw showed that straw on the soil surface in a chisel-disc treatment was wetter than standing straw (combine stubble), but that ascocarp production was much reduced by fungal predation in the wetter environment in contact with soil.[21] A similar hazard exists in spring wheat. Surface residue cover in a system with moldboard plow as the primary tillage had about the same control as 1 year of sorghum preceding winter wheat.[20] Sutton and Vyn[22] made similar observations about residue cover and preceding soybean.

Sunflower and dry edible bean preceded 23% of the wheat planted in 1990 in ND; the 12% of planted wheat after wheat after other crops is an indicator of a significant area with a rotation consisting of wheat-wheat-sunflower or dry edible bean. Potato in 1990 followed wheat or barley on 70 and 59% of potato plantings in MN and ND, respectively; corn was the preceding crop on 8 and 12% of the plantings. This distribution of preceding crop illustrates the value of crop residue for wind erosion control in the seedbed for potato. In 1991 only 14% of the land for potato received primary tillage with a moldboard plow (2% of surface covered with residue after planting), while the remaining 86% did not receive primary tillage with a moldboard plow.[23] Surface cover in the latter case ranged from 17 to 38%.

Sugar beet is a hub[11] crop economically on many of the finer textured soils in the Red River Valley, but wind erosion and direct plant injury are serious problems. Until recently, it was considered necessary to moldboard plow before sugar beet, particularly because of weed control, but sugar beet is now being no-till seeded into cover crops or chiseled small grain residues.[24,25] Several benefits are pre- and postplant wind erosion control, the prevention of twisting on the stem axis when at the four- through eight-leaf stage, and the prevention of wind erosion in the autumn after sugar beet harvest. This new technology will undoubtedly reinforce a rotation of small grain and sugar beet.

Continuous sorghum in NE constituted 38% of the sorghum planted in 1991 in NE; 20% of the sorghum was in rotation with soybean, while 12% was grown in a dryland rotation of sorghum-fallow-winter wheat.[26]

11.4. CONSERVATION TILLAGE SYSTEMS

11.4.1. Definitions and Descriptions

Definitions and descriptions of conservation tillage, as well as the adoption, have evolved dramatically since the conservation objectives of tillage were first developed in the 1930s.[3] The most general definition of conservation tillage is "a system in which either crop residues are retained on or near the soil surface, a rough soil surface is maintained, or both to control soil erosion and to achieve good soil-water relations." This general definition emphasizes a system of factors for erosion control. Even though good soil-water relations (or infiltration) is mentioned, relationship to conservation tillage is not well quantified.

Two different operational definitions are currently used.[27,28] Both place a major emphasis on crop residue cover after planting, but they differ in how the surface cover is determined. Full vs partial width of soil disturbance, as well as the primary tillage tool, are involved in the definition. In the Conservation Technology Information Center[27] definition, three types of conservation tillage-planting systems have been defined since 1989: no-till, ridge-till, and mulch-till. All have >30% residue cover at planting, and to facilitate interpretation of wind erosion the 30% residue cover is equated to "1000 pounds of small grain equivalent." Soil is not disturbed in no-till except for nutrient injection before planting; seeds are placed in a seedbed prepared by coulters, disc openers, row cleaners, and sweeps; post-plant cultivation is often used in row crops for weed control and relief of soil compaction. In mulch-till the soil is disturbed any time after harvest and before planting using chisels, sweeps, cultivator shovels, or discs. The seedbed in ridge-till usually begins with a ridge formed in the preceding crop; is undisturbed from harvest to planting except for preplant nutrient injection; and then is cut down somewhat by row cleaners ahead of the seeder.[29] A trend toward more nutrient injection in the autumn avoids disturbance of the seedbed at planting. Two nonconservation type tillage-planting systems operationally surveyed are the 15 to 30% residue cover and the 0 to 15% residue cover. The first of these two forms generally results when repeated secondary tillage follows a primary tillage other than moldboard plowing; the latter form generally results from primary tillage with a moldboard plow, in which case

Table 11.1. Conservation Tillage and Nonconservation Tillage as a Percent of Planted Cropland Tabulated by State within the NmCUS in 1991

Planted Crop	State	Percent of Planted Cropland: Conservation Tillage: No-Till	Ridge-Till	Mulch-Till	Nonconservation[a]: $30 > r > 15$	$15 > r \geq 0$
Soybean	IA	3.8	1.7	31.4	34.8	28.4
	MN	2.2	4.7	15.6	32.4	45.2
	ND	1.0	0.8	7.3	24.9	66.0
	NE	4.4	4.9	28.1	36.4	26.2
	SD	1.6	3.7	20.9	29.0	44.8
Corn	IA	5.2	1.5	19.7	32.5	41.0
	MN	1.6	4.8	16.5	25.7	51.4
	ND	3.1	2.0	22.4	26.9	45.5
	NE	9.0	13.0	33.4	28.5	16.1
	SD	2.3	3.5	23.4	36.8	34.0
Wheat[b]	MN	1.0	—[c]	15.9	39.9	43.2
	MT	3.9	—	22.2	44.6	29.4
	NE	1.8	—	26.6	41.2	30.4
	ND	3.5	—	25.7	32.4	38.4
	SD	2.4	—	25.0	33.2	39.4

Source: "National Survey of Conservation Tillage Practices, Including Other Tillage Types," NACD Conservation Technology Information Center, West Lafayette, IN, 1991.

[a] r is percent of surface covered with crop residue.
[b] Spring wheat in all except NE, which is winter wheat.
[c] — is none reported.

secondary tillage has little impact on the residue distribution.[30] The primary tillage is not, however, defined in these two forms of tillage.

Four tillage-planting forms have been defined by the Economic Research Service since 1989:[28] conv/w mbd. plow (any system that uses a moldboard plow for primary tillage and has <30% residue cover at planting), conv/wo mbd. plow (any system that does not use a moldboard plow for primary tillage and has <30% residue cover at planting), mulch-till, and no till. The mulch-till and no-till are conservation tillage systems because they provide the 30% cover at planting. The no-till has no residue incorporating tillage postharvest and preplanting, and the mulch-till includes all tillage forms other than no-till that give >30% residue cover at planting. In this system ridge-tillage is not identified separately. The systems are identified by measured yield of grain, a harvest index, the recorded operations each with a standard ratio of incorporated residue, and standard algorithms relating residue cover and weight.

11.4.2. Surveys of Adoption

Both systems[27,28] of tillage definition are used for surveys of conservation tillage adoption, and they are a somewhat complimentary source of information. Both corn and soybean show greater use of no-till in IA and NE than in the more northern states (Table 11.1), but MN and NE are the states with the greater use of ridge-till. Nearly equal corn and soybean cropland planted with ridge-till in MN reflects a stronger corn-soybean rotation, but the greater use in corn than soybean in NE reflects a significant continuous corn planting with ridge-till to facilitate furrow irrigation. These ridge-till cropland percentages are the highest in the nation.[3] The northernmost states in the NmCUS show much higher use of nonconservation (sum of the systems with residue cover <30%) than conservation tillage systems both in the soybean and corn plantings. The percentage ranges from 45 for corn in NE to 90 for soybean in ND. The greater percentage of mulch-till for soybean than corn in IA reflects the smaller amount of available soybean than corn residue before tillage in the corn-soybean rotation. Lower adoption of tillage methods with >30% surface reflects the adverse influence of residue cover on early growth of corn.[31] Although the developmental morphology of the soybean seedling[32] is not inherently as sensitive as the corn seedling[33,34] to reduced soil temperature, there are difficulties of seedling population due to surface residue (especially from corn stalks) interference with planters; this is one explanation for the preferred use of a nonconservational system, especially that with 15 to 30% cover.

Conservation tillage adoption using the second operational definition[28] is summarized in Tables 11.2 to 11.4. The various forms of tillage for wheat planting give similar estimates among states with about 70 to 75% considered to be nonconservational. The use of a chisel, disc, or sweep for primary tillage but with a final surface residue cover <30% is remarkably higher than use of the moldboard plow system, i.e., an average of 63 vs 8%, respectively.

For soybean planting there is a high use of moldboard plow for primary tillage in IA and especially in MN (Table 11.3), which agrees with Table 11.1. Again, the system with a chisel, sweep, or disc primary tillage and with <30% cover has an adoption ranging from 40 to 55%. If one adds this category to the mulch-till category, the use of chisel, sweep, or disc for primary tillage for soybean ranges from 53 to 97% of the planted cropland. This same criteria ranges higher for both planted wheat (Table 11.2) and corn (Table 11.4). The range among states for the various tillage categories for planted corn is less than for other crops. From 50 to 75% of the corn crop is planted in a system without moldboard tillage and <30% cover at planting. The use of the four tillage systems for sorghum in 1991 in NE is nearly the same as with corn in 1989.[26] No-till is a relatively minor form of tillage (Tables 11.1 through 11.4) in the NmCUS area; both survey methods show the same results, i.e., an adoption ranging from 1 to 9% of the planted cropland where reported.

The computed surface residue cover (Tables to 11.4) are characteristic of the system, i.e., 2 to 3% for the conv/w mbd. plow, 13 to 19% for the conv/wo mbd. plow, 36 to 43% for mulch-till, and >50% for no-till. The number of passes (Tables 11.2 to 11.4) did not differ significantly between the two conventional systems but exceeded the number in the mulch-till and no-till systems.

Table 11.2. Tillage Practices Used for Wheat Production in Four States within the NmCUS in 1989

Tillage System[a]	Characteristic[b]	MN	MT	ND	SD
Conv/w mbd. plow					
	Planted cropland (%)	12	NR[c]	10	10
	Surface covered (%)	2	NR	3	2
	Trips	4.6	NR	2.9	2.8
Conv/wo mbd. plow					
	Planted cropland (%)	64	73	56	58
	Surface covered (%)	13	17	15	19
	Trips	4.3	4.5	4.3	3.0
Mulch-till					
	Planted cropland (%)	24	27	32	33
	Surface covered (%)	39	43	40	39
	Trips	2.7	2.7	2.8	2.6

Source: Data from Bull, L., *Agricultural Resources: Inputs Situation and Outlook Report, AR-17,* U.S. Department of Agriculture, Washington, D.C., 1990, pp. 19–27.

[a] Conv/w mbd. plow is any tillage system that uses a moldboard plow and has <30% residue cover at planting; conv/wo mbd. plow is any tillage system that does not use a moldboard plow and has <30% residue cover at planting; mulch-till has >30% residue cover at planting and is not no-till; and no-till has no residue incorporating tillage operation prior to planting.

[b] Number of trips includes planting and such operations as stalk chopping. No data were given for no-till.

[c] NR = None reported.

Table 11.3. Tillage Practices Used for Soybean Production in Three States within the NmCUS in 1989

Tillage System[a]	Characteristic[b]	IA	MN	NE
Conv/w mbd. plow				
	Planted cropland (%)	20	41	NR
	Surface covered (%)	2	3	NR
	Trips	4.5	4.4	NR
Conv/wo mbd. plow				
	Planted cropland (%)	53	39	55
	Surface covered (%)	19	18	18
	Trips	4.1	5.0	3.8
Mulch-till				
	Planted cropland (%)	24	14	42
	Surface covered (%)	37	37	37
	Trips	3.4	3.5	2.8
No-till				
	Planted cropland (%)	2	7	2
	Surface covered (%)	—	57	—
	Trips	—	1.8	—

Source: Data from Bull, L., *Agricultural Resources: Inputs Situation and Outlook Report, AR-17,* U.S. Department of Agriculture, Washington, D.C., 1990, pp. 19–27.

[a] Tillage system explained in Table 11.2.

[b] — , insufficient data; NR, none reported.

Table 11.4. Tillage Practices Used for Corn Production in Four States within the NmCUS in 1989

Tillage System[a]	Characteristic	IA	MN	NE	SD
Conv/w mbd. plow					
	Planted cropland (%)	14	9	8	18
	Surface covered (%)	2	3	2	3
	Trips	3.9	4.0	3.2	3.7
Conv/wo mbd. plow					
	Planted cropland (%)	65	76	54	61
	Surface covered (%)	17	14	19	16
	Trips	3.5	3.7	3.8	3.5
Mulch-till					
	Planted cropland (%)	20	13	32	21
	Surface covered (%)	36	38	38	38
	Trips	3.0	2.8	2.4	3.0
No-till					
	Planted cropland (%)	1	3	6	NR[b]
	Surface covered (%)	—	—	70	NR
	Trips	1.0	1.2	1.0	NR

Source: Data from Bull, L., *Agricultural Resources: Inputs Situation and Outlook Report, AR-17,* U.S. Department of Agriculture, Washington, D.C., 1990, pp. 19–27.

[a] See Table 11.2 for explanation of tillage systems.
[b] — , insufficient data; NR, none reported.

11.4.3. Agronomic Considerations

The 1985 Food Security Act mandates an approved farming system for highly erodible lands (HEL) after 1994. The approved plan usually includes conservation tillage (30% surface cover at planting) for cropland in the whole farm field if it includes more than one third HEL. On a national basis the use of mulch-till and no-till increased somewhat[28] for winter wheat, soybean, corn, and spring wheat planted on HEL lands in 1989 compared to those lands not declared highly erodible. The same is true for the conv/wo mbd. plow category; both increased at the expense of the moldboard based system.

Tillage and residue management for control of water erosion can be estimated from the C (cropping practice) factor[35] in the revised universal soil loss equation (RUSLE);[36] the three factors are surface random roughness, surface covered with residue, and crop residue (including root) buried in the 1 to 10 cm depth. Allmaras et al.[3] used the algorithms of Laflen et al.[35] to demonstrate the component value of each factor for several tillage systems, including the effect of primary and secondary tillage starting with harvest of wheat/corn/soybean. While the surface residue cover had the greatest effect, random roughness and shallow buried residue are important factors for control of water erosion.

Wind erosion is controlled by anchored crop residue and nonerodible clods on the surface.[37] The array of nonerodible clods on the surface is qualitatively related to random roughness. A shallow incorporated residue band may concentrate crop residue sufficiently to increase soil aggregation or reduce soil crust formation during high intensity but short duration rainfall — both may influence wind erosion.[37,38]

The overwhelming use (Tables 11.2 to 11.4) of chisels, sweeps, and discs for primary tillage no deeper than 15 cm is significant because of the potential change in infiltration. Staricka et al.[39] have shown that crop residue burial patterns are specific for the tillage tools: moldboard plow, chisel, and disc. Secondary tillage tools also have little effect on the 80% of the crop residue buried deeper than 10 cm with the moldboard plow.[30] A shallow layer of enriched crop residue should immediately become biologically active. Rasiah et al.[40] found that water-stable aggregates increased and dispersible clay decreased as organic matter increased due to changes in tillage and crop sequence. Bruce et al.[41] have noted these related effects of greater biological activity, more water-stable aggregates, and increased infiltration. Significant improvements in infiltration should be expected due to improved aggregate stability. This potential infiltration response to conservation tillage requires more study. Most economic analyses indicate conservation tillage acceptance when crop yields are increased, especially when soil water relations were improved and yields increased without change of input costs. Infiltration improvements could be part of such a scenario.

The switch away from moldboard plowing may somewhat reduce the need for secondary tillage because chisels, sweeps, and discs generally produce less random roughness than a moldboard plow. If the soil surface is excessively rough, a secondary tillage may be required before a preplant herbicide can be incorporated with the next secondary tillage. Roughly 40% of the planting after full-width tillage in the NmCUS is made after the herbicide is incorporated. Johnson[42] indicates that a rough surface causes an abnormally thick layer of incorporated herbicide, which reduces efficacy of the herbicide. Shallow incorporated crop residue associated with the use of discs, chisels, or sweeps for primary tillage may also reduce soil compaction due to the greater elasticity of the residue compared to mineral matter in the soil,[43] and also the enhanced biological activity, but again, these effects have not been quantified in the field.

Herbicides were used on 91, 97, and 99% of the planted wheat, corn, and soybean, respectively, in the NmCUS in 1989;[44] the number of times an application is made are, respectively, 1.2, 1.4, and 1.6. Preemergence herbicides constituted 25, 70, and 65% of the respective herbicide use for wheat, corn, and soybean. The major herbicides for wheat are foliar 2,4-D and MCPA and preplant incorporated trifluralin. For corn they are alachlor, metolachlor, and atrazine, to be applied no later than preemergence, EPTC mandatory to be preplant incorporated, and dicamba for pre- or postemergence use. For soybean they are preplant incorporated trifluralin, imazethapyr, and alachlor no later than preemergence, and bentazon for postemergence contact. About 20 to 40% of the herbicides used, respectively, for wheat and corn-soybean must be incorporated. Incorporation requires one and often two secondary tillage passes,

Table 11.5. Characteristics of Fertilizer Use for Corn, Soybean, and Wheat in 1989 in the NmCUS

		Percent of Crop Receiving Fertilizer			Fertilizer Application Rate (kg/ha)			Application Time	
Crop	Statistic	N	P	K	N	P	K	Preplant and at Planting	Pre- and Post-Plant
Corn (a)[a]	Mean	97	81	65	129	21	43	80	18
	± S.E.	2	11	32	15	5	21	6	5
Corn (b)[a]	Obs.	69	58	30	69	14	19	89	5
Soybean[b]	Mean	16	20	16	16	17	39	97	3
	± S.E.	8	7	3	0	4	19	2	2
Spring wheat[c]	Mean	56	58	23	51	13	13	95	2
	± S.E.	13	23	28	17	2	10	3	2
Winter wheat[d]	Obs.	76	13	NR[e]	41	13	NR[e]	87	2

Source: Data from Taylor, H., and Vroomen, H., *Agricultural Resources: Inputs Situation and Outlook Report, AR-17,* U.S. Department of Agriculture, Washington, D.C., 1990, pp. 4–14, 52–55.

[a] Corn (a) is a composite of statistics for MN, IA, and NE; corn (b) is that observed for SD.
[b] Composite of statistics for MN, IA, and NE.
[c] Composite of statistics for MN, ND, SD, and MT.
[d] Winter wheat for NE.
[e] NR = None reported.

thus the large adoption of the nonconservation categories of >15 and <30% surface cover (Table 11.1) and the use of conv/wo mbd. plow (Tables 11.2 to 11.4). The comparative number of passes (Tables 11.2 to 11.4) between conv/wo mbd. plow and mulch-till suggest herbicide incorporation as one cause for reduced surface cover at planting.

More than 70% of the corn in the NmCUS receives nitrogen fertilizer, whereas that for spring wheat ranges below 60% (Table 11.5). The rate of applied nitrogen fertilizer in both crops is large enough to require a special pass before planting; Table 11.5 indicates that 80% or more of the fertilization for corn and wheat is done before or during planting. Anhydrous NH_3 or a solution of urea-ammonium nitrate is usually injected below the 15-cm depth, with shanks nominally 50 to 75 cm apart. Volatilization and immobilization in the crop residue layer (0 to 15 cm) are concerns that discourage shallow incorporation.[46,47] A surface application of fertilizer with subsequent tillage for incorporation is no longer popular. Fertilization of corn and wheat with phosphorus is roughly 70% as frequent as nitrogen fertilization; it is about 50% for potassium. Soybean receives phosphorus and potassium fertilization much less frequently than corn and wheat, but when fertilized the rates are about the same; phosphorus and potassium for both corn and soybean is often applied to the corn crop. Because of nitrogen uptake and content of the shoot irrespective of nitrogen fertilization, soybean serves as a significant sink for nitrogen.[48] Most phosphorus and potassium is applied as a band or with the seed at planting; however, deep placement with the nitrogen before planting is expected to increase.[46,47]

The standard errors in Table 11.5 reflect the variation among means reported for the states. For spring wheat the high standard error reflects a more frequent use of nitrogen, phosphorus, and potassium in MN than in ND, SD, and MT; the rate of nitrogen fertilization shows the same effect. The large difference of corn fertilization in SD compared to the average for MN, NE, and IA required separate means in Table 11.5. Undoubtedly these variations among states display lower nitrogen use especially, where the available water is less.

11.5. WHEAT-FALLOW SYSTEMS

Two wheat-fallow systems predominate where precipitation is <400 mm/annum (Figure 11.1). In the winter wheat-fallow system, wheat is grown for about 10 months followed by a 14-month fallow period. In the spring wheat-fallow system, wheat is grown for 3 months followed by a 21-month fallow period. In both systems, one wheat crop is produced every 2 years. Crop residue management for erosion control and soil-water conservation is critical during the 14- and 21-month fallow periods. Wheat after summer fallow in 1989 constituted as much as 70% of the planted wheat in MT while that in ND was near 30%.[16]

Although shallow tillage (sweeps, cultivators, and rod weeders) was used for summer fallow operations before the 1960s, the number of operations required to control weeds and break up soil crusts ultimately produced a dust mulch. Although there were fine residue fragments in the dust mulch, it was dry and biologically inactive, especially during the summer. The dust mulch prevented water vapor losses during hot, dry periods, but failed overall because of poor infiltration and an excessive evaporation, which is driven by adverse radiation balances and mass transfers over a bare surface without a boundary layer (as produced when there is surface residue) to shift upward the zone of gaseous exchange. Soil water storage (SWS) averaged 20% of precipitation during the 14- and 21-month fallow periods;[49] SWS was less efficient in the southernmost locations of the Great Plains (a shorter fallow period but a higher evaporative potential, Figure 11.4). Greb et al.[50,51] demonstrated that increases of added straw up to 6730 kg/ha without a change in other aspects of the commonplace system could increase the efficiency of stored water from 20 to 37% in a 14-month fallow period north of latitude 40°N; the greatest differential of SWS among straw residue rates occurred in late spring.

Table 11.6. Soil Water Storage (SWS) and Precipitation Storage Efficiency (PSE) for No-Till (NT), Reduced-Till (RT), and Stubble Mulch (SM) in the 14-Month Fallow Period for Winter Wheat Production

Seasonal Segment	Fallow Method					
	NT		RT		SM	
	SWS (mm)	PSE (%)	SWS (mm)	PSE (%)	SWS (mm)	PSE (%)
Postharvest	83	57	71	48	53	37
Overwinter	65	83	59	78	49	63
Summer	108	37	80	27	70	24
Total	256	49	210	40	172	33

Source: Averages from Greb,[53] Greb and Zimdahl,[54] Smika,[55] and Smika and Whitfield.[56]

Table 11.7. Soil Water Storage (SWS) and Precipitation Storage Efficiency (PSE) for No-Till (NT), Reduced-Till (RT), and Stubble Mulch (SM) in the 21-Month Fallow Period for Spring Wheat Production

Seasonal Segment	Fallow Method					
	NT		RT		SM	
	SWS (mm)	PSE (%)	SWS (mm)	PSE (%)	SWS (mm)	PSE (%)
Postharvest	36	54	38	56	30	45
First overwinter	52	57	50	55	48	53
Summer	39	21	34	18	35	18
Second overwinter	16	21	12	15	14	18
Total	143	34	134	32	127	30

Source: Tanaka[57] and Tanaka and Aase.[58]

No-till and reduced-till concepts were introduced into the wheat-fallow system by using herbicides instead of cultivation.[52] Standing wheat stubble and ground cover with no soil disturbance during critical periods served to increase biological activity and infiltration and to change the factors controlling evaporation. SWS and precipitation storage efficiency (PSE) were both increased by no-till and reduced-till in both the 14- and 21-month fallow periods (Tables 11.6 and 11.7). Winter wheat residue losses during the 14-month fallow period are 25% for no-till as compared to 75% for the stubble mulch system.[59] Respective spring wheat residue losses during the 21-month fallow period are 55 and 87%.[60] Losses for the reduced-till system are between the extremes of the no-till and stubble mulch systems.

As mechanical weed control decreased, the PSE increased from 33 to 49% (Table 11.6); the greatest break between systems was the low PSE of 63% for stubble mulch compared to 80% for the no-till and reduced-till systems during the overwinter segment. This difference relates to the greater SWS afforded by herbicides without tillage after harvest. Another sharp break among systems is the PSE (49%) of no-till compared to 26% for reduced-till and stubble mulch in the summer segment. Tillage system-mediated PSE ranged much less in the

21-month period (Table 11.7) of spring wheat production than in the 14-month fallow system; the only sharp difference among tillage systems was during the post-harvest segment, in which the no-till and reduced-till systems had a PSE of 55% compared to 45% for the stubble mulch system. The PSE in Table 11.7 is low both for the summer and second overwinter segments; three possible causes are poor wintertime infiltration into a wet and frozen ground without cover, slow redistribution of infiltrated water, and excessive evaporation from a soil surface sparsely covered with crop residue. Deibert et al.[61] observed a similar low recharge efficiency for the 21-month fallow period, which was attributed to soil profile wetness and failure of spring wheat to use water from below 120cm in soils of central and western ND. The PSE in continuous wheat was 75% compared to 29% in the fallow. For this reason, Bauer[62] suggests that the summer segment of fallow be bypassed by planting a spring crop when available water exceeds 76 mm in ND. For reduced-till and no-till in fallow systems to store more soil water than stubble mulch, surface residue at the beginning of the summer segment must be at least 2500 kg/ha (estimated cover, 70%), soils cannot be at or near field capacity, and precipitation must occur in sufficient quantity and frequency to redistribute water deep enough into the soil to suppress evaporation.[57,63]

Control of soil erosion by wind is critical during the summer and after wheat planting in the 14-month system, but in the 21-month system the danger is especially great during early spring after thaw and before the spring wheat produces a controlling roughness above the soil surface. Surface residue cover in no-till is three times greater than in the stubble mulch system during this second summer.[60] Water erosion is usually most serious during late summer of the fallow season.

The increase in PSE due to greater quantities of surface residue in no-till fallow has stimulated research to reduce fallow frequency by using more intensive cropping systems.[64-66] Not only is a no-till practice used, but crop rotation also is used to utilize soil water stored in the rooting zone of cereals. As discussed above, care must be used to avoid diseases caused by inoculum in the surface residue.

11.6. COVER CROPS: THEIR POTENTIAL AND RESEARCH NEEDS

Cover crops are generally used in the cool season to augment the system for cash crop production in the warm season. Cover crops can be harvested for grain or forage, or merely used for cover. The unharvested portion of the cover crop often facilitates the use of no-till. Cover crops have been an integral part of the double-crop system in the southeastern United States, i.e., at latitudes south of 40°N and in the humid to superhumid climate zones. There are numerous objectives for a cover crop,[67] but in the southeastern United States cover crops are needed especially to sustain high infiltration, retain nutrients from leaching, and shift cultivation to a season when climate energy is least likely to produce erosion.[3] The milder cool season of the southeastern United States facilitates a wide variety of species for cover crop, but rapid residue decomposition rates require a continuous source of carbonaceous materials to maintain a soil structure suitable for infiltration.[41] Overwinter decomposition rates are estimated to be about 20% in the NmCUS, whereas a rate of 50% can be expected in the southeastern United States.

Cover crops have been used as barriers for snow trapping to control wind erosion and reduce evaporative losses in the NmCUS, but gradual implementation of practices required by the 1985 Food Security Act will encourage cover crop technology for uniform ground cover in the absence of crop residues. This is imperative for control of wind erosion after summer fallow or harvest of potato/sugar beet. Cover crops immobilize residual soil nitrates present at harvest and maintain cover for control of wind and water erosion,[68] especially when the previous crop residue provides insufficient cover (as in corn silage, soybean, sunflower, potato, sugar beet) or when there has been summer fallow. There is much interest in leguminous cover crops substituting biologically fixed nitrogen for purchased nitrogen inputs[69]. Sparse winter precipitation and high evaporative demand in the NmCUS (Figures 11.1, 11.2, and 11.4) may cause cover crops to overdeplete water held in the upper soil horizons[70] and reduce soil moisture reserves for the following summer crop.

Because of the soil disturbance necessary for cover crop establishment, cover already afforded by residues of small grains and corn, and potential for adverse depletion of stored water, cover crops are not encouraged in areas with <500 mm precipitation unless a leguminous cover crop is used to fix nitrogen biologically. There are other circumstances when a cover crop would provide both cover and nitrogen management.

Spring precipitation in the eastern part of the NmCUS is usually sufficient to recharge the upper soil horizons from which cover crops may extract water. Power et al.[71] tested a system with a hairy vetch (*Vicia villosa* Roth) cover crop in a continuous corn sequence. When the cover crop was incorporated at planting, corn grain yields in eastern NE equalled or exceeded those that received normal nitrogen fertilizer applications. Similar observations have been made in IA and elsewhere.[72]

Corn germination and emergence may be delayed sufficiently to reduce grain yields when precipitation is deficient at planting time. With irrigation in a subhumid climate, a water deficit from a cover crop can be remedied to assure germination and emergence of a following corn crop. In practice, however, this does not always occur because of the large quantities of corn residue (13 Mg/ha) after harvest. This residue cover prevents seed of the cover crop from germinating and emerging with vigor before winter freeze. Thus, stands of the cover crop are often poor.

Problems with water deficit become more common and severe in the drier, semiarid parts of the NmCUS.[73] To date, no proven cover crop management system has been developed for the winter wheat-fallow area, where the fallow period is 14 months long. The ecofallow concept is appropriate, wherein rotations other than wheat-fallow are used, usually with reduced- or no-till techniques.[74] These rotations, while not using cover crops, often use 3- or 4-year rotations which include corn, sorghum, millet (*Setaria italica* L.), and possibly annual forages in addition to wheat and fallow.

In the spring wheat-fallow area, more possibilities exist with the 21-month fallow period. Research in MT[75] and in the Canadian prairie provinces[76] has shown that a number of spring-seeded annual legumes may have potential for use during spring of the fallow year. These include faba bean (*Vicia faba* L.), lentil (*Lens culinaris* Medikus), field pea (*Pisum sativum* L.), and medic (*Medicago* spp.). When seeded early in the spring and allowed to grow 6 to 8 weeks, these species can fix enough nitrogen to meet the requirements of the subsequent spring wheat. Again, reduced- and no-tillage methods are preferred. Flax (*Linum usitatissimum* L.) or sometimes oat is seeded in strips across fallow fields in the second autumn of the 21-month fallow period for spring wheat. Spacing and azimuth orientation of the strips (with respect to wind direction) determine their effectiveness for trapping snow and controlling wind erosion. Flax has some advantage over oat because of the upright canopy structure after winter kill. Power[77] showed that soybean, as well as field pea and faba bean, performed well when seeded in June and July of the fallow year in ND. Zachariassen and Power[78] used temperature-controlled water baths in the greenhouse to show that several annual grain legumes grew more rapidly than did smaller-seeded perennial legumes for the first 42 to 63 days after emergence when grown at cool temperatures (10°C).

A lack of suitable genetic diversity for a cover crop is a major problem.[75] For example, hairy vetch is the only legume available for use as a winter annual. It is the only annual legume species that can survive winters in this region, and may not survive north of latitude 40°N. Sims et al.[79] showed that several medics had promise for spring seedlings, and have developed the cultivar George by selection. Essentially no research effort is being devoted to the improvement of germplasm sources through breeding, however. Adapted cover crop species offer a real opportunity for improving cover crop management in this region.

11.7. CONSERVATION TILLAGE UNDER IRRIGATION

Nearly 70% of the 4 million ha of irrigated land in the NmCUS is located in NE. Other areas where irrigation is practiced include lower precipitation regions (generally <600 mm), and sandy soils in more humid parts of the NmCUS. Corn is the most common irrigated crop (see Figure 11.5), although potato, dry bean, sugar beet, and alfalfa are also important. About 60% of the irrigated land is watered by sprinkler (center pivot more than line sprinkler), and the remaining 40% is surface irrigated.

Corn in 1990 after corn in 1989 after corn in 1988 made up 73% of the planted corn under irrigation, whereas the same sequence under dryland in NE made up only 22% of the corn planted in 1990.[16] The latter frequency was similar to that in IA and MN. The frequency of the various tillage systems for corn under irrigation in NE in 1989 was nearly the same as that shown for dryland (Table 11.4); comparative values for irrigated corn were 3% for conv/w mbd. plow, 55% for conv/wo mbd. plow, 23% for mulch-till, and 18% for no-till.[28] The overall number of trips was only about 10% greater. Under irrigation nearly 80% of the land received a primary tillage with other than a moldboard plow; this retention of residue at depths of 0 to 15 cm is especially significant for soil erosion control because more than 70% of the crop is corn with residues ranging above 8 Mg/ha.

Water application rates in sprinkler systems produce a significant raindrop impact, and they may greatly exceed the soil infiltration rate. Soil erosion potential, therefore, can be serious, especially where parts of the field have a slope.[80] Surface roughness due to clodiness and crop residues usually does not interfere with water conveyance in furrow irrigation and helps to maintain high infiltration rates.[80] This benefit may explain the high use of primary tillage without a moldboard plow (discussed above). Overland flow can be nearly eliminated by small dikes defined as "reservoir tillage,"[80] even when the application rate greatly exceeds the infiltration rate. However, this roughness is considered a disadvantage for harvest in NE. Another opposing development is low energy sprinklers to reduce evaporation. These sprinklers apply water at a higher rate over a smaller area (band) than applied by higher energy systems.[81]

Furrow irrigation presents a dilemma for conservation tillage with high amounts of corn residue not banded as in ridge till. Crop residues may be retained on the surface or buried no deeper than 15 cm in irrigated soils to reduce runoff and erosion potential, but this practice increases hydraulic resistance for water advance[82] and slows advance of water across the field.[83,84] Surface roughness also increases hydraulic resistance. This uneven water distribution can be counteracted by shortened furrow length, but irrigation costs are increased.

Several practices are recommended to reduce adverse effects of the hydraulic resistance in furrow irrigation caused by surface residues and greater roughness in conservation tillage.[80] The furrows may be tracked to reduce roughness; another is to use surge flow,[85] in which case the water delivery rate into the furrow is modulated to irrigate the lower end of the field without abnormally high infiltration at the head portion caused by hydraulic resistance. Bedding techniques are being used to facilitate surface retention of residues and irrigation furrows free of residue. Another practice is to use smaller and more closely spaced furrows in no-till corn;[80] soil erosion is thus drastically reduced. This research may explain the higher use of no-till for irrigated corn as discussed above.

11.8. GENERAL SUMMARY AND CONCLUSIONS

Primary tillage with a chisel, disc, or sweep was used for 76 to 91% of the planted wheat, corn, and soybean in the NmCUS in 1989. Only one exception was the 40% of the land that was moldboard plowed before soybean planting in MN. The remaining planted land was either tilled with a moldboard plow or was no-till/ridge-till planted. This is a marked change in tillage management because moldboard plowing was the major primary tillage for corn/soybean as late as 1975.[3] The sweep has been the primary tillage tool since the 1940s in the mandatory fallow-wheat agriculture, but there has been a major shift away from the moldboard plow in the more humid portion of the NmCUS where wheat is not alternated with fallow.

According to the operational definitions of conservation tillage requiring 30% of the surface covered at planting, only 20 to 30% of the plantings in the NmCUS have achieved a conservation goal that ultimately may be in all farm conservation (sustainable) plans. Yet it is of major significance that most cropland in the NmCUS now has crop residues buried within the upper 15 cm instead of below 15 cm as with a moldboard plow. Benefits from this shallow-buried residue are better nutrient retention, more water-stable aggregates, less crusting, greater infiltration, and reduced soil erosion. To use these tillage systems for surface-retained and shallow-buried residues, major necessary advances in tillage tools and associated herbicidal weed control have taken place since 1970. Economic gains from enhanced infiltration may be prompting the change; input costs may be reduced by recent improvements in machinery/herbicide/fertilizer systems, feasible where the residue is buried at a shallow depth. Research is needed to evaluate how these potential gains may control the acceptance of shallow-buried residue.

The areal distribution of corn/wheat/soybean within the NmCUS was consistent with total and warm-season precipitation, length of growing season, and evaporative potential/rainfall during the cool season. Some area distributions of crop sequence were related to control of foliar and root diseases. Disease hazard appeared not to suppress use of primary tillage tools that provide shallow burial of residue, but it may explain additional secondary tillage after a nonmoldboard primary tillage to reduce the amount of disease inoculum on or above the soil surface. Adversely low soil temperature produced by surface residues are well known, especially where there is less benefit from improved water conservation in the corn-soybean sequence.

Numerous new tillage system technologies are under development to safeguard production and environment in this region. The adoption of ridge-till is growing[3] because it offers a reduced herbicide use in combination with cultivation for weed control, maintains a residue-free cover over the germinating and emerging crop, and controls water and wind erosion through a combination of winter cover and surface residue to provide hydraulic resistance to overland flow in the furrows. Ridge-till appears to be less sensitive than other conservation tillage systems to poor drainage. No-till shows increased adoption in irrigated continuous corn in NE, possibly due to its ability to reduce the wetting perimeter and erosion in corrugates used for water conveyance. It is also being adopted along with small grain stubble and cover crop techniques to control wind erosion in sugar beet and potato. Another benefit to sugar beet is the reduced direct wind damage. Cover crops are being used to provide barriers for snow-catch (reduced wind shear on the soil), and for complete cover after crops (sugar beet, potato) and practices (summer fallow) that leave insufficient residue cover. Much of the urgency here is to comply with the conservation provisions for highly erodible lands as mandated in the 1985 Food Security Act.

Sorption and irreversible retention of herbicides may be accelerated in soils with high organic matter contents.[86] Consequently, the dominant use of tillage tools in the NmCUS that produce a shallow burial of crop residue may have an extensive influence on the binding, efficacy, and transport of incorporated herbicides. Moreover, the incorporated crop residue and herbicide are placed together in the voids.[30,39] Research is emphasizing herbicide application during and after planting to reduce residue in the soil, but moisture conditions for herbicidal action/field operations are frequently uncertain in this region.

Farming-by-soil to improve production and environmental safeguard is in the early developmental stages.[87] Most fields contain a range of soil mapping units such that one nutrient rate/pesticide application/tillage is not optimum for all mapping units. Rates are often excessive in some parts of the field, which leads to wasted input and environmental damage. Technology is being developed and applied such that inputs can be varied manually or according to an on-board computer map of the field. Tillage variation on-the-go is especially relevant to the region because of landscapes with soil mapping units that have variable internal drainage/erosion susceptibility.

11.9. ACKNOWLEDGMENT

Joint contribution of the USDA-ARS and the Minnesota Agriculture Experimental Station paper no. 20,294 of the Scientific Journal Series.

11.10. REFERENCES

1. Tanner, C. B., and T. R. Sinclair. "Efficient Water Use in Crop Production: Research or Re-Search?" in *Limitations to Efficient Water Use in Crop Production*, A. M. Taylor, W. R. Jordon, and T. R. Sinclair (Eds.) (Madison, WI: American Society of Agronomy, 1983), pp. 1–27.
2. Allmaras, R. R., and R. H. Dowdy. "Conservation Tillage Systems and Their Adoption in the United States," *Soil Tillage Res.* 5:197–222 (1985).
3. Allmaras, R. R., G. W. Langdale, P. W. Unger, and R. H. Dowdy. "Adoption of Conservation Tillage and Associated Planting Systems," in *Soil Management for Sustainability*, R. Lal, and F. J. Pierce (Eds.) (Ankeny, IA: Soil and Water Conservation Society, 1991), pp. 53–83.

4. "Climatic Data of Iowa, Minnesota, Montana, Nebraska, North Dakota, and South Dakota," (Asheville, NC: National Oceanic and Atmospheric Administration, 1992).
5. "Comparative Climatic Data Through 1977," (Asheville, NC: National Oceanic and Atmospheric Administration, 1978).
6. Soil Survey Staff. Soil Conservation Service, "Soil Taxonomy: A Basic System of Soil Classification for Making and Interpreting Soil Surveys," Agric. Hdbk. 436, (Washington, DC: U.S. Dept. of Agricuture, 1975).
7. Cosper, H. R. "Soil Suitability for Conservation Tillage," *J. Soil Water Conserv.* 38:152–155 (1983).
8. Soil Conservation Service, "Land Resource Regions and Major Land Resource Areas of the United States," Agric. Hdbk. 296, (Washington, DC: U.S. Dept. of Agriculture 1981).
9. Bureau of Census. "1987 Census of Agriculture, Vol. 2, Subject Series, Part 1, Atlas of the United States," (Washington, DC: U.S. Dept. of Commerce, 1987).
10. Stewart, B. A. (coordinator). "Control of Water Pollution from Cropland, Vol. 1, A Manual for Guideline Development," (Washington, DC: U.S. Dept. of Agriculture, 1975).
11. Pearson, L. C. "Crop Rotations, Chapter 5," in *Principles of Agronomy* (New York: Reinhold, 1967), pp. 73–84.
12. Cook, R. J., and K. F. Baker. *The Nature and Practice of Biological Control of Plant Pathogens* (St. Paul, MN: American Phytopathology Society, 1983).
13. Windels, C. E., and J. V. Wiersma. "Incidence of *Bipolaris* and *Fusarium* on Subcrown Internodes of Spring Barley and Wheat Grown in Continuous Conservation Tillage," *Phytopathology* 82:699–705 (1992).
14. Deibert, E. J. "Reduced Tillage Systems Influences on Yield of Sunflower Hybrids," *Agron. J.* 81:274–279 (1989).
15. Helm, J. L., K. F. Grafton and A. A. Schneiter. "Dry Bean Production Handbook, A-602 Revised" (Fargo, ND: Extension Service, North Dakota State University, 1990).
16. Gill, M., and S. Daberkow. "Crop Sequences Among 1990 Major Field Crops and Associated Farm Program Participation," in *Agricultural Resources: Inputs Situation and Outlook Report, AR-24* (Washington, DC: Economic Research Service, U.S. Dept. of Agriculture, 1991), pp. 39–46.
17. Wilcoxson, R. D., T. Kommedahl, E. A. Ozmon and C. E. Windels. "Occurrence of *Fusarium* Species in Scabby Wheat from Minnesota and Their Pathogenicity to Wheat," *Phytopathology* 78:586–589 (1988).
18. Windels, C. E., T. Kommedahl, W. C. Steenstra and P. M. Burnes. "Occurrence of *Fusarium* Species in Symptom-Free and Overwintered Cornstalks in Northwestern Minnesota," *Plant Dis.* 72:990–993 (1988).
19. Crookston, R. K., J. E. Kurle, P. J. Copeland, J. H. Ford and W. E. Leuschen. "Rotational Cropping Sequence Affects Yield of Corn and Soybean," *Agron. J.* 83:108–113 (1991).
20. Bockus, W. W., and M. M. Claassen. "Effects of Crop Rotation and Residue Management Practices on Severity of Tan Spot of Winter Wheat," *Plant Dis.*76:633–636 (1992).
21. Zhang, W., and W. F. Pfender. "Effect of Residue Management on Wetness Duration and Ascocarp Production by *Pyrenophora tritici-repentis* in Wheat Residue," *Phytopathology* 82:1434–1439 (1992).

22. Sutton, J. C., and T. J. Vyn. "Crop Sequences and Tillage Practices in Relation to Diseases of Winter Wheat in Ontario," *Can. J. Plant Pathol.*12:358–368 (1990).
23. Bull, L. "Tillage Systems," in *Agricultural Resources: Inputs Situation and Outlook Report, AR-28* (Washington, DC: Economic Research Service, U.S. Dept. of Agriculture, 1992), pp. 12–15.
24. Deibert, E. J., J. F. Giles, J. Enz and D. Lizotte. "Reduced Tillage in Sugarbeet Production," in *Sugarbeet Research and Extension Reports, Vol. 12* (Fargo, ND: Extension Service, North Dakota State University, 1982) pp. 123–127.
25. Stordahl, J. B., A. G. Dexter and A. W. Cattanach. "Production of Sugar Beet in Living Cover Crop," in *Sugarbeet Research and Extension Reports, Vol. 21* (Fargo, ND: Extension Service, North Dakota State University, 1991), pp. 213–215.
26. Gill, M., H. Delvo, H. Taylor and L. Bull. "Sorghum Production Practices and Input Use," in *Agricultural Resources: Inputs Situation and Outlook Report, AR-28* (Washington, DC: Economic Research Service, U.S. Dept. of Agriculture, 1992), pp. 36–39.
27. "National Survey of Conservation Tillage Practices, Including Other Tillage Types" (West Lafayette, IN: NACD Conservation Technology Information Center, 1991).
28. Bull, L. "Tillage Systems," in *Agricultural Resources: Inputs Situation and Outlook Report, AR-17* (Washington, DC: Economic Research Service, U.S. Dept. of Agriculture, 1990), pp. 19–27.
29. Griffith, D. R., S. D. Parsons and J. V. Mannering. "Mechanics and Adaptability of Ridge-Planting for Corn and Soyabean," *Soil Tillage Res.* 18:113–126 (1990).
30. Staricka, J. A., R. R. Allmaras, W. W. Nelson and W. E. Larson. "Soil Aggregate Longevity as Determined by Incorporation of Ceramic Spheres," *Soil Sci. Soc. Am. J.* 56:1591-1597 (1992).
31. Swan, J. B., E. C. Schneider, J. F. Moncrief, W. H. Paulson and A. E. Peterson. "Estimating Corn Growth and Grain Moisture from Air Growing Degree Days and Residue Cover," *Agron. J.* 79:53–60 (1987).
32. Lersten, N. R., and J. B. Carlson. "Vegetative Morphology," in *Soybeans: Improvement, Production and Uses,* 2nd ed., Agron. Monogr. 16, J. R. Wilcox (Ed.) (Madison, WI: American Society of Agronomy, 1987), pp. 49–94.
33. Larson, W. E., and J. J. Hanway. "Corn Production," in *Corn and Corn Improvement,* Agron. Monogr. 18, G. F. Sprague (Ed.) (Madison, WI: American Society Agronomy, 1977), pp. 625–699.
34. Shaw, R. H. "Climatic Requirement," in *Corn and Corn Improvement,* Agron. Monogr. 18, G. F. Sprague (Ed.) (Madison, WI: American Society of Agronomy, 1977), pp. 591–623.
35. Laflen, J. M., G. R. Foster and C. A. Onstad. "Simulation of Individual Storm Loss for Modeling the Impact of Soil Erosion on Soil Productivity," in *Soil Erosion and Conservation*, S. A. El-Swaify, W. C. Moldenhauer and A. Lo (Eds.) (Ankeny, IA: Soil and Water Conservation Society, 1985), pp. 285–295.
36. Renard, K. G., G. R. Foster, G. A. Weesies and J. R. Porter. "RUSLE: Revised Universal Soil Loss Equation," *J. Soil Water Conserv.* 46:30–33 (1991).
37. Fryrear, D. W. "Soil Cover and Wind Erosion," *Trans. Am. Soc. Agric. Eng.* 28:781–784 (1985).

38. Zobeck, T. M., and T. W. Popham. "Influence of Microrelief, Aggregate Size, and Precipitation on Soil Crust Properties," *Trans. Am. Soc. Agric. Eng.* 35:487–492 (1992).
39. Staricka, J. A., R. R. Allmaras and W. W. Nelson. "Spatial Variation of Crop Residue Incorporated by Tillage," *Soil Sci. Soc. Am. J.* 55:1668–1674 (1991).
40. Rasiah, V., B. D. Kay and T. Martin. "Variation of Structural Stability with Water Content: Influence of Selected Soil Properties," *Soil Sci. Soc. Am. J.* 56:1604–1609 (1992).
41. Bruce, R. R., G. W. Langdale, L. T. West and W. P. Miller. "Soil Surface Modification by Biomass Inputs Affecting Rainfall Infiltration," *Soil Sci. Soc. Am. J.* 56:1614–1620 (1992).
42. Johnson, R. R. "Soil Engaging-Tool Effects on Surface Residue and Roughness with Chisel-Type Implements," *Soil Sci. Soc. Am. J.* 52:237–243 (1988).
43. Soane, B. D. "The Role of Organic Matter in Soil Compactability: A Review of Some Practical Aspects," *Soil Tillage Res.* 16:179–201 (1990).
44. Delvo, H. W. "Pesticides," in *Agricultural Resources: Inputs Situation and Outlook Report, AR-17* (Washington, DC: Economic Research Service, U.S. Dept. of Agriculture, 1990), pp. 14–19, 56–59.
45. Taylor, H., and H. Vroomen. "Fertilizer," in *Agricultural Resources: Inputs Situation and Outlook Report, AR-17* (Washington, DC: Economic Research Service, U.S. Dept. of Agriculture, 1990), pp. 4–14, 52–55.
46. Randall, G. W., K. L. Wells and J. J. Hanway. "Modern Techniques in Fertilizer Application, in *Fertilizer Technology and Use, Third Edition*, O. Englestad (Ed.) (Madison, WI: Soil Science Society of America, 1985), pp. 521–560.
47. Randall, G. W., and V. A. Bandel. "Overview of Nitrogen Management for Conservation Tillage Systems: An Overview", in *Effects of Conservation Tillage on Groundwater Quality: Nitrates and Pesticides*, T. J. Logan, J. M. Davidson, J. L. Baker and M. R. Overcash (Eds.) (Chelsea, MI: Lewis Publishers, 1987), pp. 39–63.
48. Varvel, G. E., and T. A. Peterson. "Nitrogen Fertilizer Recovery by Soybean in Monoculture and Rotation Systems," *Agron. J.* 84:215–218 (1992).
49. Mathews, O. R., and T. J. Army. "Moisture Storage on Fallowed Wheatland in the Great Plains," *Soil Sci. Soc. Am. Proc.* 24:414–418 (1960).
50. Greb, B. W., D. E. Smika and A. L. Black. "Effect of Straw Mulch Rates on Soil Water Storage During Summer Fallow in the Great Plains," *Soil Sci. Soc. Am. Proc.* 31:556–559 (1967).
51. Greb, B. W., D. E. Smika and A. L. Black. "Water Conservation with Stubble Mulch Fallow," *J. Soil Water Conserv.* 25:58–62 (1970).
52. Greb, B. W. "Water Conservation: Central Great Plains," in *Dryland Agriculture*, H. E. Dregne, and W. O. Willis (Eds.) (Madison, WI: American Society Agronomy, 1983), pp. 57–70.
53. Greb, B. W. *Snowfall and Its Potential Management in the Semiarid Central Great Plains, ARM-W-18* (Washington, DC: U.S. Dept. of Agriculture, 1980), pp. 1–46.
54. Greb, B. W., and R. L. Zimdahl. "Ecofallow Comes of Age in the Central Great Plains," *J. Soil Water Conserv.* 35:230–233 (1980).
55. Smika, D. E. "Fallow Management Practices for Wheat Production in the Central Great Plains," *Agron. J.* 82:319–323 (1990).

56. Smika, D. E., and C. J. Whitfield. "Effect of Standing Wheat Stubble on Storage of Winter Precipitation," *J. Soil Water Conserv.* 21:138–141 (1966).
57. Tanaka, D. L. "Chemical and Stubble-Mulch Fallow Influences on Seasonal Soil Water Contents," *Soil Sci. Soc. Am. J.* 49:728–733 (1985).
58. Tanaka, D. L., and J. K. Aase. "Fallow Method Influences on Soil Water and Precipitation Storage Efficiency," *Soil Tillage Res.* 9:307–316 (1987).
59. Fenster, C. R., and G. A. Peterson. *Effects of No-Tillage Fallow as Compared to Conventional Tillage in a Wheat-Fallow System,* Res. Bull. 289 (Lincoln, NE: Agriculture Experimental Station, University of Nebraska, 1979).
60. Tanaka, D. L. "Wheat Residue Loss for Chemical and Stubble-Mulch Fallow," *Soil Sci. Soc. Am. J.* 50:434–440 (1986).
61. Deibert, E. J., E. French and B. Hoag. "Water Storage and Use by Spring Wheat Under Conventional Tillage and No-Till in Continuous and Alternate Crop-Fallow Systems in the Northern Great Plains," *J. Soil Water Conserv.* 41:53–58 (1986).
62. Bauer, A. "Evaluation of Fallow to Increase Water Storage for Dryland Wheat Production," *ND Agric. Exp. Stn. Farm Res.* 25(5):6–9 (1968).
63. Black, A. L., and A. Bauer. "Strategies for Storing and Conserving Soil Water in the Northern Great Plains," in *Challenges in Dryland Agriculture — A Global Perspective,* P. W. Unger, W. R. Jordon, T. V. Sneed and R. W. Jensen (Eds.) (College Station, TX: Texas Agriculture Experimental Station, 1988), pp. 137–139.
64. Black, A. L., and A. Bauer. "Sustainable Cropping Systems for the Northern Great Plains," in *Conservation Tillage, Great Plains Conservation Tillage Symposium,* Bull. 131 (Bismarck, ND: Great Plains Agricultural Council, 1990), pp. 15–22.
65. Peterson, G., and D. Westfall. "Sustainable Dryland Agroecosystems," in *Conservation Tillage, Great Plains Conservation Tillage Symposium,* Bull. 131 (Bismarck, ND: Great Plains Agricultural Council, 1990), pp. 23–30.
66. Beck, D., and R. Doerr. "Sustainable No-Till Crop Rotations," in *Conservation Tillage, Great Plains Conservation Tillage Symposium,* Bull. 131 (Bismarck, ND: Great Plains Agricultural Council, 1990), pp. 285–290.
67. Lal, R., E. Regnier, D. J. Eckert, W. M. Edwards and R. Hammond. "Expectations of Cover Crops for Sustainable Agriculture," in *Cover Crops for Clean Water*, W. L. Hargrove (Ed.) (Ankeny, IA: Soil and Water Conservation Society, 1991), pp. 1–10.
68. Gilley, J. E., J. F. Power, P. J. Reznicek and S. C. Finkner. "Surface Cover Provided by Selected Legumes," *Appl. Eng. Agric.* 5(3):379–385 (1989).
69. Power, J. F. (Ed.) *The Role of Legumes in Conservation Tillage Systems* (Ankeny, IA: Soil Conservation Society of America, 1987).
70. Power, J. F. "Green Manures in the Great Plains," in *Proc. Great Plains Soil Fertility Conference,* Vol. 3, J. L. Havlin, and J. S. Jacobsen (Ed.) (Manhattan, KS: Kansas State University, 1990), pp. 1–18.
71. Power, J. F., Doran, J. W. and Koerner, P. T. "Hairy Vetch as a Winter Cover Crop for Dryland Corn Production," *J. Prod. Agric.* 4:62–67 (1991).
72. Doran, J. W., and M. S. Smith. "Overview: Role of Cover Crops in Nitrogen Cycling," in *Cover Crops for Clean Water*, W. L. Hargrove (Ed.) (Ankeny, IA: Soil and Water Conservation Society, 1991), pp. 85–90.

73. Power, J. F., and Biederbeck, V. O. "Role of Cover Crops in Integrated Crop Production Systems," in *Cover Crops for Clean Water*, W. L. Hargrove (Ed.) (Ankeny, IA: Soil and Water Conservation Society, 1991), pp. 167–174.
74. Wicks, G. A. "Eco-Fallow: A Reduced Tillage System for the Great Plains," *Weeds Today* 7:20–23 (1976).
75. Sims, J. R., and A. E. Slinkard. "Development and Evaluation of Germplasm and Cultivars of Cover Crops, in *Cover Crops for Clean Water*, W. L. Hargrove (Ed.) (Ankeny, IA: Soil and Water Conservation Society, 1991), pp. 121–129.
76. Slinkard, A. E., V. O. Biederbeck, L. Bailey, P. Olson, W. Rice and L. Townley-Smith. "Annual Legumes as a Fallow Substitute in the Northern Great Plains of Canada," in *The Role of Legumes in Conservation Tillage Systems*, J. F. Power (Ed.) (Ankeny, IA: Soil Conservation Society of America, 1987), pp. 6–7.
77. Power, J. F. "Growth, N Accumulation and Water Use of Legume Cover Crop in a Semiarid Environment," *Soil Sci. Soc. Am. J.* 55:1659–1663 (1991).
78. Zachariassen, J. A., and Power, J. F. "Growth Rate and Water Use by Legumes Species at Three Soil Temperatures," *Agron. J.* 83:408–413 (1991).
79. Sims, J. R., S. Koala, R. L. Ditterline and L. E. Wiesner. "Registration of 'George' Black Medic," *Crop Sci.* 25:709–710 (1985).
80. Carter, D. L. "Soil Erosion on Irrigated Lands," in *Irrigation of Agricultural Lands*, B. A. Stewart, and D. R. Nielsen (Eds.) (Madison, WI: American Society Agronomy, 1990), pp. 1143–1172.
81. Gilley, J. R., L. N. Mielke and W. W. Wilhelm. "An Experimental Center-Pivot Irrigation System for Reduced Energy Crop Production Studies," *Trans. Am. Soc. Agric. Eng.* 26:1375–1379, 1385 (1983).
82. Gilley, J. E., E. R. Kottwitz and G. A. Wieman. "Roughness Coefficients for Selected Residue Materials," *J. Irrig. Drainage Eng. ASCS* 177:503–514 (1991).
83. Dickey, E. C., D. E. Eisenhauer and P. J. Jasa. "Tillage Influences on Erosion During Furrow Irrigation," *Trans. Am. Soc. Agric. Eng.* 27:1468–1474 (1984).
84. Berg, R. D., and D. L. Carter. "Furrow Erosion and Sediment Losses on Irrigated Cropland," *J. Soil Water Conserv.* 35:267–270 (1980).
85. Kemper, W. D., T. J. Trout, A. S. Humpherys and M. S. Bullock. "Mechanisms by which Surge Irrigation Reduces Furrow Infiltration Rates in a Silty Loam Soil," *Trans. Am. Soc. Agric. Eng.* 31:821–829 (1988).
86. Green, R. E., and S. W. Karickhoff. "Sorption Estimates for Modeling," in *Pesticides in the Soil Environment: Processes, Impact, and Modeling*, H. H. Cheng (Ed.) (Madison, WI: Soil Science Society of America, 1990), pp. 79–101.
87. Larson, W. E., and P. C. Robert. "Farming by Soil," in *Soil Management for Sustainability,* R. Lal, and F. J. Pierce (Eds.) (Ankeny, IA: Soil and Water Conservation Society, 1991), pp. 103–112.

CHAPTER 12

Constraints on Conservation Tillage under Dryland and Irrigated Agriculture in the United States Pacific Northwest

Robert E. Sojka and David L. Carter
U.S. Department of Agriculture; Kimberly, Idaho

TABLE OF CONTENTS

12.1. INTRODUCTION

The Pacific Northwest (PNW) supports significant areas of both irrigated and rainfed agriculture. This bimodality is also impacted by the diversity of crop and animal agriculture it supports. Drilled grain and pulse crops, row crops, vegetable and horticultural crops, grass sod, and perennial alfalfa (*Medicago sativa* L.) hay are among the choices that can appear in a farm's cropping system. Developing soil management practices and tillage systems to accommodate such diversity has been a challenge to soil conservationists. To date most research published from the region has concentrated on small grain production in the dryland areas. Another smaller body of literature has dealt with conservation tillage of irrigated field crops. The potential for development of conservation tillage in the PNW derives from the region-wide severity of erosion in both dryland and irrigated agriculture. Residue management has been the essential element of the tillage systems in both cases. Although preserving crop residues at the soil surface is a key strategy, conservation tillage in the PNW has embraced other practices as well. Furthermore, greater recognition of the extent and severity of erosion under irrigated conditions is warranted, and research on erosion and conservation tillage for irrigated systems should be a high priority.

Two areas not specifically covered in this chapter are wind erosion and plant pathology. Wind erosion is significantly abated by maintenance of soil vegetative cover, and many aspects of that strategy are dealt with at length in this chapter. Those with an interest in wind erosion in the PNW will find Vomocil and Ramig[1] a good, if somewhat dated, reference. The pathology of conservation tillage in the PNW is a voluminous topic. The aspects of straw management briefly covered herein address cultural principles relevant to pathology as well as the agronomic and soil issues focused on in this chapter. Those wanting more detail are referred to the excellent review by Cook.[2]

12.2. SOIL AND CLIMATIC CONSTRAINTS

Conservation tillage in the PNW has been greatly influenced by soil properties affecting and affected by structure and aggregation, and the climatic interactions with these physical properties. Generally, the PNW has medium-textured loessal soils containing some volcanic ash and little organic matter, with poor structure and few stable aggregates. Dryland production areas frequently utilize long uninterrupted steep slopes. Their cropping systems vary with annual precipitation (200 to 600 mm), but in general, soils under conventional tillage are worked bare in late summer following harvest to accommodate subsequent fallowing or planting of small grain or pulse crops.

Both water and wind erosion are serious problems. Erosion by water is consistently the greatest threat. Soil loss tolerances to water erosion vary from

2.2 to 11.2 t/ha, depending on soil depth and lithic contact. In addition, because of frequency of steep slopes "tillage erosion" (downslope displacement of soil by implement usage) has also been a serious component of upslope soil loss.[3,4] Annual soil loss on nonirrigated soils throughout the region ranges from 4 to 60 t/ha, of which 85% occurs during the winter,[5] compared to 20% for the 37 states east of the Rocky Mountains.[6]

Soil freezing seriously accelerates erosion in these soils by virtually eliminating infiltration and promoting runoff.[7-9] In model simulations using PNW meteorological records, the number of freeze-thaw cycles varied from 1 to 7/year, and averaged 3/year. Soils were frozen 51% of the time in December, 67% of the time in January, and 53% of the time in February.[10] Duration and depth of soil freezing was reduced by maintenance of surface residues, providing increased probability of infiltration from seasonal precipitation.[11] Coupling residue maintenance with chiseling or paraplowing* increased spring infiltration rates threefold over no-tillage alone.[12] An unconstrained soil matrix also maintains greater aggregate stability during freeze-thaw cycles.[13] Thus runoff, and hence erosion, can be minimized by reducing the duration of ice blockage of soil pores, and by increasing the proportion of macropores, which block less easily.

Poor residue coverage promotes soil freezing. Consequently, the potential amount of profile water storage also decreases, especially deep in the profile. This is exacerbated by nocturnal migration of water to the frost depth, where it is exposed to diurnal thawing and evaporation loss.[14-19] The number of diurnal freeze-thaw cycles from November through March varies from 60 to 120.[20] Over a winter season, evaporation losses can be significant. Residue maintenance reduces both the temperature gradients that drive this water movement, and the evaporative loss of soil water from the surface few centimeters of soil where water accumulates during transient diurnal frost episodes.

The effects of freezing and thawing on soil hydraulics couple with effects on soil structure to further aggravate erosion. On a macroscale, Formanek et al.[21] showed that a single freeze-thaw cycle reduced soil cohesive strength by more than half. Subsequent cycles had less effect. Similar strength reduction has been observed in the field,[22] attributed to the separation of aggregates by freezing and thawing.[23] Upon thawing, soil cohesion returns as a function of soil water tension, and throughout these episodes surface soil shear strength provides a reasonable index of erodibility.[21]

12.3. NONIRRIGATED CONSERVATION TILLAGE

PNW conservation tillage research from the 4 million ha of dry farmlands has been reviewed several times in the last 15 years.[3,24-29] These reviews

* Mention of trademarks, proprietary products, or vendors does not constitute a guarantee or warranty of the product by the U.S. Department of Agriculture, and does not imply its approval to the exclusion of other products or vendors that may also be suitable.

documented the fragility of the region's soil resource and identified technologies and strategies for reducing tillage and preserving residue. Allmaras[24] matched management systems to specific crops and environmental needs. The PNW has promoted conservation tillage more successfully than some regions because of the aggressive manner in which the technology was developed and spread in the Solutions to Environmental and Economic Problems (STEEP) program.[30,31]

12.3.1. Conservation Tillage Programs

The STEEP program, conceived in 1972 and funded since 1976, pooled the resources of Idaho, Oregon, and Washington to wage a coordinated assault on soil erosion. The program initially had five research objectives: (1) development of conservation tillage and plant management systems, (2) plant breeding to suit conservation tillage, (3) pest management for conservation tillage, (4) improved erosion and runoff prediction, and (5) evaluation of soil conservation economics and socioeconomics. In 1982 an extension program was initiated to augment the research program. STEEPs success resulted from a timely conjunction of several key factors. Producer groups were committed to program goals, were frequently consulted, and were involved in its priority setting and operation. The program employed multidisciplinary interaction among experiment stations, the Soil Conservation Service (SCS) the Agricultural Research Service, and the Extension Service components. All participants shared in federal funding, which was equally distributed among the three states. Research funds were allocated through a proposal-review system targeted at solving problems in order of priority and probability of success.

STEEP researched farmer and public attitudes and perceptions of erosion severity, program effectiveness, and needed priorities.[32,33] The specific insight of these surveys showed that the conservation ethic was a less effective motivation for adoption of conservation tillage than demonstration of economic benefit. In the early 1980s minimum tillage became recognized as a management practice that maximized net returns during an era of declining agricultural commodity prices. Farmers knowledgeable about soil erosion were found to be more likely to adopt control practices than uninformed farmers. If they perceived the problem existed on their farm they took conservation action, using available research and advice.

STEEP research results have been shared and discussed at annual meetings and promoted to users through publications, newsletters, slide sets, radio and television coverage, grower meetings, and demonstration plots. This technology transfer was accomplished by the intimate involvement of researchers, county extension agents, conservation district supervisors, and the SCS. The program has also benefitted from strong and coordinated administrative and technical leadership, and from participant commitment and *esprit de corps.*

12.3.2. Management Strategies

The most effective strategy for combatting erosion in PNW dryland systems has been conservation of crop residues at the soil surface through various systems of tillage reduction. The SCS has for many decades promoted stubble mulch farming to prevent soil loss and conserve water and soil organic matter.[34] The positive relationship between "topsoil" depth in the Palouse and the productivity of wheat (*Triticum aestivum* L.) over a range of soil organic matter contents was confirmed by Pawson et al.[35] The relationship of tillage, soil fertility, crop residue management, and organic matter for the region was recently reviewed in depth by Rasmussen and Collins.[29] They recognized the negative impact of excessive tillage and fallowing on the oxidative loss of organic matter for nitrogen mineralization. Long-term effects of conservation tillage on organic carbon and nitrogen in soil were summarized for 11 sites worldwide (Table 12.1).

Specific conservation practices vary widely to suit local needs. Allmaras et al.[26] concluded that on slopes less than 12%, tillage systems and residue management alone could significantly control erosion, but for inclines of 12 to 20% slope length also had to be interrupted through terracing. Their work suggested that for slopes greater than 20%, even combining these approaches would still result in soil loss above tolerance limits. Improved new approaches include slot mulching by placing compacted straw in trenches extending to below the frost layer. Performed on the contour and in conjunction with no-till or chemical fallow, these practices offer another method with which to improve infiltration and reduce runoff and erosion on steep slopes.[36]

Fallowing in the driest of the nonirrigated cropped areas of the PNW is a major contributor to erosion. Stubble left standing over the winter months can increase net soil water storage (SWS) by as much as 90 mm through better snow capture and prevention of soil freezing.[37] The effectiveness of this practice is enhanced by deep chiseling.[19,38] Where surface mulching is practiced and soil water retention is increased, deep chiseling also provides drainage to prevent saturation of surface soil, which can otherwise cause overwinter oxygen stress and denitrification.[39] In all but the most marginal situations, e.g., where shallow soils limit SWS capacity, these increases (especially if coupled with no-till cropping), and/or delayed spring tillage and early maturing varieties make annual cropping more economical than summer fallowing in most years.[28]

Managing previous crop residues significantly impacts conservation tillage success. Straw yields of PNW winter wheat are typically double the grain yield. This can amount to 10 to 15 Mg/ha from a well-managed crop. The once-prevalent practice of burning has been largely discredited and is discouraged both on agronomic merits and air quality considerations. Short-term weed, nutrient, and disease benefits have been shown to be less certain than the long-term reduction of soil organic matter and immediate impact on erosion.[40]

Table 12.1. Change in Soil Organic Carbon (C) and Nitrogen (N) Levels Resulting from Conservation Tillage as Compared to Conventional Tillage

Location and Soil	Annual Precipitation (mm)	Soil Depth (cm)	Length of Study (years)	Tillage System[a]	Increase (%/year) C	Increase (%/year) N	Ref.
South Africa							
Haploxeralf	412	10	10	TT	5.6	3.4	85
Haploxeralf	412	10	10	NT	7.3	5.1	85
Germany							
Podsol		30	5	NT	3.2	1.4	86
Podsol		30	5	NT	2.4	1.6	86
Podsol		30	6	NT	1.3	1.3	86
Australia							
Western							
Psamment	345	15	9	NT	1.6	—	87
Alfisol	307	15	9	NT	0.7	—	87
Alfisol	389	15	9	NT	1.4	—	87
Queensland							
Pellustert	698	10	6	NT	1.2	1.3	88
Saskatchewan, Canada							
Chernozem		15	6	NT	6.7	2.8	89
United States							
North Dakota							
Haploboroll	375	45	25	SM	1.8	1.3	90
Haploboroll	375	45	25	SM	–0.1	0.1	90
Argiboroll	375	45	25	SM	0.5	0.4	90
Kansas							
Haplustoll		15	11	NT	0.7	0.6	91
Nebraska							
Haplustoll	446	9	15	NT	2.8	2.4	92
	446	10	15	NT	1.2	1.0	92
Oregon							
Haploxeroll	416	15	44	SM	0.3	0.4	93
Washington							
Haploxeroll	560	5	10	NT	1.9	2.0	94
Mean					2.2	1.7	
Minimum					–0.1	0.1	
Maximum					7.3	5.1	

Source: From Rasmussen and Collins, *Adv. Agron.*, 45:101, 1991. With permission.

[a] TT, tine-till; NT, no-till; SM, stubble mulch.

Despite the benefits of straw retention, however, it must be managed. Straw kept upright for snow capture and soil protection should be laid down by planting time to assure radiation penetration into developing wheat canopies and for soil warming and maximum photosynthesis.[41]

Combining should generally cut only enough straw to ensure no escape of grain beneath the cutter bar, and may require minor modification of chaff spreaders to prevent an uneven distribution of chaff behind the combine.[42-44] Failure to abide by these precautions can result in greater soil freezing and erosion potential from uncovered areas, and greater disease potential in high chaff areas. Uneven straw decomposition and variable nutrient availability (complicating subsequent fertilization), and uneven implement performance in subsequent tillage and planting operations also result from uneven spreading of residues.

Successful conservation tillage requires development of a sound soil fertility program to meet yield goals and to accommodate changes in nutrient cycling and organic matter retention in the presence of high residue levels.[29,40] Conservation tillage changes both crop nutrient requirements and system dynamics affecting conservation tillage success. With stubble mulching, nitrogen additions can offset reduced mineralization,[45] but the practice can encourage grassy weed competition[46] and crop water use,[47] both of which can limit yield.

The general requirement of no-till drill openers-fertilizer banders was reviewed by Erbach et al.[48] A test of various designs was reported for PNW conditions by Wilkins et al.[49] They stated that the best emergence was produced with a deep furrow opener which placed seeds in contact with soil containing sufficient soil water to allow germination and emergence. Subsequent evaluations[50,51] have shown particular promise for a strip till seeder and for the New Zealand style Cross Slot™ opener (see Chapter 8).

Experience has shown that fertilizer can be optimally placed near wheat roots to favor wheat uptake, and to limit uptake by competing weeds.[52] This concept can be expanded to include twin-row planting of grain (one row on each side of the fertilizer band) to "hide" fertilizer from competing weeds between pairs of wheat rows.[53]

12.4. IRRIGATED CONSERVATION TILLAGE

The irrigated areas of the PNW are generally flatter, occur at lower elevation, and receive less precipitation than adjacent nonirrigated croplands. Many irrigated areas are in river valleys and the soils are commonly alluvial deposits along the floodplain. Over 3.2 million ha are irrigated in the PNW (Table 12.2). About 1.85 million ha are sprinkler irrigated, and about 1.35 million ha are surface irrigated.[54] The conversion from surface to sprinkler irrigation and the development of new sprinkler irrigated lands has taken place mostly during the past 30 years. Drilled field crops are produced on both irrigated and nonirrigated lands, but nearly all row crops and high value cash crops in the PNW are grown under irrigation. The number of different crops grown under irrigation is three or four times greater than in rainfed agriculture, resulting in more diverse and often greater amounts of residue to manage under irrigated agriculture. For

Table 12.2. Summary of Irrigated Farmland in the Pacific Northwest

	Irrigated Area (thousands of hectares)			
Irrigation Type	ID	OR	WA	Total
Center pivot	221	99	185	505
Other sprinkler	605	312	414	1331
Gravity	826	336	191	1353
Drip/trickle	<1	2	12	15
Crop types				
Small grains	646	89	101	836
Row crop/Vegetable	325	97	166	588
Hay/grass seed/pasture	575	494	417	1486
Tree and other horticulture	8	48	113	169

Source: Compiled from *Irrig. J.*, 41(1):23–34, 1991.

example, alfalfa is commonly grown in rotation with other crops on irrigated land in the PNW. This crop has an extensive, deep taproot system, and these roots perform much like buried residue when the alfalfa is killed to allow planting of the next crop in the rotation. The traditional approach has been to kill crowns with herbicides or sweep tillage or both, followed by discing and moldboard plowing to bury taproot residues. To accomplish what has been perceived as necessary for a satisfactory seedbed, an average of ten tillage operations has been used for row crops following alfalfa in rotation.[55]

Irrigation-induced erosion was first recognized as a problem in the 1940s.[56-60] Early research on the subject related slope and stream size to sediment loss, and early researchers warned irrigators against irrigating land that was too steep, cautioning them to use streams as small as possible. These warnings were largely unheeded until Public Law 92-500, the Water Quality Act of 1972, focused attention on the water pollution problems associated with irrigation runoff. Ironically, federal funds to combat erosion from irrigated farmland do not reflect the severity of the problem because conservation funds are distributed in relation to legally recognized estimates of erosion. In the past this has relied heavily on the universal soil loss equation (USLE). The USLE generates erosion estimates based on climate data (rainfall), and current legislation does not make allowances for adjustments of the production to take irrigation into account. Thus, soil loss from irrigated arid land (for conservation funding purposes) has been based on unrealistically low estimates of runoff and erosion.

Tillage for soil conservation on irrigated row crop culture may not always mean no-till, or even maintenance of residues on the soil surface. Subsoiling in furrow or sprinkler irrigation and basin or reservoir tillage under sprinkler irrigation are examples of tillage operations that may take place in otherwise conventional systems to improve infiltration, reduce runoff, and prevent soil erosion. Only in the past 8 to 10 years have no-tillage systems been introduced to irrigated land.[55,61,62]

12.4.1. Furrow Irrigation

From 5 to 50 tons of soil per hectare can be lost per year from typical surface irrigated fields, with three or more times that amount lost near furrow inlets.[63,64] Mech[65] reported the loss of 50.9 t/ha from a single 24-hr irrigation. Figure 12.1 illustrates the erosion that occurs near the inlet ends of furrows. Even on low sloping fields this type of hydraulic leveling proceeds rapidly enough to completely denude the topsoil from upper reaches of some fields in only a few decades.[62,66] Because many PNW soils are underlain with subsurface horizons rich in calcium carbonates, their exposure or mixing with surface soil results in reduced productivity.[66] This productivity loss cannot be restored except by returning topsoil to the denuded area.[67] Figure 12.2 illustrates how erosion, combined with plowing, has mixed white subsoil with surface soil, resulting in a lighter color on the inlet ends of fields.

In most published papers on irrigation-induced erosion, sediment loss from the lower end of the field is referred to as erosion. There must be erosion for sediment loss to occur, but there can be extensive erosion within a field without sediment loss from the field as a result of the deposition of sediment eroded from upper reaches of a furrow at the lower reaches of the furrow before being carried away with the runoff. Upper reach erosion with simultaneous lower reach depostion occurs because irrigation furrows serve as both conveyance channels and infiltrating surfaces for water to enter the soil. This supplies water to satisfy the infiltration needs for the crop over the entire furrow length.[62,67] Water flow rates at the upper reaches, therefore, are significantly greater than at the lower reaches of the furrow because of the cumulative downstream effect of infiltration. The size of the furrow stream required to overcome the cumulative stream size reduction resulting from infiltration is generally large enough to be erosive near the inlet ends of furrows, but the sediment may be deposited before the water reaches the outlet ends of the furrows. Hence, erosion can occur in the upper reaches of the field without sediment loss from the field.

Typical sediment losses for major crops grown with traditional tillage are presented in Table 12.3. These data were mean values from measurements made on more than 100 fields. Sediment loss values vary severalfold at the same slope; therefore, caution should be exercised when applying these data. Sediment loss can be reduced by a variety of approaches, including vegetative filters,[62] settling ponds,[68] minibasins, and buried pipe runoff control systems.[69,70]. Erosion and sediment loss can be reduced by field incorporation of residue and reduced tillage,[71-76] permanent furrow sodding,[77] no-tillage systems,[55,78] selection of furrow spacing, irrigation set duration and plant proximity to furrows,[79] zone-subsoiling beside furrows,[80,81] and with the introduction of flocculating polymers at dilute concentrations in the furrow streams.[82]

In conventional tillage, Sojka et al.[79] found that applying equal amounts of water in shorter duration irrigation sets by using narrower row spacings could improve infiltration and reduce erosion. This effect was both the systematic

Figure 12.1. Erosion near the inlet ends of irrigation furrows.

result of decreasing runoff time and growing plants in closer proximity to the irrigated furrows. This allowed plants to stabilize furrows with roots and vegetative debris (e.g., flowers shed by dry beans). Yield and quality of corn (*Zea mays* L.) and sugar beet (*Beta vulgaris* L.) were unaffected and yield of dry beans (*Phaseolus* spp.) was slightly improved by narrower rows. Another study[80,81] showed that zone subsoiling decreased runoff and erosion, increased infiltration and yield, and improved tuber grade in furrow-irrigated Russet Burbank potatoes (*Solanum tuberosum* L.) (Table 12.4). Subsoiling was under beds alongside furrows, using the Tye Paratill (i.e., paratilling) in otherwise conventional culture. Recent work has shown an almost complete reduction in both erosion and sediment loss for furrow-irrigated systems in which the water advance was treated with 5 to 10 ppm of polyacrylamide.[82]

No-tillage systems for furrow-irrigated land were developed and evaluated by Carter and Berg[55] and by Carter et al.[78] They showed that cereal or corn can be easily grown following alfalfa, corn following cereal, or corn and cereal following corn without tillage using the same furrows for irrigating the subsequent no-tillage crop as the original. Both erosion and sediment loss were greatly reduced and in many cases completely eliminated. These crops yielded as well and were of equal quality without tillage as with traditional tillage. Not only did no-tillage conserve soil by reducing erosion and sediment loss, but net income increased more than $125/ha each year over a 5-year cropping sequence[78] as a result of reduced tillage costs.

The recent work by Carter and Berg[55] and Carter et al.[78] demonstrated that conservation tillage can be successful on furrow-irrigated land and that it is

Figure 12.2. White upper ends of irrigated fields caused by loss of topsoil as a result of irrigation furrow erosion and subsoil mixing with topsoil by plowing.

currently the best approach for soil and water conservation on these lands. The data in Table 12.5 illustrate the effectiveness of conservation tillage in reducing sediment loss for dry bean and corn production. The primary difference between the traditional and conservation tillage treatments was burial of crop residues by moldboard plowing in traditional tillage treatments, whereas plowing was not done in any of the conservation tillage treatments.

Table 12.3. Estimated Sediment Losses (t/ha) from Fields of Different Crops Furrow Irrigated from Concrete-Lined Ditches with Siphon Tubes

Crop	Average Field Slope (%)			
	0.5–1	1–2	2–3	>3
Alfalfa	0.0	1.6	5.2	12.6
Cereal grain or pea	2.5	7.2	14.3	23.3
Dry bean or corn	5.6	19.5	41.2	62.8
Sugar beet	7.2	27.1	59.2	98.6

Source: From Carter, D. L., *Irrig. Agric. Lands Agron.*, 30:1148, 1990. With permission.

Note: Run length was 200 m.

Conservation tillage systems generally reduce erosion and sediment loss from 50 to nearly 100% compared to traditional tillage systems for row crop production.[62] The reader will note that the sediment losses in Table 12.5 for dry beans differ substantially from Table 12.3. Table 12.3 presents average sediment losses from nearly 100 fields. Table 12.5 uses data from a limited number of sites in which conservation tillage was compared. The higher erosion rates reflect the choice of these particular highly erosive sites to study conservation tillage.

12.4.2. Sprinkler Irrigation

Sprinkler-irrigated systems generally allow most of the same conservation tillage practices used in dryland farming,[83] particularly no-till, mulch-tillage, deep chiseling, or subsoiling. A major difference is the frequency and intensity of water application, particularly in center pivot systems. Even on relatively shallow slopes, the outer portions of most center pivots apply water at rates that may cause runoff and erosion. Many center pivot systems cover areas of highly variable slope. Therefore, it is almost impossible to design a system to adequately supply water to the growing crop over the entire irrigated area without causing runoff and attendant erosion on part of that area. Actually, more flexibility exists for residue management under sprinkler irrigation than with either rainfed or surface-irrigated areas. More residue can be tolerated on the soil surface under sprinkler irrigation than with surface irrigation because sprinklers apply water more evenly over the irrigated area, whereas excess residue can inhibit water flow with surface irrigation. The advantage of sprinkler irrigation over rainfed culture is the ability to apply water when needed, eliminating the need to sow with deep seeding drills or other specialized drills that still may not work as well with high crop residues. Furthermore, with irrigation, plant emergence is less likely to be restricted on PNW fields if planted shallow.

The erosion processes under sprinkler irrigation, although similar to those under rainfall, do exhibit some differences. For example, once streams begin to flow in sprinkler-irrigated areas, they increase in size as runoff water

Table 12.4. Summary of Water Infiltration and Cumulative Sediment Loss for 1989 and 1990 as Affected by Zone Subsoiling

	Traffic Furrows			Nontraffic Furrows			All Furrows		
Treatment	Sediment Loss (kg/ha)	Infiltration (mm)	Sediment Infiltration[a] (kg/mm/ha)	Sediment Loss (kg/ha)	Infiltration (mm)	Sediment Infiltration[a] (kg/mm/ha)	Sediment Loss (kg/ha)	Infiltration (mm)	Sediment Infiltration[a] (kg/mm/ha)
					1989				
Mean −ZS	1154	281	15.0	315	439	2.31	1469	721	6.90
Mean +ZS	871	306	9.5	297	426	2.27	1168	732	5.24
					1990				
Mean −ZS	8450	254	41.2	976	377	3.21	9428	631	18.52
Mean +ZS	2604	321	10.3	771	398	2.37	3388	718	5.84

Source: Adapted from Sojka et al., *Soil Tillage Res.*, 25:358, 1993. With permission.

[a] Infiltration occurs only during sediment monitoring.

Table 12.5. Sediment Losses from Experimental Plots Where Traditional and Conservation Tillage Treatments were Compared on the Same Fields under Furrow Irrigation

Crop and Previous Crop	Slope (%)	Sediment Loss (t/ha) Traditional Tillage	Conservation Tillage
Dry beans following wheat	1.3	114	29.4
Dry beans following wheat	3.3	11.0	2.7
Dry beans following wheat	0.6	30.3	14.1
Sweet corn following alfalfa	1.1	11.0	5.8
Silage corn following wheat	0.6	12.6	0.4
Silage corn following corn	1.4	12.1	1.8

Source: From Carter and Berg, *J. Soil Water Conserv.*, 46:140, 1991.

increases. However, once the stream exits the sprinkled area, the stream size diminishes as infiltration removes water from the stream. This latter phase is similar to furrow irrigation erosion processes. Residues on the soil surface or various tillage practices reduce erosion and sediment loss with sprinkler irrigation in the same manner as under rainfall. Effects beyond the zone being sprinkled are the same as for furrow irrigation. In the latter case, more residue can be tolerated than with furrow irrigation, and the reduction of erosion is greater.[62]

An excellent tillage method for combating erosion and sediment loss under sprinkler irrigation is reservoir tillage. It is a process of making small catchment basins, 50 cm or less in length, 20 to 25 cm in width, and 15 to 20 cm deep (Figure 12.3). These small reservoirs trap runoff water when the sprinkler application rate exceeds the infiltration rate. The water is held until it infiltrates the soil, sometimes after irrigation has ceased. Kincaid et al.[84] have developed successful cropping systems using reservoir tillage under sprinkler irrigation. These systems are widely used for potato production and various other crops.

12.5. GENERAL SUMMARY AND CONCLUSIONS

Crop and water constraints on conservation tillage on dryland in the PNW are reasonably well understood because many years of research and technology transfer have identified problems and provided solutions or alternative approaches to most of them. The STEEP program has been particularly effective in encouraging the application of new conservation tillage technology to rainfed agriculture. In contrast, crop and water constraints to conservation tillage of irrigated land are less well understood. Most of the conservation tillage research on irrigated land is recent. Research results are promising, but the application of these results is just beginning. Conservation tillage can be highly successful for conserving water and soil and can increase net income through tillage cost savings. A rapid expansion of conservation tillage technology to irrigated land is encouraged because the conservation impact has great

Figure 12.3. Reservoir tillage basins in a potato field under sprinkler irrigation.

potential. Educational programs, conservation tillage demonstration projects, and incentive programs have been shown to be effective means of accelerating conservation practice implementation.

12.6. REFERENCES

1. Vomocil, J. A., and R. E. Ramig. "Wind Erosion Control on Irrigated Columbia Basin Land, a Handbook of Practices," Spec. Rep. 466 (Corvallis: Oregon State University Extension Service, 1976).
2. Cook, R. J. "Diseases Caused by Root-Infecting Pathogens in Dryland Agriculture," *Adv. Soil Sci.* 13:215–239 (1990).
3. Papendick, R. I., and D. E. Miller. "Conservation Tillage in the Pacific Northwest," *J. Soil Water Conserv.* 32:49–56 (1977).
4. Mech, S. J., and G. R. Free. "Movement of Soil During Tillage Operations," *Agric. Eng.* 23:379–382 (1942).
5. McCool, D. K., R. I. Papendick and F. L. Brooks. "The Universal Soil Loss Equation as Adapted to the Pacific Northwest," in *Proc. 3rd Federal Inter-Agency Sedimentation Conference* (Washington, DC: Water Resources Council, 1976), pp. 135–147.
6. Wischmeier, W. H. "Storms and Soil Conservation," *J. Soil Water Conserv.* 17:55–59 (1962).
7. Zuzel, J. F., R. R. Allmaras and R. Greenwalt. "Runoff and Soil Erosion on Frozen Soil in Northeastern Oregon," *J. Soil Water Conserv.* 37:351–354 (1982).
8. Zuzel, J. F., and J. L. Pikul. "Infiltration Into a Seasonally Frozen Agricultural Soil," *J. Soil Water Conserv.* 42:447–450 (1987).

9. Pikul, J. L., Jr., J. F. Zuzel and D. E. Wilkins. "Water Infiltration Into Frozen Soil: Field Measurements and Simulation," paper presented at National Symposium on Preferential Flow, December 16 to 17, Chicago (1991).
10. Zuzel, J. F., J. L. Pikul, Jr. and R. N. Greenwalt. "Point Probability Distributions of Frozen Soil," *J. Clim. Appl. Meteorol.* 25:1681–1686 (1986).
11. Pikul, J. L., Jr., J. F. Zuzel and R. N. Greenwalt. "Formation of Soil Frost as Influenced by Tillage and Residue Management," *J. Soil Water Conserv.* 41:196–199 (1986).
12. Pikul, J. L., Jr., J. F. Zuzel and R. E. Ramig. "Effect of Tillage-Induced Soil Macroporosity on Water Infiltration," *Soil Tillage Res.* 17:153–165 (1990).
13. Lehrsch, G. A., R. E. Sojka, D. L. Carter and P. M. Jolley. "Freezing Effects on Aggregate Stability Affected by Texture, Mineralogy, and Organic Matter," *Soil Sci. Soc. Am. J.* 55:1401–1406 (1991).
14. Cary, J. W., R. I. Papendick and G. S. Campbell. "Water and Salt Movement in Unsaturated Frozen Soil: Principles and Field Observation," *Soil Sci. Soc. Am. J.* 43:3–8 (1979).
15. Harlan, R. L. "Analysis of Coupled Heat Fluid Transport in Partially Frozen Soil," *Water Resource Res.* 9:1314–1323 (1973).
16. Fuchs, M., G. S. Campbell and R. I. Papendick. "An Analysis of Sensible and Latent Heat Flow in a Partially Frozen Unsaturated Soil," *Soil Sci. Soc. Am. J.* 42:379–385 (1978).
17. Pikul, J. L., Jr., and R. R. Allmaras. "Hydraulic Potential in Unfrozen Soil in Response to Diurnal Freezing and Thawing of the Soil Surface," *Trans. ASAE* 28:164–168 (1985).
18. Pikul, J. L. Jr., L. Boersma and R. W. Rickman. "Temperature and Water Profiles during Diurnal Soil Freezing and Thawing: Field Measurements and Simulation," *Soil Sci. Soc. Am. J.* 53:3–10 (1989).
19. Pikul, J. L., Jr., and J. F. Zuzel. "Heat and Water Flux in a Diurnally Freezing and Thawing Soil," in *Proc. Int. Symp. Frozen Soil Impacts on Agricultural, Range and Forest Lands* Spec. Rep. No. 90-1 (Spokane, WA: Cold Regions Research and Engineering Laboratory, 1990).
20. Hershfield, D. M. "The Frequency of Freeze-Thaw Cycles," *J. Appl. Meteorol.* 13:348–354 (1974).
21. Formanek, G. E., D. K. McCool and R. I. Papendick. "Freeze-Thaw and Consolidation Effects on Strength of a Wet Silt Loam," *Trans. ASAE* 27:1749–1752 (1984).
22. Bradford, J. M., and R. B. Grossman. "In-Situ Measurement of Near-Surface Soil Strength by the Fall-Cone Device," *Soil Sci. Soc. Am. J.* 46:685–688 (1982).
23. Chamberlain, E. J. 1981. *Frost Susceptibility of Soil,* CRREL Monogr. 81-2. (Hanover, NH: Cold Regions Research and Engineering Laboratory, 1981).
24. Allmaras, R. R. "Soil Conservation: Using Climate, Soils, Topography, and Adapted Crops Information to Select Conserving Practices," in *Dryland Agriculture,* H. E. Dregne and W. O. Willis (Eds.), Agron. Monogr. No. 23 (Madison, WI: American Society of Agronomy/Crop Science Society of America/Soil Science Society of America, 1983), pp. 139–153.
25. Allmaras, R. R., S. C. Gupta, J. L. Pikul, Jr. and C. E. Johnson. *Effects of Tillage and Crop Residue Removal on Erosion, Runoff, and Plant Nutrients,* Spec. Publ. 25 (Ankeny, IA: Soil Conservation Society of America, 1979), pp. 15–20.

26. Allmaras, R. R., S. C. Gupta, J. L. Pikul, Jr. and C. E. Johnson. "Soil Erosion by Water as Related to Management of Tillage and Surface Residues, Terracing, and Contouring," Western Ser. No. 10 (Washington, DC: U.S. Dept. of Agriculture, 1980).
27. Jennings, M. D., B. C. Miller, D. F. Bedzicek and D. Granatstein. "Sustainablilty of Dryland Cropping in the Palouse: An Historical View," *J. Soil Water Conserv.* 45:75–80 (1990).
28. Papendick, R. I., D. K. McCool and H. A. Krauss. "Soil Conservation: Pacific Northwest," in *Dryland Agriculture,* H. E. Dregne and W. O. Willis (Eds.), Agron. Monogr. No. 23, (Madison, WI: American Society of Agronomy/Crop Science Society of America/Soil Science Society of America, 1983), pp. 273–279.
29. Rasmussen, P. E., and H. P. Collins. "Long-Term Impacts of Tillage, Fertilizer, and Crop Residue on Soil Organic Matter in Temperate Semiarid Regions," *Adv. Agron.* 45:93–134 (1991).
30. McDole, R. E., and S. A. Reinertsen. "STEEP, and Interagency, Multidisciplinary Approach to Soil Conservation," *J. Soil Water Conserv.* 38:244–245 (1983).
31. Oldenstadt, D. L., R. E. Allan, G. W. Bruehl, D. A. Dillman, E. L. Michalson, R. I. Papendick and D. J. Rydrych. "Solutions to Environmental and Economic Problems (STEEP)," *Science* 217:904–909 (1992).
32. Boersma, L., R. G. Mason and D. Faulkenberry. "Soil Erosion in Oregon Volume I: Perceptions and Attitudes of Farmers and Non-Farmers," Spec. Rep. 764 (Corvallis: Oregon State University, 1986).
33. Boersma, L., R. G. Mason and D. Faulkenberry. "Use of Minimum Tillage in Oregon," Spec. Rep. 807. (Corvallis: Oregon State University, 1987).
34. Duley, F. L. "Stubble-Mulch Farming to Hold Soil and Water," Farmer's Bull. 1997. (Washington, DC: U.S. Government Printing Office, 1948).
35. Pawson, W. W., O. L. Brough, J. P. Swanson and G. M. Horner. "Economics of Cropping Systems and Soil Conservation in the Palouse," Tech. Bull. No. 2, Agric. Exp. Stn., Idaho, Oregon, and Washington and ARS (Washington, DC: U.S. Dept. of Agriculture, 1961).
36. Saxton, K. E., D. K. McCool and R. I. Papendick. "Slot Mulch for Runoff and Erosion Control," *J. Soil Water Conserv.* 36:44–47 (1981).
37. Massee, T. W., and H. McKay. "Improving Dryland Wheat Production in Eastern Idaho with Tillage and Cropping Methods," Idaho Agric. Exp. Stn. Bull. No. 581 (1979).
38. Massee, T. W., and F. H. Siddoway. "Fall Chiseling for Annual Cropping of Spring Wheat in the Intermountain Dryland Region," *Agron. J.* 61:177–182 (1969).
39. Rickman, R. W., and B. L. Klepper. "Wet Season Aeration Problems beneath Surface Mulches in Dryland Winter Wheat Production," *Agron. J.* 72:733–736 (1980).
40. Rasmussen, P. E., and C. R. Rohde. "Stubble Burning Effects on Winter Wheat Yield and Nitrogen Utilization under Semiarid Conditions," *Agron. J.* 80:940–942 (1988).
41. Wilkins, D. E., B. L. Klepper and P. E. Rasmussen. "Management of Grain Stubble for Conservation-Tillage Systems," *Soil Tillage Res.* 12:25–35 (1988).

42. Allmaras, R. R., C. L. Douglas, Jr., P. E. Rasmussen and L. L. Baarstad. "Distribution of Small Grain Residue Produced by Combines," *Agron. J.* 77:730–734 (1985).
43. Douglas, C. L., Jr., R. R. Allmaras, P. E. Rasmussen, R. E. Ramig and N. C. Roager, Jr. "Wheat Straw Composition and Placement Effects on Decomposition in Dryland Agriculture of the Pacific Northwest," *Agron. J.* 44:833–837 (1980).
44. Douglas, C. L., Jr., P. E. Rasmussen and R. R. Allmaras. "Cutting Height, Yield Level, and Equipment Modification Effects on Residue Distribution by Combines," *Trans. ASAE* 32:1258–1262 (1989).
45. Elliott, L. F., V. L. Cochran and R. I. Papendick. "Wheat Residue and Nitrogen Placement Effects on Wheat Grown in the Greenhouse," *Soil Sci.* 131:48–52 (1981).
46. Gill, G. S., and W. M. Blacklow. "Effect of Great Brome (*Bromus diandrus* Roth.) on the Growth of Wheat and Great Brome and Their Uptake of Nitrogen and Phosphorous," *Aust. J. Agric. Res.* 35:1–8 (1984).
47. Ramig, R. E., R. R. Allmaras and C. M. Smith. "Nitrogen-Sulfur Relations in Soft White Winter Wheat. I. Yield Response to Fertilizer and Residual Sulfur," *Agron. J.* 67:219–224, 1975.
48. Erbach, D. C., J. E. Morrison and D. E. Wilkins. "Equipment Modification and Innovation for Conservation Tillage," *J. Soil Water Conserv.* 38:182–185 (1983).
49. Wilkins, D. E., G. A. Muilenburg, R. R. Allmaras and E. C. Johnson. "Grain-Drill Opener Effects on Wheat Emergence," *Trans. ASAE* 26:651–655 (1983).
50. Wilkins, D. E., F. E. Bolton and K. E. Saxton. "Evaluating Seeders for Conservation Tillage Production of Peas," ASAE Pap. No. 90-1532, (St. Joseph, MI: ASAE, 1990).
51. Baker, C. J., and K. E. Saxton. "The Cross-Slot Conservation-Tillage Grain Drill Opener," ASAE Pap. 88-1568 (St. Joseph, MI: ASAE, 1988).
52. Klepper, B. L., P. E. Rasmussen and R. W. Rickman. "Fertilizer Placement for Cereal Root Access," *J. Soil Water Conserv.* 38:250–252 (1983).
53. Veseth, R., R. McDole, C. Engle and J. Vomocil. "Fertilizer Band Location for Cereal Root Access," Pacific Northwest Extension Publication. ID, WA, OR. PNW283 (1986).
54. "1990 Irrigation Survey," *Irrig. J.* 41(1):23–34 (1991).
55. Carter, D. L. and R. D. Berg. "Crop Sequences and Conservation Tillage to Control Irrigation Furrow Erosion and Increase Farmer Income," *J. Soil Water Conserv.* 46:139–142 (1991).
56. Gardner, W., and C. W. Lauritzen. "Erosion as a Function of the Size of the Irrigating Stream and the Slope of the Eroding Surface," *Soil Sci.* 62:233–242 (1946).
57. Gardner, W., J. H. Gardner and C. W. Lauritzen. "Rainfall and Irrigation in Relation to Soil Erosion," *Utah Agric. Exp. Stn. Bull.* 326 (1946).
58. Israelson, O. W., G. D. Clyde and C. W. Lauritzen. "Soil Erosion in Small Furrows," *Utah Agric. Exp. Stn. Bull.* 320 (1946).
59. Mech, S. J. "Effect of Slope and Length of Run on Erosion under Irrigation," *Agric. Eng.* 30:379–383 (1949).
60. Taylor, C. A. "Transportation of Soil in Irrigation Furrows," *Agric. Eng.* 21:307–309 (1940).

61. Carter, D. L., C. E. Brockway and K. K. Tanji. "Controlling Erosion and Sediment Loss from Irrigated Cropland," *Am. Soc. Civ. Eng., Irrig. Drain. Div.* (in press).
62. Carter, D. L. "Soil Erosion on Irrigated Lands," *Irrig. Agric. Lands Agron.* 30:1143–1171 (1990).
63. Berg, R. D., and D. L. Carter. "Furrow Erosion and Sediment Losses on Irrigated Cropland," *J. Soil Water Conserv.* 35:267–270 (1980).
64. Kemper, W. D., T. J. Trout, M. J. Brown and R. C. Rosenau. "Furrow Erosion and Soil Management," *Trans. ASAE* 28:1564–1572 (1985).
65. Mech, S. J. "Soil Erosion and Its Control under Furrow Irrigation in the Arid West," Info. Bull. 184, (Washington, DC: USDA-ARS, 1959).
66. Carter, D. L., R. D. Berg and B. J. Sanders. "The Effect of Furrow Irrigation Erosion on Crop Productivity," *Soil Sci. Soc. Am. J.* 49:207–211 (1985).
67. Carter, D. L. "Furrow Irrigation Erosion Lowers Soil Productivity," *Am. Soc. Civ. Eng., Irrig. Drain. Div.* (in press).
68. Brown, M. J., J. A. Bondurant and C. E. Brockway. "Ponding Surface Drainage Water for Sediment and Phosphorous Removal," *Trans. ASAE* 24:1478–1481 (1981).
69. Carter, D. L., and R. D. Berg. "A Buried Pipe System for Controlling Erosion and Sediment Loss on Irrigated Land," *Soil Sci. Soc. Am. J.* 47:749–752 (1983).
70. Carter, D. L., and R. D. Berg. "A Buried Drain Erosion and Sediment Loss Control System," *Univ. Idaho Curr. Info. Ser.* No. 760 (1985).
71. Miller, D. E., and J. S. Aarstad. "Furrow Infiltration Rates as Affected by Incorporation of Straw or Furrow Cultivation," *Soil Sci. Soc. Am. Proc.* 35:492–495 (1971).
72. Aarstad, J. S., and D. E. Miller. "Effects of Small Amounts of Residue on Furrow Erosion," *Soil Sci. Soc. Am. J.* 45:116–118 (1981).
73. Berg, R. D. "Straw Residue to Control Furrow Erosion on Sloping, Irrigated Cropland," *J. Soil Water Conserv.* 39:58–60 (1984).
74. Brown, M. J. 1985. "Effect of Grain Straw and Furrow Irrigation Stream Size on Soil Erosion and Infiltration," *J. Soil Water Conserv.* 40:389–391 (1985).
75. Evans, R. G., J. S. Aarstad, D. E. Miller and M. W. Kroeger. "Crop Residue Effects on Surge Furrow Irrigation Hydraulics," *Trans. ASAE* 30:424–429 (1987).
76. Miller, D. E., J. S. Aarstad and R. G. Evans. "Control of Furrow Erosion with Crop Residues and Surge Flow Irrigation," *Agron. J.* 51:421–425 (1987).
77. Cary, J. W. "Irrigating Row Crops from Sod Furrows to Reduce Soil Erosion," *Soil Sci. Soc. Am. J.* 50:1299–1302 (1986).
78. Carter, D. L., R. D. Berg and B. J. Sanders. "Producing No-Till Cereal or Corn Following Alfalfa on Furrow-Irrigated Land," *J. Prod. Agric.* 4:174–179 (1991).
79. Sojka, R. E., M. J. Brown and E. C. Kennedy-Ketcheson. "Reducing Erosion from Surface Irrigation using Furrow Spacing and Plant Position," *Agron. J.* 84:668–675 (1992).
80. Sojka, R. E., E. C. Kennedy-Ketcheson, M. J. Brown and D. T. Westermann. "Erosion and Infiltration of Furrow Irrigated Potato Fields as Affected by Zone Subsoiling," in *Proc. International Erosion Control Assoc. Conf.* (Steamboat Springs, CO: International Erosion Control Association, 1991) pp. 301–315.
81. Sojka, R. E., D. T. Westermann, M. J. Brown and B. D. Meek. "Zone Subsoiling Effects on Infiltration, Runoff, Erosion, and Furrow Irrigated Potatoes," *Soil Tillage Res.* 25:351–368 (1993).

82. Lentz, R. D., I. Shainberg, R. E. Sojka and D. L. Carter. "Preventing Irrigation Furrow Erosion with Small Application of Polymers," *Soil Sci. Soc. Am. J.* 56:1926–1932 (1992).
83. Miller, D. E., and J. S. Aarstad. "Preplant Tillage Effects on Irrigated Corn Grown on a Sandy Soil," *J. Prod. Agric.* 3:606–609 (1990).
84. Kincaid, D. C., I. McCann, J. R. Busch and M. Hasheminia. "Low Pressure Center Pivot Irrigation and Reservoir Tillage," in *Proc. 3rd National Irrigation Symposium — Visions of the Future* (St. Joseph, MI: American Society of Agricultural Engineers, 1990), pp. 54–60.
85. Agenbag, G. A., and P. C. J. Maree. "The Effect of Tillage on Soil Carbon, Nitrogen and Soil Strength of Simulated Surface Crusts in Two Cropping Systems for Wheat (*Triticum aestivum*)," *Soil Tillage Res.* 14:53–65 (1989).
86. Fleige, H., and K. Bauemer. "Effect of Zero Tillage on Organic Carbon and Total Nitrogen Content and Their Distribution in Different N-Factions in Loessal Soils", *Agroecosystems* 1:19–29 (1974).
87. White, P. F. "The Influence of Alternative Tillage Systems on the Distribution of Nutrients and Organic Carbon in some Common Western Australia Wheat Belt Soils," *Soil Water Manage. Conserv.* 28:95–116 (1990).
88. Saffinga, P. G., D. S. Powlson, P. C. Brookes and G. A. Thomas. "Influence of Sorghum Residues and Tillage on Soil Organic Matter and Soil Microbial Biomass in an Australian Vertisol," *Soil Biol. Biochem.* 21:759–765 (1989).
89. Campbell, C. A., V. O. Biederbeck, M. Schnitzer, F. Selles and R.P. Zenter. "Effect of 6 Years of Zero Tillage and N Fertilizer Management on Changes in Soil Quality of an Orthic Brown Chernozem in Southwest Saskatchewan," *Soil Tillage Res.* 14:39–52 (1989).
90. Bauer, A., and A. L. Black. "Soil Carbon, Nitrogen, and Bulk Density Comparisons in Two Cropland Tillage Systems after 25 Years and in Virgin Grassland," *Soil Sci. Soc. Am. J.* 45:1166–1170 (1981).
91. Havlin, J. L., D. E. Kissel, L. D. Maddux, M. M. Claasen and J. H. Long. "Crop Rotation and Tillage Effects on Soil Organic Carbon and Nitrogen," *Soil Sci. Soc. Am. J.* 54:448–452 (1990).
92. Doran, J. W. "Soil Microbial and Biochemical Changes Associated with Reduced Tillage," *Soil Sci. Soc. Am. J.* 44:765–771 (1980).
93. Rasmussen, P. E., and C. R. Rohde. "Long-Term Tillage and Nitrogen Fertilzation Effects on Organic Nitrogen and Carbon in a Semiarid Soil," *Soil Sci. Soc. Am. J.* 52:1114–1117 (1988).
94. Granatstein, D. M., D. F. Bezdicek, V. L. Cochran, L. F. Elliot and J. Hammel. "Long-Term Tillage and Rotation Effects on Soil Microbial Biomass, Carbon, and Nitrogen," *Biol. Fertil. Soils* 5:265–270 (1987).

CHAPTER 13

Tillage Systems for Soil and Water Conservation on the Canadian Prairie

Francis J. Larney and C. Wayne Lindwall
Agriculture Canada Research Station; Lethbridge, Alberta, Canada

R. César Izaurralde
University of Alberta; Edmonton, Alberta, Canada

Alan P. Moulin
Agriculture Canada Research Station; Melfort, Saskatchewan, Canada

TABLE OF CONTENTS

0-87371-571-3/94/$0.00+$.50

13.1. INTRODUCTION

Conservation tillage is a farming concept that has been constantly evolving world wide for more than 50 years. Its main objective is to provide protection against erosion mainly through the presence of surface residues. In the context of western Canada, soil erosion remains the dominant threat against the long-term sustainability of farming.[1-3] Erosion has an impact on long-term soil productivity via its effects on soil quality. Increased adoption of conservation tillage systems is seen by many as one of the few options to assure long-term sustainability and economic viability of farms across the Canadian prairie provinces.[4-6] The currently accepted view on conservation tillage design and use is that a whole systems approach is needed. In using such systems, one must consider elements such as crops to be grown, soil type, equipment, fertilizer, herbicides, and interactions like soil × tillage and tillage × genotype. The objective of this chapter is to provide an overview of conservation tillage systems as they are practiced or developed under Canadian prairie conditions. Specific emphasis is placed on the constraints that prevent massive adoption of these systems as well as on the environmental implications of their use.

13.2. DESCRIPTION OF AGROECOSYSTEM

13.2.1. Soil Types

Soils of the Canadian prairie provinces range in type from Brown to Dark Brown Chernozemic in southern Saskatchewan and Alberta, to Black Chernozemic in Manitoba, Dark Gray Chernozemic in the central prairies, and Gray Luvisolic in the north (Figure 13.1). The native vegetation of the dry Brown and Dark Brown soils consists mainly of xerophytic and mesophytic grasses and forbs. The Black soils occur in the fescue prairie-aspen grove (parkland) and true prairie grasslands, whereas the Dark Gray soils are located in the transitional areas between grasslands and forests. The Gray soils developed under mixed deciduous and evergreen forests, on highly basic, mineral parent materials in subhumid to humid, mild to very cold climates.

The Brown soil zone occupies the south-central region of the prairies. The average depth of the Ap horizon is 12 cm and the organic matter content of the surface 30 cm is about 2%. About 80% of these soils are glaciated and are of medium loam texture. Topography varies from nearly level to very hilly. In this soil zone, moisture deficit is the major drawback to sustained agricultural

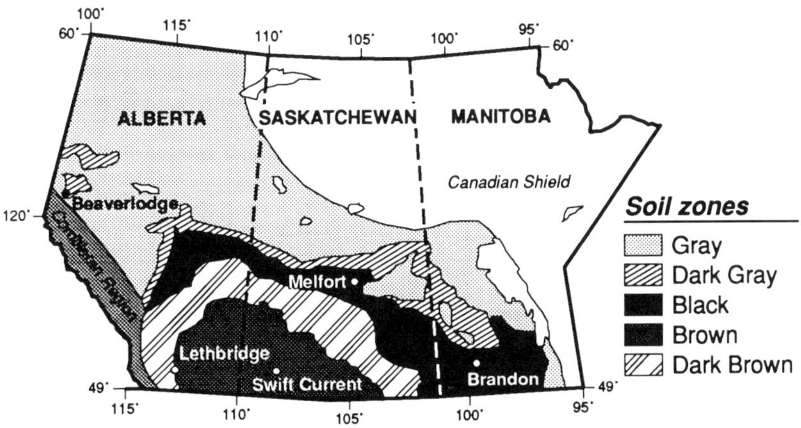

Figure 13.1. Generalized soil map of the Canadian prairie provinces.

production. The so-called Palliser Triangle comprises a major portion of the Brown soil zone of the Canadian prairies. It was deemed unfit for agricultural settlement by surveyor Capt. John Palliser in the 1850s.

In the Dark Brown soil zone, the depth of the Ap horizon is about 18 cm. Organic matter content of the surface 30 cm is about 4%. The lower moisture deficit in this zone as compared to the Brown soil zone allows greater flexibility in the choice of crop rotations.

In the Black soil zone the depth of the surface horizon averages 20 to 25 cm and the soil contains about 7% organic matter in the surface 30 cm. Because precipitation is higher in this zone, moisture deficits are less of a problem in cropping systems.

The Dark Gray and Gray Luvisolic forest soils occur north of the Black soil zone and include the Peace River region of Alberta and British Columbia. The soils generally possess a thin layer (about 5 cm) of dark-colored humus overlying a gray colored B horizon. The organic matter content of this grayish layer is generally low. With sufficient application of chemical fertilizers these soils have good productivity. Their major limitation is a short growing season because of their northern location.

13.2.2. Climatic Factors

Throughout the arable prairies the warmest month is July and the coldest is January. June has the greatest precipitation and July the highest potential evaporation. Differences between the warmest and coldest months vary between 29 and 39°C. Mean annual temperatures are generally higher in the Brown and Dark Brown soil zones than in the Black and Gray zones. About 50% of the annual precipitation falls between May and September, with about 30% falling as snow in the winter months. Snow provides a source of soil water, insulation for crops such as winter wheat (*Triticum aestivum* L.), and protection from wind erosion. Potential evaporation generally exceeds precipitation from April to September, resulting in a moisture deficit. In general, annual precipitation increases from <350 mm in the Brown soil zone to >475 mm in parts of the Black and Gray soil zones; however, potential evaporation decreases from the Brown to the Gray zones. Hence, annual moisture deficits decrease from about 400 mm in the Brown zone to negligible in the Gray zone.

Lethbridge, in southern Alberta, has an average of 1689 growing degree days (GDD; >5°C, 1951 to 1980) while Beaverlodge, in the Peace River area of northwestern Alberta, has 1221. Brandon, Manitoba, close to the eastern edge of the prairies, is slightly higher with 1705 GDD. Annual frost-free days on the prairies varies from about 93 in the Melfort area of north-central Saskatchewan to 117 in southern Alberta and the Swift Current area of southwestern Saskatchewan.

The chinook phenomenon is confined to southern Alberta and southwestern Saskatchewan. Chinook winds bring mild Pacific air over the Rocky Mountains in the winter period. Warmer temperatures cause rapid snowmelt and as the surface soil is generally frozen, infiltration is poor. Instead of contributing to moisture recharge, snowmelt becomes runoff. Chinooks also dry the soil surface and erosion with wind speeds gusting to 130 km/hr may occur. In addition they increase the number of freeze-thaw cycles in the winter period, which breaks down soil aggregates, leaving them more susceptible to wind erosion.

13.2.3. Cropping Systems

Cultivation of the Canadian prairie is a relatively recent phenomenon, beginning in the late 19th century. The uncertainty in the amount and frequency distribution of precipitation on the Canadian prairies historically has limited the choice of options used in rotations to drought-hardy spring wheat and summer fallow.[7] Fallow or summer fallow is a system in which the land is not cropped and weeds are controlled with tillage or herbicides for a 20-month period (September through April). Average fallowing frequency ranges from once every 2 years in the Brown soil zone to once every 4 years in the Black zone, in direct relation to moisture availability. Summer fallowing is a means of reducing the risk of poor yields and minimizing input costs in the drier regions and many producers are reluctant to move toward continuous cropping. By controlling weed growth and increasing moisture reserves (although it conserves only 20 to 30% of the precipitation received) summer fallowing reduces the dependency for optimum yields on seasonal rainfall.[8] Thus, year to year variation in yield is much lower when crops are grown on fallow rather than on stubble. Plowless summer fallow, whereby crop residue is kept on the soil surface as a protection against evaporation and wind erosion, has been practiced in western Canada since 1915.[9] This technique is also called trash cover farming or stubble mulching.

Although spring wheat is the major crop, barley (*Hordeum vulgare* L.), canola (*Brassica campestris* L. and *B. napus* L.), and flax (*Linum usitatissimum* L.) occupy significant proportions of the seeded area in the Brown and especially the Dark Brown soil zones. Mixed farming is prevalent in the Black, Dark Gray Chernozeimic, and Gray Luvisolic zones, and rotations there include a high proportion of forage legumes and grasses for pasture and hay. Irrigation is utilized on approximately 500,000 ha of land in southern Alberta where the two main crops are soft white spring wheat and alfalfa (*Medicago sativa* L.). Irrigation is also practiced in parts of southern Saskatchewan and the Red River Valley of Manitoba. Crops grown under irrigation include sugar beet (*Beta vulgaris* L.), potatoes (*Solanum tuberosum* L.), corn (*Zea mays* L.), field peas (*Pisum sativum* L.), beans (*Phaseolus* spp.), and other specialty crops. In British Columbia, only about 4% of the total land area is suitable for agriculture. Most of this is in the Lower Fraser Valley, with a smaller area in the Dawson Creek region of northeastern British Columbia which has agricultural practices that are similar to those of the Peace River area of northwestern Alberta.

Of approximately 30 million ha of cultivated lands in the three prairie provinces, about 50% is located in the Black and Gray zones, 27% in the Dark Brown zone and 23% in the Brown zone.[7] Fallow land peaked in the early 1970s at 11.5 million ha, but decreased to about 8.5 million ha in 1991. From 1957 to 1984 the area of cultivated land increased by 5 million ha. Of this increase, 80% resulted from the breaking of new land and the cultivation of pasture. Most of the increase (60%) has occurred in the Black and Gray zones on the northern edge of the parkland.

13.3. SOIL AND CLIMATIC CONSTRAINTS FOR CONSERVATION TILLAGE

There are few soil constraints for the adoption of conservation systems in western Canada. No-till has been successfully adopted on a wide range of soil textures (sands to clays), provided crop residues are managed appropriately. Excess precipitation in spring, coupled with poor drainage, has been a constraint in the adoption of conservation tillage, particularly no-till, on some heavier clay soils in more humid locations of North America. In western Canada, naturally poorly drained soils are generally not under cultivation and above normal precipitation in spring time is looked upon as beneficial rather than constraining.

Soil physical property measurements are useful for interpreting and explaining yield differences between rotation, tillage, and seed drill treatments.[10-13]

However, because the so-called conventional tillage systems in southern Alberta (wide-blade cultivator, heavy-duty cultivator) would be considered minimum tillage in more humid climates because of the shallow depth of tillage (8 to 12 cm) and the lack of soil inversion, differences in most physical properties between conventionally tilled (wide-blade) soil and no-till soil are not great.[14-16] Differences in soil physical properties are reported when comparing conventional tillage (moldboard plowing to 20 to 24 cm) with no-till in studies in more humid climates. Soil physical conditions (e.g., degree of packing) in the narrow seedbed zone of conservation tillage systems are probably more important in terms of their effect on seed germination, emergence, and establishment, than with conventional tillage systems. If plant establishment is poor or nonuniform then yield potential of the crop is reduced from the outset.

Climatic constraints in the adoption of conservation tillage in western Canada are related to soil temperature effects. No-tilled soil with its extra crop residue, as compared with conventionally tilled soil (autumn and spring tillage), may be slower to warm up in spring and crop emergence may be somewhat reduced. To some extent, this can be compensated for by shallower seeding with no-till systems. Climate may be more constraining for conventional tillage than conservation tillage in the western Canadian scenario. For example, with no-till, stubble standing over the winter period has a greater capacity to trap snow and increase soil moisture reserves than stubble flattened or buried by conventional autumn tillage.[17] Also, winter wheat seeded into standing stubble in a no-till system has a better chance of winter survival because of the snow-trapping capability of the stubble. The trapped snow insulates the crop and protects against low temperatures and potential winterkill.

In more humid regions of North America, the adoption of conservation tillage is often hampered by excess residue generated by optimum soil moisture conditions. Excess or windrowed crop residue may cause blockage of tillage implements and seeding equipment. Minimum tillage, with one preseeding tillage operation, may be used to improve seedbed conditions under these constraints. However, in semiarid regions such as the western Canadian prairie, excess crop residue is rarely a problem except in wet growing seasons and under irrigation. In these cases, it can be baled and removed for livestock without a detrimental effect on long-term organic matter levels.

13.4. CRITICAL FACTORS RELATED TO ADOPTION OF CONSERVATION TILLAGE

13.4.1. Conventional Summer Fallow

The use of summer fallow on the Canadian prairies is often advanced as the number one cause of various types of soil degradation. Excessive use of summer fallow is associated with reduced soil organic matter levels, soil erosion, dryland salinity, deterioration in soil tilth, loss of nitrogen and less efficient crop use of available water. Strip cropping, whereby adjacent strips are alternately cropped and fallowed, was introduced mainly for erosion control in the 1920s. However, in recent decades, the width of summer fallow strips has been increased or strip cropping has been abandoned altogether to accommodate increased machinery size, a trend that has increased the area exposed to wind erosion. Fallow moisture conservation efficiency is usually quite low. With no-till it averages about 50%, while conventional "black" fallow (no residue conservation) averages about 20%; i.e., only 20% of the moisture received during the fallow period is actually stored in the soil for subsequent crop use. Many producers derive short-term advantages from summer fallow. The main reasons for its popularity are (1) to accumulate soil moisture during the 20 months the land lies fallow; (2) to accelerate decomposition of crop residue and soil organic matter, thereby increasing available nutrients; (3) to control weeds, insects, and soil-borne diseases; and (4) to produce a more stable income (reduced risk of crop failure) compared to other rotations, continuous cropping, or flexcropping. A major review of historical rotation data from the prairies concluded that summer fallow remains a legitimate option in the cropping systems of western Canada, though its role and recommended use vary with edaphic and climatic factors.[7] In the Black and Gray soil zones, where moisture deficits are relatively small, summer fallowing can only be justified for the control of unmanageable pests or in the event of potential drought. In Brown and Dark Brown soils, where moisture stress is the primary yield-limiting factor, the replenishment of soil moisture during summer fallow reduces economic risks and warrants the inclusion of some summer fallow in crop rotations.

The goal of research in fallow frequency is to increase soil water storage (SWS) and to decrease the soil's vulnerability to wind erosion.[18] Although frequent inclusion of summer fallow enhances soil degradation, this effect can be minimized by using conservation tillage techniques.[19,20] Soil degradation by erosion also may be reduced by the use of partial fallows such as green manure plowdowns or by reducing the frequency of fallow in the rotation.[21-23] Economic analyses indicate that the optimum fallowing frequency is about once in 3 years, although that value varies depending on soil and economic variables.[24]

The replacement of some or all tillage operations with chemicals (chemical fallow) is also an alternative to conventional summer fallow.[18,25-27] In southern Alberta, an economic feasibility study of substituting herbicides for mechanical tillage showed that substantial savings in resources (labor, fuel, oil, machinery repairs) could be achieved by using herbicides to replace some or all of the tillage operations.[28] A possible limitation to chemical fallow is the perceived harmful effects on the environment over the longterm. However, there were no deleterious effects on soil biota after 40 years of 2,4-D use in Swift Current, Saskatchewan.[29] In Lethbridge, Alberta, 2,4-D and glyphosate applied at ten times the field rate caused no detectable effects on measured soil biological factors.[30]

13.4.2. Benefits of Stubble Management

In the semiarid regions of the Canadian prairies, water availability is the main factor limiting plant growth.[31-34] Here, unlike humid areas, increased moisture conservation is probably on a par with erosion control as a reason for development of conservation tillage systems.[35,36] An integral part of the development of conservation tillage systems is residue management.[37,38] In addition to increasing SWS, good crop residue management also reduces erosion risk. Maintaining adequate crop residue levels can become a major concern in semiarid climates where periodic droughts are common.[39] Differences in surface residue incorporation by various tillage implements have been documented.[40,41]

In a drought year, straw yields as well as crop grain yields are depressed. This means less residue is available for moisture conservation and erosion control. If the following year is also dry, the problem is exacerbated. For most soils on the Canadian prairies, the critical level of crop residues needed to maintain an adequate level of organic matter, soil fertility, and soil physical properties is not well defined. Speaking generally, as crop residue levels decline, annual wind erosion-related soil losses increase.

Soil water conservation under conservation tillage systems has been studied.[42-45] Maintenance of straw mulches in no- and minimum tillage generally increases water stored in the soil profile when compared to that with conventional tillage (Table 13.1). Other practices aimed at improving moisture conservation include stubble sculpting at harvest to improve snow-trapping, and fall subsoiling to improve infiltration of snowmelt in spring.[17,46,47]

13.4.3. Conservation Cropping Systems

Spring wheat is the most common crop in western Canada and has been included in several long-term rotation studies.[48-52] Fallow-spring wheat-spring wheat generally provided the highest economic returns as compared to fallow-spring wheat and continuous spring wheat. It also resulted in reduced losses of organic matter, and contributed less to soil degradation; yield variability and risk was greater than fallow-spring wheat, however. Continuous spring wheat, even when properly fertilized, provided the lowest economic return, but it was superior from a soil quality perspective. The conservation benefits of continuous cropping was also supported by soil organic carbon concentrations for a continuous spring wheat rotation, which were 19% higher than those for a fallow-spring wheat rotation at Lethbridge, Alberta.[53] Appreciable declines in aggregate stability with increasing frequency of fallow in spring wheat rotations have also been documented.[54,55]

Winter wheat, though it often outyields spring wheat (by up to 20%), has not been favored by producers because of its lower market value, lower protein content, and susceptibility to winterkill.[56] Introduction of winter wheat rather than spring wheat into a rotation can give some measure of erosion control by providing adequate cover in autumn and early spring when soils are particularly

Table 13.1. Effects of Tillage Systems on Total Spring Water (cm) under Stubble Cropping Conditions at Indian Head, Saskatchewan

Tillage System	Soil Depth (cm) 0–30[a]	30–60	60–120	0–60[a]	0–120[b]	% Full Profile
No-Till	11.3 a	10.8 a	20.3 a	22.1 a	42.4 a	88
Minimum	11.4 a	10.3 a	19.9 a	21.7 a	41.6 a	88
Conventional	10.5 b	10.1 a	19.6 a	20.6 b	40.2 b	84

Source: From Lafond, G. P. et al, *Can. J. Plant Sci.,* 72, 109, 1992. With permission.

Note: Values are averaged over a 4-year rotation (1987–1990) of peas, spring wheat, winter wheat, spring wheat. Means followed by the same letters are not significantly different from each other.

[a] Significant at the 1% level.
[b] Significant at the 5% level.

prone to wind erosion. If seeded at the appropriate time (late August to early September) and in the appropriate manner to ensure early spring growth, it is better suited to take advantage of seasonal precipitation distribution than is spring wheat. These beneficial effects can be enhanced with no-till, which maintains more available water in the root zone and conserves more crop residue on the soil surface.[18] Also, the benefits to soil quality should be increased, because switching from spring wheat to winter wheat reduces the fallow phase from 20 to 13 months.

Despite these apparent advantages, winter wheat accounted for less than 6% of the total area of wheat grown in the province of Alberta in the mid-1980s.[57] Prior to the 1970s, winter wheat was limited to the chinook areas of southern Alberta, but with the introduction of no-till "stubbling-in" techniques the potential areas of production have grown to include most of western Canada.[58,59] However, the general adoption of winter wheat in rotation on the Canadian prairies has been hindered by severe winterkill in 1984 and 1985, disease, and 3 years with severe drought during the preanthesis stage.[60]

A recent review of crop rotation studies on the Canadian prairies concluded that winter wheat has not been used for a sufficient time in rotations to allow proper assessment of its potential.[7] The introduction of crops such as canola and flax into rotations also warrants investigation. These crops would replace the fallow phase of a rotation, thus controlling erosion, and may use less water than continuous cropping to cereals. In a 12-year study in the dark brown soil zone of west-central Saskatchewan, rotations that included canola and flax showed good potential to provide consistently high net returns.[61] There has been a recent renewed interest in incorporating annual legume crops such as faba beans (*Vicia faba* L.), lentils (*Lens culinaris* Medic.), and field peas into cropping systems as a way to increase the biological diversity of agroecosystems and consequently augment their resilience. The research work has concentrated on the use of legumes as partial fallow replacements, cereal-legume intercrops, adaptation of legumes to no-till, and on the functioning of annual legumes in short-term rotations under no-till and low-input systems.[21-23,62-64]

It is difficult to compare annual crop yields in conservation and conventional tillage systems for western Canada due to climatic variability. Conventional tillage may outyield no-till in some years, but the reverse may be true in others. If tillage systems are compared on an annual basis or over a short period, crop yields may not be higher in minimum and no-till than conventional tillage in all years. Although reducing or eliminating fallow tillage may save 5 to 8% more soil water prior to planting (the equivalent of 10 to 15 mm of soil water), the benefit of no-till relative to conventional fallow is a function of growing season precipitation.[65] If rainfall received during the growing season is adequate and timely for crop growth little or no difference will be seen between yields under no-till, minimum, or conventional tillage.

The effects of conservation vs conventional tillage have been compared at locations throughout the Canadian prairies.[59,65-69] Long-term studies in the black soil zone of Manitoba showed that yields under no-till were higher than under conventional tillage (Table 13.2). No-till and conventional tillage under various cropping systems were compared in north-central Alberta in the 1980s (Table 13.3). A 6-crop-year study conducted in 1987 and 1988 comparing conventional and no-till production of barley, barley-field pea intercrop, and faba bean showed no yield difference for the cropping systems studied, and thus suggested the feasibility of growing these crops without tillage.[63] A Brown soil in southwestern Saskatchewan showed a less favorable response to no-till in some years (Table 13.4). On Black soils at Melfort, Saskatchewan, yields of cereals (barley, wheat), oilseeds (canola, flax), and pulses (faba bean, lentil, and field pea) were higher under no-till than conventional tillage (Table 13.5). Other tillage comparisons for noncereal crops such as canola, field pea, and sugar beet have been made,[71-73] and the economics of various tillage practices have been reported.[74-76] Results of conservation tillage and soil management research are often reported mainly in terms of effects on crop yield with less emphasis on effects on soil chemical and physical properties both over the short and long term. The effects of conservation tillage on soil quality parameters (organic matter, microbial biomass) has been documented.[77-80] The general conclusion from all these studies, whether on yield, economics, or soil properties, is that conservation tillage is a feasible and viable alternative for crop production in western Canada.

The acceptance of conservation tillage has depended on the development of herbicides for suitable weed control. Weed populations are the result of many factors including crop, climate, tillage, and herbicide history. With minimum tillage, especially no-till (by definition), cultivation for weed control is not always an alternative. This places increased demand or dependence on herbicides to provide consistent broad-spectrum weed control, both preplanting and in-crop or during the fallow period.[81] Changes in weed populations accompany changes in tillage management systems and often influence the adoption of conservation tillage techniques.[82] Most reports suggest that annual grass weeds will increase under minimum or no-till. In Manitoba, an increase in wild oats (*Avena fatua* L.) under no-till was reported.[83] At Lethbridge, no-till resulted in

Table 13.2. Yield Comparisons Between No-Till and Conventional Tillage from Trials Conducted on Black Chernozemic Soils Located at the University of Manitoba

Crop	Year	Station Years	Yield (kg/ha) No-Till	Yield (kg/ha) Conventional Till	Difference (%)
Wheat	1969–1989	31	2580 a[a]	2450 b	5.2
Barley	1969–1989	25	4010 a	3910 a	2.6
Canola	1969–1989	23	1590 a	1400 b	14.0
Flax	1969–1971	8	790 a	730 a	8.6

Source: Lafond, G. P. et al, *Crop Management for Conservation, Proc. Soil Conserv. Symp. (Yorkton)* (Saskatchewan, SK: Division of Extension, University of Saskatchewan, 1990). With permission.

[a] Means followed by the same letter are not significantly different from each other ($p = 0.05$).

Table 13.3. Comparison of No-Till vs Conventional Tillage under Various Cropping Systems, Soil Types, and Fertilizer Practices in North-Central Alberta

Soil	Cropping System	Straw Treatment	Fertilizer Nitrogen (kg/ha)	Crop Yield (Mg/ha) Conventional Tillage	Crop Yield (Mg/ha) No-Till
Orthic Black Chernozemic[a]	Continuous barley	Removed	0	2.13	1.23
		Left	0	1.88	0.76
		Removed	112	3.33	3.24
		Left	112	3.61	3.04
Orthic Gray Luvisol[a]	Continuous barley	Removed	0	2.04	1.20
		Left	0	1.65	1.48
		Removed	112	3.72	3.34
		Left	112	3.72	3.75
Orthic Black Chernozemic[b]	Barley	Left	70	3.73	3.36
	Barley–pea intercrop	Left	70	3.96	3.81
	Faba bean	Left	0	3.12	2.87
Orthic Black Chernozemic[c]	Continuous barley	Left	0	4.64[d]	5.12
	Continuous barley	Left	50	5.92	6.22
	Barley after faba bean	Left	0	4.41	6.05
	Barley after barley–pea intercrop	Left	0	4.36	5.34
	Barley within intercrop	Left	0	2.51	3.18

[a] Nyborg and Mahli.[67]
[b] Izaurralde et al.[63]
[c] Choudhary et al.[64]
[d] Yields are above-ground dry matter.

Table 13.4. Yield Comparisons Between No-Till and Conventional Tillage from Trials Conducted for a Wheat-Fallow Rotation on a Brown Chernozemic Silt Loam Soil Located in Southwestern Saskatchewan

Year	Yield (kg/ha)		
	No-Till	Minimum Tillage	Conventional Tillage
1982	2852 a[a]	2930 a	2979 a
1983	2448 ab	2362 b	2585 a
1984	1248 b	1468 a	1558 a
1985	1293 a	878 b	949 b
1986	2925 ab	3024[b]	3141 a
1987	2447 a	2167 b	2424 a
1988	1661 a	1434 b	1447 b

Source: From Lafond, G. P. et al, *Crop Management for Conservation, Proc. Soil Conserv. Symp. (Yorkton)* (Saskatchewan, SK: Division of Extension, University of Saskatchewan, 1990). With pemission.

[a] Means within years followed by the same letters are not significantly different from each other ($p = 0.01$).
[b] Data excluded from analysis as no-till carried out on minimum tillage fallow treatment the previous year.

Table 13.5. Yield (kg/ha) of Cereals, Oilseeds, and Pulses Sown into Cereal Stubble Left Standing Over the Winter on Black Chernozemic Soils Located at Melfort, Saskatchewan

Crop	Spring Tillage	No-Till
Cereals and pulses[a]		
Barley	3220	3390
Wheat	2040	2280
Faba bean	2120	2260
Lentil	1200	1510
Field Pea	1890	2200
Cereals and oilseeds[b]		
Barley	3140	3540
Wheat	2010	2210
Canola	1830	2070
Flax	1300	1390

Source: From Wright, T., *Crop Management for Conservation, Proc. Soil Conserv. Symp. (Yorkton)* (Saskatchewan, SK: Division of Extension, University of Saskatchewan, 1990). With permission.

[a] Mean of 12 station years.
[b] Mean of 10 station years.

fewer annual broadleaf weeds and more annual grass weeds such as wild oats and green foxtail (*Setaria viridis* (L.) Beauv.) in the following crop as compared to conventional fallow.[18] Nearly all workers agree that perennial weeds increase in severity when the intensity of tillage is reduced. Canada thistle

(*Cirsium arvense* (L.) Scop.), quackgrass (*Elytrigia repens* (L.) Nevski), foxtail barley (*Hordeum jubatum* L.), and wild rose (*Rosa acicularis* Lindl.) are the species most often mentioned in reports from western Canada.[84]

The effectiveness of rotations in controlling crop disease has been documented. However, little information is available with regard to the effect of conservation tillage on crop disease.[85] A study at Lethbridge reported the influence of minimum tillage on the severity of common root rot in wheat.[86] The risk of root rot was often less with minimum tillage because of shallower seeding and less moisture stress.

13.4.4. Seeding Technology

Another critical factor in the adoption of conservation tillage systems is the use of proper seeding equipment.[87,88] With the introduction of conservation tillage systems that leave large amounts of residue on the soil surface, seed drill design needed to be modified to prevent mechanical blockage but still place the seed into moisture, with reduced or no seedbed preparation.[89-92] In addition to the general features to consider when buying seeding equipment (cost effectiveness, horsepower requirements, ease of maintenance, etc.) are the following factors: the ability to cut through and clear heavy trash; even penetration in hard soil conditions; good depth control on openers to provide proper seed placement and good seed-soil contact; adequate packing system; and minimum stubble knock-down. Traditionally, farmers have used preseeding tillage in spring to obtain a firm, moist seedbed and provide some weed control. In addition, extra tillage operations may be applied in autumn to incorporate herbicides and/or reduce crop residues. However, if a change is made to conservation tillage the benefits of preseeding tillage are called into question. Excessive or unnecessary preseeding tillage operations can result in the loss of soil moisture, resulting in a dry, cloddy seedbed and subsequent poor stand. This is especially true when recropping stubble land, when deeper planting is required to reach soil moisture. Seedbed preparation on summer fallow land usually has little effect on crop yield.

The development of new seeding technology (high-clearance hoe drills, no-till drills, air seeders) in the 1980s provided drills that could effectively penetrate heavy crop residue and more compact soils to place seed at the appropriate depth in moist soil. However, despite the availability of this equipment, tradition and misconceptions about the need for cultivation still limits the widespread adoption of direct seeding or no-till.[93] Farmers are reluctant to drop proven practices. Greater economic and management inputs and perceived risks associated with maintaining surface crop residues also limit adoption. Just as there is no one simple recipe for the best tillage system, there is no simple answer to the question of what is the best minimum tillage seeder, as no single machine will meet the needs of all farmers. Long-term trials at Lethbridge have shown that differences among drills averaged over several years are small, but depending on the particular year and seedbed conditions, one drill may have an advantage over another.[26,87] In dry conditions, hoe drills provided the most consistent performance and highest yields. In years when moisture conditions were more favorable and crop residue levels were not excessive, disc drills proved superior. They cause less soil disturbance and this usually results in less stimulation of weed growth and volunteer grain. Some disc drills may be susceptible to plugging when crop residues are high. Unfortunately, the seed drill is often blamed for problems of inadequate placement or poor plant stands when the actual problem may have been related to poor trash distribution or inadequate weed control.

Fertilizer placement is also important in conservation tillage systems. Field studies of soil nitrogen transformations and crop growth under conventional and no-till systems showed that conserving both seedbed moisture and total soil water had beneficial effects on winter wheat yield at Lethbridge and Vauxhall, Alberta.[94] However, restricted nitrogen availability resulting from fertilizer nitrogen immobilization reduced grain nitrogen concentration in no-till relative to conventional tillage. The study also suggested that fertilizer nitrogen should be placed deeper in the soil, below the straw mulch layer, to minimize nitrogen immobilization. A recently completed 5-year study documented the relative merits of fertilizer banding with various commercial and prototype seed drills for no-till.[95] The optimum design criteria, application parameters, and potential benefits for point injection of liquid nitrogen fertilizer for winter wheat have been established.[96-99] Timing and placement of fertilizer for no-till winter wheat production has been studied in Saskatchewan and Manitoba.[100-103]

Nitrogen immobilization and loss under conservation cropping systems, particularly under chemical fallow, has been suggested as a possible factor limiting the maximization of wheat yields associated with these systems, and hence their more general adoption.[104] Providing additional nitrogen in the form of ammonium nitrate to chemical fallow plots increased wheat yield by up to 37% at Lethbridge.[105]

13.5. ENVIRONMENTAL ISSUES IN CONSERVATION TILLAGE ADOPTION

13.5.1. Use of Agrochemicals

Environmental issues have become headline news in recent years, particularly with respect to chemicals in the environment.[106,107] Several recent public surveys have indicated that the majority of Canadians are very concerned about the impact that agricultural practices have on water quality and more specifically about the increased dependence on pesticides.[108] Public perceptions can have a greater impact on government policy than scientific fact, and this is particularly true for issues related to the impact of pesticides on food safety and water quality.

The widespread adoption of minimum or no-till systems may be limited because of the apparent increased dependence on pesticides, particularly herbicides, with these systems. Some believe that the potential negative environmental effects associated with pesticides used in minimum tillage systems may be more serious than the sustainability concerns associated with traditional tillage systems. That is to say that the side effects of the potential cure for soil degradation may be more damaging than the problem itself. For this reason it is vital that the effects and fate of applied pesticides be monitored and well documented to ensure that recommended soil conservation practices are environmentally sustainable and that food safety is not compromised.

Since 1986, the Alberta Department of Environment, in cooperation with the Food Production and Inspection Directorate of Agriculture Canada, has been monitoring pesticide residues in surface and groundwater.[109] Samples taken from each of the six major rivers in Alberta were analyzed for all of the major pesticides used in agriculture and in all cases none were found at the detection limit of 1 ppb. Similarly, since 1988 more than 300 samples taken from farm wells throughout the province revealed no evidence of pesticide levels that exceeded the acceptable concentration in water permitted by Health and Welfare Canada. Contrary to some of the findings in the U.S. corn belt and other regions where intensive management practices are used, these findings suggest that Alberta groundwaters are relatively free of pesticides.

Farmers are looking for less expensive chemical inputs for crop production and some are abandoning chemical-intensive agriculture. For example, legumes in crop rotations can substitute for synthetic fertilizer nitrogen and animal manures also serve as nutrient sources. Many farmers, both organic and nonorganic, now use these inputs to maintain soil productivity and prevent soil degradation. However, the benefits of manure to maintain soil productivity is dependent on its availability and haulage costs.[110]

13.5.2. Soil Erosion

Agricultural practices that contribute to soil erosion (both water and wind) have substantial negative effects on air and water quality and contribute to environmental degradation.[106,111,112] The costs of soil erosion are not only directly felt by the producer but also by society in general.[113,114] Management of crop residue is a critical factor in erosion control systems for western Canada. In the Brown soil zone excessive tillage reduces residues below levels required to control erosion. Many coarse- and fine-textured soils are in danger of eroding with conventional fallow, particularly if wheat yields are low. For example, a wheat yield of 1750 kg/ha produces approximately 2600 kg/ha of crop residue, which results in a soil cover of 1750, 1300, and 725 kg/ha for no-till, minimum, and conventionally tilled fallow. Cover of flat crop residue required to control wind erosion is 1000, 1500, and 2000 kg/ha for loamy, clay, and sandy soils, respectively. Consequently, fine and coarse soils are susceptible to erosion under minimum and conventional fallow tillage.[65]

Crop yields and residue are higher in the Dark Brown, Black, Dark Gray, and Gray soils due to higher precipitation and lower potential evapotranspiration. Under continuous cropping with no-till or minimum tillage, crop residue levels, except those following a crop failure, are adequate to control erosion. However, residue cover may not be adequate to control erosion in fields conventionally fallowed after low-residue crops such as canola.

Wind erosion on irrigated land is a problem in southern Alberta. Soils are often pulverized into erodible aggregates in preparation for seeding of sugar beets, potatoes, and pulse crops. These crops in turn produce small amounts of residue, as compared with cereals, and this may exacerbate a potential problem.

13.6. GENERAL SUMMARY AND CONCLUSIONS

The principles of soil conservation have been well established through many years of applied research and successful adoption in western Canada. Numerous documents and symposia have focused on the extent and severity of the region's soil degradation problems. The main reason for the adoption of conservation tillage practices such as no-till is to reverse or at least slow down the various soil degradation processes and to reduce input costs, rather than to increase yields in the short term.[115,116] This has been a major factor in the reluctance of some producers to adopt conservation practices in the short term. However, interest in conservation tillage has increased dramatically in recent years as a result of major research and extension efforts. The 1991 Census of Agriculture reported that 33% of all land seeded to annual crops on the Canadian prairies was under conservation tillage or no-till.[117] Conventional practices were utilized the least in Saskatchewan (64%), followed by Manitoba (66%), Alberta (73%), and the Peace River region of British Columbia (83%). Use of no-till was highest in Saskatchewan (10%) and lowest in Alberta (3%). However, many real or perceived constraints to more widespread adoption of conservation tillage systems remain. These constraints will continue to be overcome, aided by interdisciplinary research and coordinated education and technology transfer efforts.

13.7. REFERENCES

1. Acton, D. F. "Concepts and Criteria of Soil Quality in the Context of Western Canada," in *Soil Quality in the Canadian Context — 1988 Discussion Papers*, Tech. Bull. 1991-1E, LRRC Contr. No. 89-12, S. P. Mathur, and C. Wang (Eds.) (Ottawa: Research Branch, Agriculture Canada, 1991), pp. 2–4.
2. Dumanski, J. "Towards a Soil Conservation Strategy for Canada," in *In Search of Soil Conservation Strategies in Canada,* D.W. Anderson (Ed.) (Ottawa: Agricultural Institute of Canada, 1987), pp. 1–10.

3. Dumanski, J., D. R. Coote, G. Luciuk, and C. Lok. "Soil Conservation in Canada," *J. Soil Water Conserv.* 41:204–210 (1986).
4. Nowland, J. L. "Canadian Concerns for Soil Quality," in *In Search of Soil Conservation Strategies in Canada,* D.W. Anderson (Ed.) (Ottawa: Agricultural Institute of Canada, 1987), pp. 11–22.
5. Van Kooten, G. C., and W. H. Furtan. "A Review of Issues Pertaining to Soil Degradation in Canada," *Can. J. Agric. Econ.* 35:33–54 (1987).
6. Van Kooten, G. C., and W. H. Furtan. "Is Soil Worth Saving? The Economics and Ethics of Soil Deterioration," in *In Search of Conservation Strategies in Canada,* D.W. Anderson (Ed.) (Ottawa: Agriculture Institute of Canada, 1987), pp. 67–82.
7. Campbell, C. A., R. P. Zentner, H. H. Janzen and K. E. Bowren. "Crop Rotation Studies on the Canadian Prairies," Agric. Can. Publ. 1841/E (Ottawa: Agriculture Canada, 1990).
8. Baier, W. "An Agroclimatic Probability Study of the Economics of Fallow-Seeded and Continuous Spring Wheat in Southern Saskatchewan," *Agric. Meteorol.* 9:305–321 (1972).
9. Hopkins, E. S., A. E. Palmer and W. S. Chepil. "Soil Drifting Control in the Prairie Provinces," Dept. of Agric. Publ. 568, Farmer's Bull. 32., (Ottawa: Department of Agriculture, 1946).
10. Arshad, M. A., and J. L. Dobb. "Tillage Effects on Soil Physical Properties in the Peace River Region: Implications for Sustainable Agriculture," *Proc. 28th Annu. Alberta Soil Sci. Workshop,* (Lethbridge, AB: Faculty of Extension, University of Alberta, Edmonton, 1991), pp. 190–199.
11. Gauer, E., C. F. Shaykewich and E. H. Stobbe. "Soil Temperature and Soil Water under Zero Tillage in Manitoba," *Can. J. Soil Sci.* 62:311–325 (1982).
12. Grant, R. F., R. C. Izaurralde and D. S. Chanasyk. "Soil Temperature under Conventional and Minimum Tillage: Simulation and Experimental Verification," *Can. J. Soil Sci.* 70:289–304 (1990).
13. Mahli, S. S., and P. A. O'Sullivan. "Soil Temperature, Moisture and Penetrometer Resistance under Zero and Conventional Tillage in Central Alberta," *Soil Tillage Res.* 17:167–172 (1990).
14. Chang, C., and C. W. Lindwall. "Effect of Long-Term Minimum Tillage Practices on Some Physical Properties of a Chernozemic Clay Loam," *Can. J. Soil Sci.* 69:443–449 (1989).
15. Chang, C., and C. W. Lindwall. "Comparison of the Effect of Long-Term Tillage and Crop Rotation on Physical Properties of a Soil," *Can. Agric. Eng.* 32:53–55 (1990).
16. Chang C., and C. W. Lindwall. "Effects of Tillage and Crop Rotation on Physical Properties of a Loam Soil," *Soil Tillage Res.* 22:383–389 (1992).
17. Maulé, C. P., and D. S. Chanasyk. "The Effects of Tillage upon Snow Cover and Spring Soil Water," *Can. Agric. Eng.* 32:25–31 (1990).
18. Lindwall, C. W., and D. T. Anderson. "Agronomic Evaluation of Minimum Tillage Systems for Summer Fallow in Southern Alberta," *Can. J. Plant Sci.* 61:247–253 (1981).
19. Anderson, D. T. "Surface Trash Conservation with Tillage Machines," *Can. J. Soil Sci.* 41:99–114 (1961).

20. Johnson, W. E. "Conservation Tillage in Western Canada," *J. Soil Water Conserv.* 32:61–65 (1977).
21. Biederbeck V. O., and Looman, J. "Growth of Annual Legumes for Soil Conservation under Severe Drought Conditions," in *Conservation for the Future, Proc. Soils Crops Workshop* (Saskatoon: University of Saskatchewan, 1985), pp. 298–311.
22. Biederbeck, V. O., and A. E. Slinkard. "Effect of Annual Legume Green Manures on Yield and Quality of Wheat on a Brown Loam," in *Proc. Soils Crop Workshop,* (Saskatoon: University of Saskatchewan, 1988), pp. 345–361.
23. Slinkard, A. E., and V. O. Biederbeck. "Indian Head Lentil as a Fallow Substitute in the Dark Brown Soil Zone," in *Proc. Soils and Crops Workshop,* (Saskatoon: University of Saskatchewan, 1987), pp. 216–221.
24. Zentner, R. P., D. W. Campbell, C. A. Campbell and D. W. Read. "Energy Considerations of Crop Rotations in Southwestern Saskatchewan," *Can. Agric. Eng.* 26:25–29 (1984).
25. Anderson, C. H. "Comparison of Tillage and Chemical Summer Fallow in a Semi-Arid Region," *Can. J. Soil Sci.* 51:397–403 (1971).
26. Foster, R. K., and C. W. Lindwall. "Minimum Tillage and Wheat Production in Western Canada," in *Wheat Production in Canada — A Review, Proc. Canadian Wheat Production Symp.,* A. E. Slinkard, and D. B. Fowler, (Eds.) (Saskatoon: University of Saskatoon, 1986), pp. 354–366.
27. McConkey, B. G., R. P. Zentner and W. Nicholaichuk. "Comparison of a Spring Wheat-Mechanical Fallow Rotation with a Winter Wheat-Chemical Fallow Rotation in Southwestern Saskatchewan," in *Proc. Soils Crop Workshop* (Saskatoon: University of Saskatchewan, 1989), pp. 66–72.
28. Zentner, R. P., and C. W. Lindwall. "Economic Evaluation of Minimum Tillage Systems for Summer Fallow in Southern Alberta," *Can. J. Plant Sci.* 62:631–638 (1982).
29. Biederbeck, V. O., C. A. Campbell and A. E. Smith. "Effects of Long-Term 2,4-D Field Applications on Soil Biochemical Processes," *J. Environ. Qual.* 16:257–262 (1987).
30. Olson, B. M. "Effect of 2,4-D and Glyphosate on Some Soil Biological Properties under Chemical Fallow," *Research Highlights, 1989,* L. J. Sears (Ed.) (Lethbridge, AB: Agriculture Canada Research Station, 1990), pp. 40–41.
31. de Jong, R., and A. Bootsma. "Estimated Long-Term Soil Moisture Variability on the Canadian Prairies," *Can. J. Soil Sci.* 68:307–321 (1988).
32. de Jong, R., J. A. Shields and W. K. Sly. "Estimated Soil Water Reserves Applicable to a Wheat-Fallow Rotation for Generalized Soil Areas Mapped in Southern Saskatchewan," *Can. J. Soil Sci.* 64:667–680 (1984).
33. Lehane, J. J., and W. J. Staple. "Influence of Soil Texture, Depth of Soil Moisture Storage and Rainfall Distribution on Wheat Yields in Southwestern Saskatchewan," *Can. J. Soil Sci.* 45:207–219 (1965).
34. Johnson, W. E. "Soil Conservation: Canadian Prairies," in *Dryland Agriculture,* Agron. No. 23, H.E. Dregne, and W.O. Willis (Eds.) (Madison, WI: American Society of Agronomy, 1983) pp. 259–272.
35. Bowren, K. E., and R. D. Dryden. "Effect of Fall and Spring Treatment of Stubble Land on Yield of Wheat in the Black Soil Region of Manitoba and Saskatchewan," *Can. Agric. Eng.* 13:32–35 (1971).

36. Juma, N. G., and R. C. Izaurralde. "Soil Conservation Strategies using Conventional and Alternative Management Systems," in *Proc. 27th Annu. Alberta Soils Sci. Workshop* (Edmonton, AB: Faculty of Extension, University of Alberta, 1990), pp. 247–256.
37. McAndrew D. W., and P. Nelson. "Conservation Tillage Systems in Northeastern Alberta," in *Research Highlights, 1989* (Lacombe, AB: Agriculture Canada Research Station, 1989), pp. 45–46.
38. Nuttall, W. F., K. E. Bowren, and C. A. Campbell. "Crop Residue Management Practices and N and P Fertilizer Effects on Crop Response and on Some Physical and Chemical Properties of a Black Chernozem over 25 Years in a Continuous Wheat Rotation," *Can. J. Soil Sci.* 66:159–171 (1986).
39. Lal, R., and H. Steppuhn. "Minimizing Fall Tillage on the Canadian Prairies — A Review," *Can. Agric. Eng.* 22:101–106 (1980).
40. Anderson, D. T., and G. G. Russell. "Effects of Various Quantities of Straw Mulch on the Growth and Yield of Spring and Winter Wheat," *Can. J. Soil Sci.* 44:109–118 (1964).
41. Anderson, D. T., and A. D. Smith. "Seeding Mechanisms for Trash Cover Farming," *Can. Agric. Eng.* 8:33–36 (1966).
42. Lafond, G. P., H. Loeppky and D. A. Derksen. "The Effects of Tillage Systems and Crop Rotations on Soil Water Conservation, Seedling Establishment and Crop Yield," *Can. J. Plant Sci.* 72:103–115 (1992).
43. de Jong, E. "Water use Efficiency under Conservation Tillage Systems," in *Crop Management for Conservation, Proc. Soil Conserv. Symp. (Yorkton),* G. P. Lafond and D. B. Fowler (Eds.) (Saskatchewan, SK: Division of Extension, University of Saskatchewan, 1990), pp. 102–126.
44. Entz, M. H., and D. B. Fowler. "Critical Stress Periods Affecting Productivity of No-Till Winter Wheat in Western Canada," *Agron. J.* 80:987–992 (1988).
45. Grevers, M. C., J. A. Kirkland, E. de Jong and D. A. Rennie. "Soil Water Conservation under Zero- and Conventional Tillage Systems on the Canadian Prairies," *Soil Tillage Res.* 8:265–276 (1986).
46. Kirkland, K. J., and C. H. Keys. "The Effect of Snow Trapping and Cropping Sequence on Moisture Conservation and Utilization in West-Central Saskatchewan," *Can. J. Plant. Sci.* 61:241–246 (1981).
47. McConkey, B. G., Steppuhn, H. and Nicholaichuk, W. "Effects of Fall Subsoiling and Snow Management on Water Conservation and Continuous Spring Wheat Yields in Southwestern Saskatchewan," *Can. Agric. Eng.* 32:225–234 (1990).
48. Campbell, C. A., D. W. L. Read, R. P. Zentner, A. J. Leyshon and W. S. Ferguson. "First 12 Years of a Long-Term Crop Rotation Study in Southwestern Saskatchewan — Yields and Quality of Grain," *Can. J. Plant Sci.* 63:91–108 (1983).
49. Campbell, C. A., R. P. Zentner and H. Steppuhn. "Effect of Crop Rotations and Fertilizers on Moisture Conserved and Moisture use by Spring Wheat in Southwestern Saskatchewan," *Can. J. Soil Sci.* 67:457–472 (1987).
50. Campbell, C. A., R. P. Zentner and P. J. Johnson. "Effect of Crop Rotation and Fertilization on the Quantitative Relationship between Spring Wheat Yield and Moisture use in Southwestern Saskatchewan," *Can. J. Soil Sci.* 68:1–16 (1988).
51. Zentner, R. P., and C. A. Campbell. "First 18 Years of a Long-Term Crop Rotation Study in Southwestern Saskatchewan — Yields, Grain Protein and Economic Performance," *Can. J. Plant Sci.* 68:1–21 (1988).

52. Campbell, C. A., R. P. Zentner, J. F. Dormaar and R. P. Voroney. "Land Quality Trends and Wheat Production in Western Canada," in *Wheat Production in Canada — A Review, Proc. Canadian Wheat Production Symp.,* A. E. Slinkard and D. B. Fowler (Eds.) (Saskatoon, SK: University of Saskatoon, 1986), pp. 318–353.
53. Janzen, H. H. "Soil Organic Matter Characteristics After Long-Term Cropping to Various Spring Wheat Rotations," *Can. J. Soil Sci.* 67:845–856 (1987).
54. Dormaar, J. F. "Chemical Properties of Soil and Water Stable Aggregates after 67 Years of Cropping to Spring Wheat," *Plant Soil* 75:51–61 (1983).
55. Dormaar, J. F., and U. J. Pittman. "Decomposition of Organic Residues as Affected by Various Dryland Spring Wheat-Fallow Rotations," *Can. J. Soil Sci.* 60:97–106 (1980).
56. Fowler, D. B. "The Potential for Winter Wheat," in *Soil Erosion and Land Degradation, Proc. 2nd Annu. Western Provincial Conf., Rationalization of Water and Soil Research and Management* (Saskatoon: Saskatchewan Institute of Pedology, 1984), pp. 303–336.
57. *Census Canada 1986. Agriculture. Alberta* (Ottawa, ON: Statistics Canada, 1987).
58. Lafond, G. P. "Evaluation of Winter Wheat Management Practices under Semi-Arid Conditions," *Can. J. Plant Sci.* 72:1–12 (1992).
59. Zentner, R. P, C. W. Lindwall and J. M. Carefoot. "Economics of Rotations and Tillage Systems for Winter Wheat Production in Southern Alberta," *Can. Farm Econ.* 22:3–13 (1988).
60. Fowler, D. B., M. H. Entz, G. P. Lafond and D. K. Tompkins. "Crop Diversification through Reduced Tillage: the Low Input, Environmentally Friendly Winter Cereal Option," in *Crop Management for Conservation, Proc. Soil Conserv. Symp. (Yorkton),* G. P. Lafond and D. B. Fowler (Eds.) (Saskatchewan, SK: Division of Extension, University of Saskatchewan, 1990), pp. 202–251.
61. Zentner R. P., S. A. Brandt, K. J. Kirkland, C. A. Campbell and G. J. Sonntag. "Economics of Rotation and Tillage Systems for the Dark Brown Soil Zone of the Canadian Prairies," *Soil Tillage Res.* 24:271–284 (1992).
62. Izaurralde, R. C., N. G. Juma and W. B. McGill. "Plant and Nitrogen Yield of Barley-Field Pea Intercrop in Cryoboreal-Subhumid Central Alberta," *Agron. J.* 82:295–301 (1990).
63. Izaurralde, R. C., W. B. McGill, N. G. Juma, D. S. Chanasyk, S. Pawluk and M. J. Dudas. "Performance of Conventional and Alternative Cropping Systems in Cryoboreal Sub-Humid Central Alberta," *J. Agric. Sci. Cambr.* 120:33–41 (1993).
64. Choudhary, M., R. C. Izaurralde and N. G. Juma. "Comparison of Conventional and Alternative Cropping Systems for Central Alberta," *Proc. 29th Annu. Alberta Soil Science Workshop (Lethbridge),* (Edmonton, AB: Department of Soil Science, University of Alberta, 1992), pp. 307–313.
65. Lafond, G. P., S. Brandt, D. W. McAndrew, E. Stobbe and S. Tessier. "Tillage Systems for Crop Production," in *Crop Management for Conservation, Proc. Soil Conserv. Symp. (Yorkton),* G. P. Lafond and D. B. Fowler (Eds.) (Saskatchewan, SK: Division of Extension, University of Saskatchewan, 1990), pp. 155–201.

66. Mahli, S. S., P. A. O'Sullivan and C. D. Caldwell. "Preliminary Results from a Feasibility Study of Zero Tillage in Central Alberta," in *Proc. Alberta Soil Science Workshop* (Edmonton, AB: Faculty of Extension, University of Alberta, 1984), pp. 137–148.
67. Nyborg, M., and S. S. Mahli. "Effect of Zero and Conventional Tillage on Barley Yield and Nitrate Nitrogen-Content, Moisture and Temperature of Soil in North-Central Alberta," *Soil Tillage Res.* 15:1–9 (1989).
68. Mahli, S. S., and M. Nyborg. "Improving Barley Yield under Zero Tillage," in *Soil Quality in Semiarid Agriculture,* Vol. 2, J. W. B. Stewart (Ed.) (Saskatoon, SK: University of Saskatchewan, Saskatchewan Institute of Pedology, 1990), pp. 237–242.
69. Brandt, S. A. "Zero vs. Conventional Tillage with Two Rotations: Crop Production over the Last 10 Years," *Proc. Soils and Crops Workshop,* (Saskatoon, SK: University of Saskatchewan, 1989), pp. 330–338.
70. Wright, T. "General Overview of Conservation Tillage," in *Crop Management for Conservation, Proc. Soil Conserv. Symp. (Yorkton),* G. P. Lafond and D. B. Fowler (Eds.) (Saskatchewan, SK: Division of Extension, University of Saskatchewan, 1990), pp. 5–15.
71. Poppe, S., and M. H. Entz. "The Comparative Response of Canola, Field Pea and Wheat to Reduced Tillage," *Proc. 28th. Annu. Alberta Soil Sci. Workshop (Lethbridge),* (Edmonton, AB: Faculty of Extension, University of Alaska, 1991), pp. 232–240.
72. Wright, A. T. "Seedbed Preparation for Rapeseed Grown on Fallow and Stubble," *Can. J. Plant Sci.* 69:805–814 (1989).
73. Bergen, P. "Sugar Beet Response to Reduced Tillage Seedbed Preparation," *Proc. 28th Annu. Alberta Soil Sci. Workshop (Lethbridge),* (Edmonton, AB: Faculty of Extension, University of Alaska, 1991), pp. 224–231.
74. Mahli, S. S., G. Mumey, P. A. O'Sullivan and K. N. Harker. "An Economic Comparison of Barley Production under Zero and Conventional Tillage," *Soil Tillage Res.* 11:159–166 (1988).
75. Zentner, R. P., and C. W. Lindwall. "An Economic Assessment of Zero Tillage in Wheat Fallow Rotations in Southern Alberta," *Can. Farm Econ.* 13(6):1–6 (1978).
76. Zentner, R. P., S. Tessier, M. Peru, F. B. Dyck and C. A. Campbell. "Economics of Tillage Systems for Spring Wheat Production in Southwestern Saskatchewan (Canada)," *Soil Tillage Res.* 21:225–242 (1991).
77. Arshad, M. A., M. Schnitzer, D. A. Angers and J. A. Ripmeester. "Effects of Till vs. No-Till on the Quality of Soil Organic Matter," *Soil Biol. Biochem.* 22:595–599 (1990).
78. Dormaar, J. F., and C. W. Lindwall. "Chemical Differences in Dark Brown Chernozemic Ap Horizons under Various Conservation Tillage Systems," *Can. J. Soil Sci.* 69:481–488 (1989).
79. Campbell, C. A., V. O. Biederbeck, M. Schnitzer, F. Selles and R. P. Zentner. "Effect of 6 Years of Zero Tillage and N Fertilizer Management on Changes in Soil Quality of an Orthic Brown Chernozem in Southwestern Saskatchewan," *Soil Tillage Res.* 14:39–52 (1989).

80. Carter, M. R., and D. A. Rennie. "Changes in Soil Quality under Zero Tillage Farming Systems: Distribution of Microbial Biomass and Mineralizable C and N Potentials," *Can. J. Soil Sci.* 62:587–597 (1982).
81. Bowren, K. E. "Wild Oat Control with Minimum and Zero Tillage," in *Proc. Wild Oat Symp.* A. E. Smith (Ed.), (Regina, SK: Canadian Plains Research Centre, 1983), pp. 53–58.
82. Holm, F. A. "Fallow Weed Control in a Conservation Tillage System," in *Crop Management for Conservation, Proc. Soil Conserv. Symp. (Yorkton),* G.P. Lafond and D.B. Fowler, (Eds.) (Saskatchewan, SK: Division of Extension, University of Saskatchewan, 1990), pp. 37–58.
83. Donaghy, D. I. "Zero Tillage Weed Control," *Weed Facts* (Winnipeg, MB: Manitoba Agriculture, 1981).
84. Thomas, A. G., and D. Derksen. "Changes in Weed Spectrum with Changes in Tillage Practices," in *Crop Management for Conservation, Proc. Soil Conserv. Symp. (Yorkton),* G. P. Lafond and D. B. Fowler (Eds.) (Saskatchewan, SK: Division of Extension, University of Saskatchewan, 1990), pp. 59–74.
85. Mortensen, K., K. L. Bailey and W. McFadden. "Plant Diseases in a Conservation Tillage System," in *Crop Management for Conservation, Proc. Soil Conserv. Symp. (Yorkton),* G. P. Lafond and D. B. Fowler (Eds.) (Saskatchewan, SK: Division of Extension, University of Saskatchewan, 1990), pp. 91–101.
86. Conner, R. L., C. W. Lindwall and T. G. Atkinson. "Influence of Minimum Tillage on Severity of Common Root Rot in Wheat," *Can. J. Plant Pathol.* 9:56–58 (1987).
87. Lindwall, C. W., and D. T. Anderson. "Effects of Different Seeding Machines on Spring Wheat Production under Various Conditions of Stubble Residue and Soil Compaction in No-Till Rotations," *Can. J. Soil Sci.* 57:81–91 (1977).
88. Lindwall, C. W., R. P. Zentner and D. T. Anderson. "Conservation Characteristics of Minimum Tillage Systems," Paper No. 79-1019. (Winnipeg, MB: Joint Meeting ASAE and CSAE, 1979).
89. Dyck, F. B., R. P. Zentner, M. Peru, C. A. Campbell and S. Tessier. "Review of Equipment and Research Results for Conservation Tillage at Swift Current, Saskatchewan," in *Conservation Tillage, Proc. Great Plains Conservation Tillage Symp. (Bismarck, ND)* (Great Plains Agricultural Council Bulletin No. 131, 1990), pp. 97–108.
90. Loeppky, H., G. P. Lafond and D. B. Fowler. "Seeding Depth in Relation to Plant Development, Winter Survival and Yield of No-Till Winter Wheat," *Agron. J.* 81:125–129 (1989).
91. Tessier, S. M. Peru, F. B. Dyck, R. P. Zentner and C. A. Campbell. "Conservation Tillage for Spring Wheat Production in Semi-Arid Saskatchewan," *Soil Tillage Res.* 18:73–89 (1990).
92. Austenson, H. M., R. Ashford, F. W. Bigsby, K. E. Bowren, W. B. Reed, D. J. Warnock, A. Wenhardt and E. H. Wiens. "A Comparison of Methods of Direct Seeding Wheat on Stubble Land in Saskatchewan," *Can. J. Plant Sci.* 58:739–743 (1978).
93. Macartney, L. K. "The Impact of Education on Summer Fallow use in the Canadian Prairies," *J. Soil Water Conserv.* 42:114–117 (1987).

94. Carefoot, J. M., M. Nyborg and C. W. Lindwall. "Tillage-Induced Soil Changes and Related Grain Yield in a Semi-Arid Region," *Can. J. Soil Sci.* 70:203–214 (1990).
95. Lindwall, C. W. "Furrow Openers for Seed and Fertilizer Placement in Minimum Tillage Systems," Final Report, Project 87-0014, *Farming for the Future,* Vol. 1, (Edmonton, AB: Alberta Agriculture, 1989).
96. Janzen, H. H., and C. W. Lindwall. "Optimum Application Parameters for Point Injection of Nitrogen in Winter Wheat," *Soil Sci. Soc. Am. J.* 53:1878–1883 (1989).
97. Janzen, H. H., C. W. Lindwall and C. J. Roppel. "Relative Efficiency of Point-Injection and Surface Applications for N Fertilization of Winter Wheat," *Can. J. Soil Sci.* 70:189–201 (1990).
98. Lindwall, C. W., H. H. Janzen and T. L. Roberts. "Fertilizer Injection in Conservation Tillage Systems," in *Conservation Tillage, Proc. Great Plains Conservation Tillage Symp., (Bismarck, ND)* (Great Plains Agricultural Council Bulletin No. 131, 1990), pp. 181–188.
99. Janzen, H. H., T. L. Roberts and C. W. Lindwall. "Uptake of Point-Injected Nitrogen by Winter Wheat as Influenced by Time of Application," *Soil Sci. Soc. Am. J.* 55:259–264 (1991).
100. Fowler, D. B., and J. Brydon. "No-Till Winter Wheat Production on the Canadian Prairies: Placement of Urea and Ammonium Nitrate Fertilizers," *Agron. J.* 81:518–524 (1989).
101. Fowler, D. B., and J. Brydon. "No-Till Winter Wheat Production on the Canadian Prairies: Timing of Nitrogen Fertilization," *Agron. J.* 81:817–825 (1989).
102. Fowler, D. B., J. Brydon and R. J. Baker. "Nitrogen Fertilization of No-Till Winter Wheat and Rye. I. Yield and Agronomic Responses," *Agron. J.* 81:66–72 (1989).
103. Grant, C. A., E. H. Stobbe and G. J. Racz. "The Effect of Fall Applied N and P Fertilizer and Timing of N Application on Yield and Protein Content of Winter Wheat Grown on Zero-Tilled Land in Manitoba," *Can. J. Soil Sci.* 65:621–628 (1985).
104. Aulakh, M. S., and D. A. Rennie. "Gaseous Nitrogen Losses from Conventional and Chemical Summer Fallow," *Can. J. Soil Sci.* 65:195–203 (1985).
105. Lindwall, C. W., B. L. Sawatsky and T. L. Jensen. "Zero Tillage in Southern Alberta," *Proc. 21st Alberta Soil Science Workshop* (Edmonton, AB: Faculty of Extension, University of Alberta, 1984), pp. 128–136.
106. McEwen, F. L., and M. H. Miller. "Environmental Effects and Strategies to Deal with Them," in *In Search of Conservation Strategies in Canada,* D.W. Anderson (Ed.) (Ottawa: Agriculture Institute Canada, 1987) pp. 157–171.
107. "A survey and Review of Information Pertaining to Chemical Use in Association with Conservation Tillage on the Canadian Prairies," Report CP (EP)WNR90-91-11 (Edmonton, AB: Environment Canada, 1991).
108. "Canadians and the Environment: The Perspective of Adults and School Children" (Toronto: Angus Reid Group, Inc., 1989).
109. Shaw, G. G. "Pesticides in the Aquatic Environment" (Edmonton, AB: Agriculture Canada, Food Production and Inspection Branch, 1991).

110. Webber, C. A., and C. W. Lindwall. "Manure Makes Money on Moderately Eroded Land," Weekly Letter No. 2954 (Lethbridge, AB: Agriculture Canada Research Station, 1990).
111. Coote, D. R. "The Extent of Soil Erosion in Western Canada," in *Soil Erosion and Land Degradation, Proc. 2nd Annu. Western Provincial Conf., Rationalization of Water and Soil Research and Management* (Saskatoon: Saskatchewan Institute of Pedology, 1984), pp. 34–48.
112. Beke, G. J., C. W. Lindwall, T. Entz and T. C. Channappa. "Sediment and Runoff Water Characteristics as Influenced by Cropping and Tillage Practices," *Can J. Soil Sci.* 69:639–647 (1989).
113. Van Kooten, G. C., W. P. Weisenel and E. de Jong. "Estimating the Costs of Soil Erosion in Saskatchewan," *Can. J. Agric. Econ.* 37:63–75 (1989).
114. Sparrow, H. O. *Soil at Risk: Canada's Eroding Future,* Report of the Standing Committee on Agriculture, Fisheries and Forestry (Ottawa, ON: Supply and Services Canada, 1984).
115. Campbell, C. A., J. S. J. Tessier and F. Selles. "Challenges and Limitations to Adoption of Conservation Tillage — Soil Organic Matter, Fertility, Moisture and Soil Environment," in *Land Degradation and Conservation Tillage, 34th Annu. Meet. Can. Soc. Soil Sci.,* Calgary, AB (1988), pp. 140–185.
116. McRorie, H. D. "Farm Production and Its Relationship to Soil Conservation," in *In Search of Soil Conservation Strategies in Canada,* D. W. Anderson (Ed.) (Ottawa: Agricultural Institute of Canada, 1987), pp. 91–106.
117. *1991 Census, Agricultural Profile of Canada, Part 1* (Ottawa: Statistics Canada, 1992).

CHAPTER 14

Conservation Tillage in the Southern United States Great Plains

Paul W. Unger
U.S. Department of Agriculture, Agriculture Research Station; Bushland, Texas

Edward L. Skidmore
U.S. Department of Agriculture, Kansas State University; Manhattan, Kansas

TABLE OF CONTENTS

4.1. INTRODUCTION

The southern Great Plains (SGP) is the southern portion of the Great Plains of the United States, a vast midcontinent region that early explorers called the "Great American Desert"[1] because precipitation was limited, there were few springs or streams, and the landscape was relatively flat and treeless. The explorers considered the region undesirable and wholly uninhabitable for people from the humid and forested eastern regions of the United States. This view of the "Great American Desert" remained in the public mind until after the end of the Civil War in 1865.

Early human inhabitants of the Great Plains were the Plains Indians, who were nomadic and nonagricultural. They were hunters, mainly of the bison (*Bison* spp.) that roamed freely on the wide, grassy expanses of the Plains. The grasses were mainly short types on the arid to semiarid western part and tall types on the subhumid eastern part of the region. Settlers from the eastern United States and Europe, both ranchers and farmers, inhabited the region during the latter decades of the 1800s. The farmers brought with them the implements and methods of farming that they had used in their region of origin. The grasses were turned under and the land was used mainly for growing grain crops. The methods of farming involved clean tillage (total incorporation of crop residues), and crop production generally was satisfactory in years of above-average precipitation. Overall, however, both the implements and methods of farming brought by the settlers were unsuitable for the harsh environment of the Great Plains that is characterized by low precipitation and generally high winds. The methods of farming were especially unsuitable when precipitation was below normal for several succeeding years. The major drought of the 1930s,[2] coupled with intense wind storms, clearly demonstrated the urgent need for developing farming practices adaptable to the Great Plains' environment. An outcome of the severe devastation of the land that occurred during the 1930s was the development of the stubble mulch tillage system, presently known as a type of conservation tillage. A widely accepted definition of

Table 14.1. MLRAs of the SGP

MLRA	Name	Total Area (km^2)	Area in Cropland (%)	Elevation (m)	Precipitation (mm)
70	Pecos-Canadian Plains and Valleys	84,830	3	1200–2100	300–400
77	Southern High Plains	126,780	60	800–2000	375–550
78	Central Rolling Red Plains	130,370	35	500–900	500–750
80A	Central Rolling Red Prairies	52,700	40	300–500	625–900
80B	Texas North-Central Prairies	25,500	15	200–700	550–750

Source: Agriculture Handbook 296.[4]

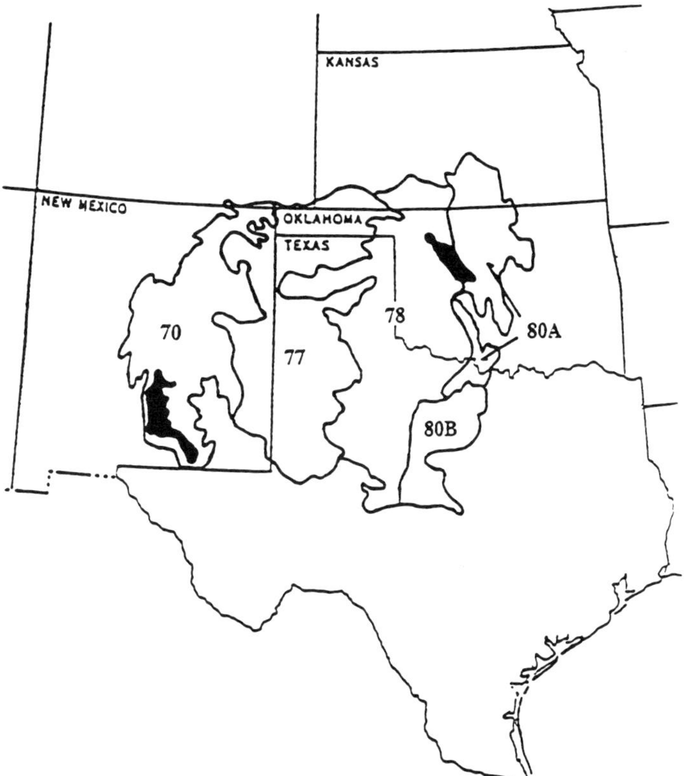

Figure 14.1. MLRAs of the SGP of the United States.

conservation tillage is any tillage or planting system that maintains at least 30% of the soil surface covered by residues after crop planting to reduce soil erosion by water. Where wind erosion is the primary concern, residues or plants of other crops equivalent to at least 1.1 Mg/ha of flat, small grain residue must be maintained on the surface during the critical erosion period.[3]

Presently, large portions of the region are extensively farmed, with cropland comprising up to twothirds of the land in some major land resource areas (MLRAs) (Table 14.1, Figure 14.1).[4] The remaining land is used mainly for livestock production; large ranches are common in some MLRAs. Relatively small areas are used for cities, roads, parks, etc. With current practices, the SGP as well as the entire Great Plains is a major crop-producing region, especially of grain crops.

14.2. FEATURES OF THE AGROECOSYSTEM

The boundaries of the SGP are imprecise. For the purposes of this chapter, the SGP is limited to MLRAs 70, 77, 78, 80A, and 80B (Table 14.1).[4] These MLRAs lie mainly in New Mexico, Oklahoma, and Texas, but portions of MLRAs 77, 78, and 80A extend into Kansas (Figure 14.1).[4] Kansas generally is considered part of the central Great Plains, but this chapter includes complete information for all MLRAs listed, including the parts that extend into Kansas.

14.2.1. Soil Types

Soils of the SGP range widely in surface texture. Major areas of sandy soils are in the northwestern and southern portions of MLRA 77 and throughout much of MLRA 78, usually having favorable water infiltration rates but low water-holding capacities, which limits the amount of water that can be stored for later use by crops. Soils high in clay predominate in the central and northeastern portions of MLRA 77. Pullman (torrertic paleustolls), the dominant series of clayey soils, has been mapped on about 1.53 million ha.[5] Other major clayey soil series are Sherm (torrertic paleustolls), Olton (aridic paleustolls), and Acuff (aridic paleustolls). In contrast to sandy soils, clayey soils have good water-holding capacity, but often are not filled to capacity because water infiltration into them is low. Thus, water availability for crops again is limited.

Surface slopes of most soils used for crop production are less than 5% in MLRAs 70 and 77, and, in many cases, the slopes are less than 1%. In the other MLRAs, slopes usually range from 5 to 10%.

All SGP soils are highly subject to erosion. Wind erosion is dominant on cropland in MLRAs 70 and 77. In MLRAs 78, 80A, and 80B, water erosion dominates, but wind erosion is a major problem in localized areas. The potential for wind erosion is especially great on sandy soils throughout the SGP.

The clayey soils of MLRAs 70 and 77 usually have adequate fertility so that a response to fertilizer is not obtained under dryland conditions.[6] With irrigation, crops respond to nitrogen and, in some areas, to phosphorus applications. The sandy soils in MLRAs 70 and 77 and most soils in MLRAs 78, 80A, and 80B are less fertile, and crops usually respond to fertilization.

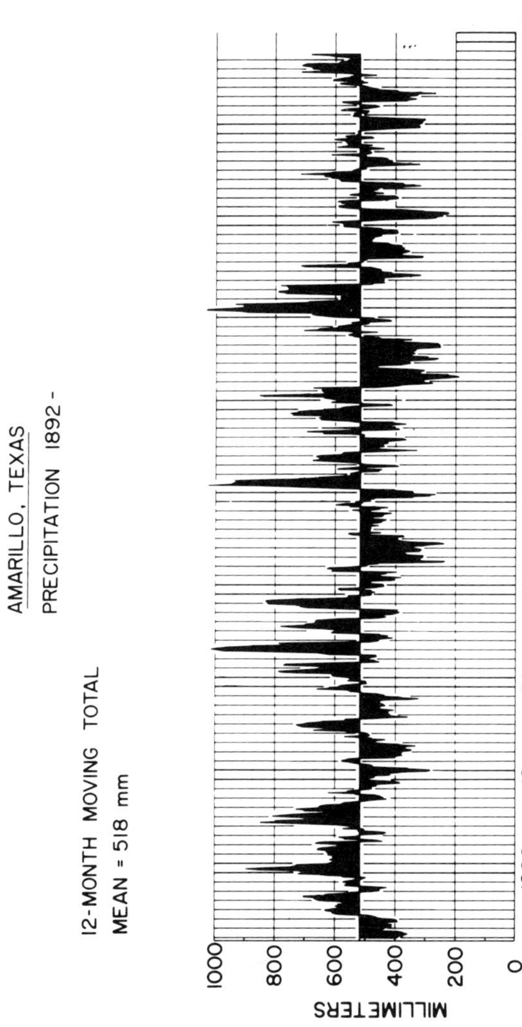

Figure 14.2. Long-term precipitation at Amarillo, TX, plotted as a 12-month moving total to show above- and below-average periods. Long-term average is 518 mm. Points on the curve represent deviation from the average for the past 12 months.

Table 14.2. Analysis of Precipitation by Storm Size, Bushland, TX, 1960–1979

Storm Size (mm)	Storms (no.)	Cumulative Storms (%)	Total Precipitation (mm)	Cumulative Precipitation (%)
0.00–1.27	561	36.8	226	2.4
1.28–2.54	229	51.9	459	7.4
2.55–6.35	321	73.0	1370	22.2
6.36–12.7	195	85.8	1757	41.2
12.8–25.4	143	95.2	2554	68.8
25.5–50.8	62	99.3	2133	91.8
50.9–76.2	8	99.8	422	96.9
76.3–101.6	2	99.9	167	98.9
101.7–127.0	1	100.0	120	100.0

Source: Adapted from Stewart, B. A., Proc. Workshop on Planning and Management of Water Conservation Systems in the Great Plains States (Lincoln, NE: U.S. Department of Agriculture, Soil Conservation Service, 1985).

14.2.2. Climate

The climate of the SGP ranges from subhumid in the eastern part to semiarid in the west. Precipitation has an east-west gradient, and lines of equal precipitation have a north-south alignment. Mean annual precipitation ranges from about 900 mm in the eastern portion of MLRA 80A to about 300 mm in MLRA 70. A major characteristic of SGP precipitation is its high variability within and among years. This variability is illustrated in Figure 14.2,[7] which depicts the long-term precipitation for Amarillo, TX. A second major characteristic is that much of the precipitation occurs in small storms, commonly less than 25 mm per storm (Table 14.2).[8] This often results in limited soil wetting and, consequently, major water losses by evaporation. High summer temperatures, low atmospheric humidities, and high winds contribute to high evaporative water losses.

Irrigation is used to supplement water from precipitation on about 45% of the cropland in MLRA 77, with the water obtained from the Ogallala Aquifer. Irrigation is less common in other MLRAs where it is used only in localized areas, and the water is obtained mainly from ponds, reservoirs, or streams.

Air temperatures in the SGP are influenced both by latitude, which ranges from about 32° to 38°N, and altitude, which ranges from about 300 to about 1370 m above mean sea level. Maximum temperatures usually approach or exceed 38°C several days each summer throughout the region. Minimum temperatures of –15 to –20°C occur during most winters in MLRAs 70 and 77, where altitudes are greatest. Temperatures limit the frost-free growing season to 180 to 190 days at the higher altitudes and from 210 to 220 days at the lower altitudes.[5] Winter wheat (*Triticum aestivum* L.), a cold-tolerant crop, is widely grown throughout the region, but is damaged by cold temperatures under some conditions, mainly when extremely cold temperatures follow periods of relatively mild temperatures.

14.2.3. Crops and Cropping Systems

The major crop in all MLRAs of the SGP, except MLRA 70, is winter wheat, which is grown for grain and for grazing by cattle. Under fully irrigated conditions (mainly in MLRA 77), grain yields of around 6.0 Mg/ha are achieved. Grain yields on dryland are highly variable and range from about 1.0 Mg/ha at the drier western locations to about 2.5 Mg/ha at the more humid eastern locations.

Grain sorghum (*Sorghum bicolor* (L.) Moench) is widely grown in MLRAs 77, 78, 80A, and 80B. With full irrigation (mainly in MLRA 77), grain yields often exceed 8.0 Mg/ha. Dryland sorghum grain yields are highly variable and usually average less than 3.0 Mg/ha.

Important crops in MLRA 70, where only about 3% of the land area is farmed, are forages and feed grains (various species). A major crop in the southern half of MLRA 77 is cotton (*Gossypium hirsutum* L.), where some is irrigated if groundwater is available. Other important crops in portions of MLRA 77 are corn (*Zea mays* L.), sugar beets (*Beta vulgaris* L.), and various vegetables. Cotton also is important in MLRAs 78, 80A, and 80B, and usually is not irrigated. Other important crops are alfalfa (*Medicago sativa* L.) and peanut (*Arachis* spp.) in MLRA 78; alfalfa, peanut, and soybean (*Glycine max* L.) in MLRA 80A; and oat (*Avena sativa* L.) in MLRA 80B.

Where adequate water is available, either from precipitation in the more humid eastern portion of the region or from irrigation, crops usually are grown annually (one crop per year). Often, crops such as wheat, sorghum, cotton, and peanut are grown continually (the same crop in the same field each year), unless weeds or other pests (insects or diseases) warrant crop rotations. Where such problems occur, another adaptable crop may be grown or a crop rotation may be implemented. Crop rotations involving a winter and a summer crop, for example, winter wheat and grain sorghum, are highly effective for controlling problem weeds. Such rotation is especially appropriate where water for irrigation is limited because fallow periods between crops provide time for additional water storage in soil for use by the next crop.[9,10] Sugar beet usually is rotated with other crops, mainly to help control sugar beet diseases.

On dryland, especially on the drier portions of the region, crops may be grown annually, but crop-fallow systems are more common. The main systems involving fallow on dryland are wheat-fallow (one crop in 2 years) and wheat-sorghum-fallow (two crops in 3 years). In some cases, grain sorghum and cotton are grown in rotation.[11]

Cotton often is grown continuously, both with irrigation and on dryland. On dryland, skip-row planting patterns often are used. Typical patterns are two rows planted and one row skipped, two rows planted and two rows skipped, and four rows planted and two rows skipped. In such cases, plant roots extend into and extract water from the nonplanted rows, thus resulting in yields not greatly different from those in which all rows are planted. Advantages of skip-row planting include lower production costs because less seed is planted and the harvested area is smaller; easier weed control; wheel traffic can be confined to the skipped rows, thus minimizing soil compaction near the planted row; and high-residue crops can be planted in skipped rows to provide vegetative barriers for wind erosion control.[11]

14.3. ADOPTION OF CONSERVATION TILLAGE

14.3.1. Conventional Tillage Systems

As in other regions, tillage that is considered or termed "conventional" changes with advances in technology. This was illustrated by Greb et al.[12] for winter wheat production on dryland in Colorado in the central Great Plains (CGP). During the early years, bare fallow achieved by plowing and harrowing was common. Later, SM tillage became the "conventional" system. Presently, a limited tillage system involving a combination of tillage and herbicides for weed control and land preparation for the next crop is being widely used in the CGP and could appropriately be called the conventional system for that region.

Early tillage systems in the SGP closely paralleled those in other Great Plains regions for crops such as winter wheat. Clean tillage for which all crop residues were plowed under was the primary method used by early settlers in the SGP. Such tillage continued to be used during the early decades of the 1900s, and was a major factor leading to the devastating dust storms that plagued the SGP during the drought of the 1930s.[2] During that drought, crop growth was extremely limited. Not enough plant material from either growing plants or crop residues was available to help control erosion. The irrigation methods that could have increased crop yields and stabilized the soil had not been developed in the SGP by the 1930s.

The SM tillage system was initially developed at Lincoln, NE in the late 1930s. Within a few years, research and development work on SM tillage was done throughout the Great Plains, including Bushland, TX.[13] With this system, large sweeps or blades undercut the surface at a depth of about 5 to 10 cm while retaining most residues on the surface. SM tillage is effective for controlling erosion, especially that caused by wind, provided adequate residues are available. It is generally considered to be *the* conventional tillage system, especially for dryland wheat and grain sorghum production in MLRA 77.[11] If the requirement of 30% surface cover or of 1.1 Mg/ha of small-grain-equivalent amount of residues on the surface after crop planting is satisfied, SM tillage can appropriately be called conservation tillage. Chisel plows are also widely used for dryland wheat production, but tillage systems involving the chisel plow seldom retain enough residues on the surface in parts of the region for the system to be classified as conservation tillage. For irrigated wheat, mainly in MLRA 77, the land usually is disced, chiseled, disced again, and furrowed to control irrigation water flow. Moldboard plowing is used by some producers, but its use has declined sharply in the last 10 to 20 years.

At more humid locations in the SGP where wheat grain yields are greater, surface residue amounts may exceed 3.0 Mg/ha and cause difficulties in seeding the next crop. Under such conditions, the first postharvest tillage may be offset discing. Subsequent tillage may employ a SM implement. In cases in which cheatgrass (*Bromus secalinus* L.) is a serious problem, moldboard plowing at approximately 4-year intervals is one effective control measure.[11]

Tillage practices for grain sorghum are similar to those for wheat, except that lister plowing (ridge tillage) is used for sorghum where a major potential for wind erosion exists. Ridge tillage helps control wind erosion by providing surface roughness in the direction perpendicular to prevailing winds.

Much of the SGP cotton is grown on soils having relatively high sand contents. Because cotton residues have a low value in controlling erosion, practices other than surface residue management must be relied upon for doing so. Terracing and ridge tillage on the contour are effective for controlling water erosion. To help control wind erosion, the land is moldboard plowed, listed, or chiseled to reduce compaction and to bring erosion-resistant clods to the surface.[11,14] Other practices for controlling wind erosion include mulching with cotton gin trash or stover of other crops — for example, pearl millet (*Pennisetum americanum* (L.) K. Schum);[15] growing high-residue-producing crops that serve as a wind barrier on the skipped rows in which cotton is planted in a skip-row pattern;[16,17] or planting a small grain cover crop between cotton rows before the cotton is harvested. In most cases water storage was greater with a mulch than with bare soil.[15,18]

Dryland cotton is planted in lister furrows or on ridges formed by lister plowing. In either case, the ridges help control wind erosion.[11] Cotton to be irrigated commonly is planted on previously formed ridges.

14.3.2. Development of Conservation Tillage Systems

Russel[19] showed the tremendous value of surface residues for reducing stormwater runoff and for increasing soil water storage (SWS), thus providing additional water for use by subsequent crops. Both also have a major impact on soil erosion. Reducing runoff minimizes water flow across the surface, thus also minimizing soil transport across the surface. Increased water storage increases the potential for improved plant growth, thus providing more vegetative matter by either growing plants or residues postharvest for greater protection of the surface against erosion by wind or water. Water conservation and erosion control are prime requisites for successful crop production in the SGP.

The early research by Russel[19] and others, which led to the development of SM tillage, also showed that runoff and evaporation decreased and SWS increased with increasing amounts of surface residues (Table 14.3).[19] Although the potential for increased water conservation with no-tillage was illustrated by this early research, no-tillage systems of crop production were not practical at that time because of weed control and equipment limitations.

Table 14.3. Water Storage, Runoff, and Evaporation from Field Plots at Lincoln, NB, 10 April to 27 September 1939

Treatment	Storage	Runoff (mm)	Evaporation	Loss (%)[a]
Straw, 2.2 Mg/ha, normal subtillage	30	26	265	83
Straw, 4.5 Mg/ha, normal subtillage	29	10	282	88
Straw, 4.5 Mg/ha, extra loose subtillage	54	5	262	82
Straw, 9.0 Mg/ha, normal subtillage	87	Trace	234	73
Straw, 17.9 Mg/ha, no-tillage	139	0	182	57
Straw, 4.5 Mg/ha, disced	27	28	266	83
No straw, disced	7	60	254	79
Contour basin listing	34	0	287	89

Source: Adapted from Russel, J. C., *Soil Sci. Soc. Am. Proc.,* 4, 65, 1939.

[a] Based on total precipitation, which was 321 mm for the period.

Herbicides for weed control in field crops became available in the early 1950s. Soon thereafter, no-tillage research was initiated for both dryland winter wheat and grain sorghum in the SGP at Bushland, TX.[20-22] Major limitations to no-tillage in the early studies were limited effectiveness of herbicides for controlling weeds and low residue production by dryland crops. The amounts of residue produced were not adequate to appreciably enhance water infiltration or suppress soil water evaporation.

Major improvements in herbicides were made during the late 1950s and in the 1960s, which led to widespread interest in developing no-tillage cropping systems. These studies showed that the no-tillage system of crop production was highly effective for controlling soil erosion, both wind and water, and for improving water conservation. The need for improved soil and water conservation in the SGP led to further conservation tillage (including no-tillage) research at several SGP locations in the 1960s.

14.3.2.1. Southern High Plains

In 1968, Unger et al.[23] applied atrazine [2-chloro-4-(ethylamino)-6-(isopropylamino)-*s*-triazine]* at 3.4 kg of active ingredient per hectare (a.i./ha) and 2,4-D [2,4-dichlorophenoxy)acetic acid] at 1.1 kg a.i./ha to some plots at Bushland, TX (MLRA 77) soon after harvest of irrigated winter wheat. The herbicides were used to control weeds and volunteer wheat during the ensuing fallow period (about 11 months later) before planting grain sorghum. On other plots, weeds were controlled by disc or sweep tillage as needed, or by a combination of one sweep tillage operation and herbicides as indicated above. The irrigated wheat produced about 11 Mg/ha of residues. Soil water contents

* Names are necessary to report factually on available data; however, the USDA neither guarantees nor warrants the standard of the product, and the use of the name by the USDA implies no approval of the product to the exclusion of others that also may be suitable.

Table 14.4. Effect of Tillage during Fallow after Winter Wheat Harvest in 1968 on Soil Water Content at Sorghum Planting in 1969, Bushland, TX

Treatment	Soil Water Content (mm)
Tandem disc tillage	145 c[a]
Tandem disc plus sweep tillage	135 c
Sweep tillage	163 b
Sweep tillage plus herbicides	196 a
Herbicides only (no-tillage)	203 a

Source: Adapted from Unger P. W. et al, *J. Soil Water Conserv.*, 26, 147, 1971.

[a] Values followed by the same letter are not significantly different at the 5% level (Duncan multiple range test).

at sorghum planting in May 1969 (Table 14.4)[23] were similarly high for the herbicides-only and the sweep tillage-herbicides combination treatments and similarly low with disc tillage only and disc plus sweep tillage treatments. At sorghum planting, about 4.6 Mg/ha of residues remained on the surface of herbicide-treated plots, but <0.2 Mg/ha on disc tillage plots.

In the foregoing study,[23] sorghum yields were not obtained, but in subsequent residue management studies after irrigated winter wheat from 1974 to 1981, plant-available soil water contents at sorghum planting averaged 213, 174, and 155 mm, and dryland sorghum grain yields averaged 3.23, 2.62, and 2.12 Mg/ha with no-, sweep, and disc tillage treatments, respectively.[9,10] Yield responses were not as great when irrigated wheat was followed by dryland sunflower (*Helianthus annuus* L.) [24] or irrigated corn.[25] However, Musick et al.[26] obtained major water storage and sorghum yield increases with no-tillage as compared to clean (disc) tillage when both the wheat and grain sorghum were irrigated.

Baumhardt et al.[27] evaluated the effects of disc and no-tillage management of wheat residues on SWS and on dryland and irrigated grain sorghum yields on Pullman soils at Bushland and Lubbock, TX. SWS was greater with no-tillage than with discing at Bushland where wheat produced 11 Mg/ha of residues. At Lubbock, where the residue amount was only 2 Mg/ha, water storage differences were slight. Average sorghum grain yields on dryland for 2 years were greater (1.19 Mg/ha increase) with no-tillage than with disc tillage at Bushland, but only in 1 of 2 years at Lubbock (0.96 Mg/ha increase). With irrigation, grain yields were not affected by tillage treatments at Bushland, but were 1.11 Mg/ha greater with no-tillage than with disc tillage at Lubbock.

Water storage and sorghum yields increased with increases in amounts of wheat residues placed on the soil surface at Bushland (Table 14.5).[28] This study clearly illustrated the value of surface residues for enhancing SWS, which is of major importance in stabilizing crop yields in the semiarid portions of the SGP.

Table 14.5. Straw Mulch Effects on Soil Water Storage during Fallow,[a] Water Storage Efficiency, and Grain Sorghum Yield at Bushland, TX, 1973–1976

Mulch Rate (Mg/ha)	Water Storage[b] (mm)	Water Storage Efficiency[b] (%)	Grain Yield (Mg/ha)
0	72 c[c]	22.6 c[c]	1.78 c[c]
1	99 b	31.1 b	2.41 b
2	100 b	31.4 b	2.60 b
4	116 b	36.5 b	2.98 b
8	139 a	43.7 a	3.68 a
12	147 a	46.2 a	3.99 a

Source: Adapted from Unger, P. W., *Soil Sci. Soc. Am. J.*, 42, 486, 1978.

[a] Fallow duration of 10 to 11 months.
[b] Water storage determined to 1.8 m depth. Precipitation averaged 318 mm.
[c] Column values followed by the same letter are not significantly different at the 5% level (Duncan multiple range test).

Annual cropping of dryland winter wheat using conservation tillage practices can usually be achieved without difficulty. SM or chisel tillage is used most often. The major limitation is the amount of residue available. Often, 30% of the surface is not covered with residues at planting for the practice to qualify as conservation tillage. Use of no-tillage for continuous wheat on dryland has shown promise.[29]

For annually cropped irrigated winter wheat, usually large amounts of residue are produced. Even if the land is disced one or two times between crops, which is a common practice, adequate residues can be retained on the surface to qualify as conservation tillage. Most attempts to grow irrigated wheat annually using a no-tillage system were not successful because of a surface residue buildup that caused planting, seedling vigor, and weed control problems.[30,31] For annual wheat, alternating between no-tillage and limited tillage was satisfactory.

Cotton residues (stalks) have little value for controlling erosion and they are shredded before the land is plowed in preparation for the next crop, usually cotton. Moldboard plowing, listing, or chiseling produces a rough, cloddy surface that helps control wind erosion.[11,14] However, the sandy soils on which much of the cotton is grown in the SGP are highly erosive and severe erosion frequently occurs.

14.3.2.2. Central Rolling Red Plains

Most crops in the central rolling red plains (MLRA 78) are not irrigated. Groundwater in most of the MLRA is scarce, and water for irrigation is derived from ponds or reservoirs that have limited storage capacity. While irrigation is limited in MLRA 78, some research involving conservation tillage for irrigated crops has been conducted in the region. At Munday, TX, irrigated winter wheat grain yields averaged 4.13 Mg/ha with clean and 3.48 Mg/ha with reduced tillage during a 5-year study. For one crop, yields were significantly lower with reduced tillage (4.46 vs 5.86 Mg/ha). Gerard and Bordovsky[32] attributed the lower yields to fewer heads associated with a lower plant population because of planting problems, in which large amounts of residues were present, and possibly because of reduced tillering. For this and the following studies, reduced tillage involved weed control with herbicides between crops and cultivation during the growing season.

For annually cropped, irrigated grain sorghum at Munday, differences in grain yield, water use, and water-use efficiency due to reduced and clean tillage treatments were not significant. For clean tillage, plots were disced twice, bedded (ridge-tilled), and cultivated before and after planting. Use of herbicides provided additional growing season weed control.[33] Clark[34] evaluated the effects of conventional and reduced tillage systems and furrow diking treatments on dryland cotton production on Abilene clay loam (fine, mixed, thermic pachic argiustoll) at Chillicothe, TX. Treatments involved diking all, alternate, or no furrows. Cotton lint yields were not affected by tillage system, but furrow diking significantly increased yields. Average yield increases over the undiked treatment were 16 and 36% with alternate and every furrow diking, respectively. This study showed that in the absence of surface residues, furrow diking can be an effective water conservation practice, at least on some soils. On the Pullman soil at Bushland, furrow diking in conjunction with no-tillage wheat residue management (dikes installed before planting wheat) did not increase SWS or sorghum yields over those obtained with no-tillage alone.[35] Surface residue amounts and soil slopes undoubtedly contributed to the different responses. Residue amounts were greater and slopes were less at Bushland than at Chillicothe.

In another study at Chillicothe, Clark et al.[36] determined the production potential and practicality of reduced, conventional, and narrow row tillage systems for dryland cotton production. Furrow diking was a key operation for each system. The reduced tillage system significantly reduced runoff and increased water storage, lint yield, water use efficiency, and profitability as compared to the conventional and narrow row systems. Net return to land, management, and risk was $244/ha with reduced tillage, which was 50 and 114% greater than with the conventional and narrow row systems, respectively.

For dryland grain sorghum production at Munday, conventional tillage involved discing twice, bedding, and cultivating before and after planting. Reduced tillage involved using glyphosate (*N*-(phosphonomethyl) glycine) to control weeds between harvest and planting, and cultivation to control weeds during the growing season. The treatments did not significantly affect grain yields, water use, or water use efficiency.[33]

Much of the wheat in Oklahoma in MLRA 78 is annually cropped. SM tillage was used there in the 1960s. Although generally successful, common problems involved weed control and poor plant populations that resulted from planting in stubble with less than adequate drills.[37]

Scientific and technological advances since the 1960s have made reduced or no-tillage systems feasible for wheat production in western Oklahoma where the systems are known as "Lo-Till." Lo-Till is any of several systems that rely on herbicides alone or in combination with tillage to provide adequate weed control during the noncrop period and to provide a favorable seedbed for the next crop. With the large amounts of residues usually produced by wheat in the region, maintaining adequate residues on the surface is not a major problem. The systems still qualify as conservation tillage, even when some tillage is performed. Use of conservation tillage increased SWS in some cases.[38] Use of a no-tillage system often resulted in the greatest soil water content.[39]

In comparisons involving Lo-Till and cooperators' practices, early results indicated a 0.66 Mg/ha average grain yield increase with Lo-Till (3.76 vs 3.11 Mg/ha). Use of Lo-Till systems provided better soil water conditions that allowed earlier and/or more timely planting than was possible with conventional tillage. Earlier planting provided more forage for grazing by cattle, thus offsetting the greater cost of herbicides used with the Lo-Till system.[37] Although early results were favorable, growing wheat annually without tillage resulted in major weed (mainly cheatgrass) problems and reduced plant vigor after 3 or 4 years at some locations. Major tillage at 3- or 4-year intervals helps overcome these problems.[40] At other locations, wheat has been grown successfully by using no-tillage continuously,[41] but delayed planting or use of a contact herbicide can also avoid the cheatgrass problem, thus avoiding the destruction of improvements in soil properties achieved through the use of no-tillage.[42]

Epplin et al.[43] compared conservation and mechanical tillage systems for weed control where winter wheat was grown annually in Oklahoma. The additional cost of herbicides, where they were relied upon for weed control, exceeded the value of fuel and labor saved by using mechanical weed control methods. However, because of less investment in equipment, some conservation tillage systems were competitive with mechanical systems on a total cost basis.

14.3.2.3. Other Land Areas

The MLRAs discussed in this section are small compared to MLRAs 77 and 78. Of the three, MLRA 70 is the largest (Table 14.1).[4] However, only about 3% of it is used for cropland. This MLRA is at a higher elevation and receives less precipitation than MLRA 77, which is east of MLRA 70. In general, tillage practices applicable to MLRA 77 are also applicable to MLRA 70. MLRAs 80A and 80B, which are east of MLRA 78, are located at a lower elevation and generally receive more precipitation than MLRA 78. Wheat, grain sorghum, and cotton are major crops in all three MLRAs, and tillage practices in MLRAs 80A and 80B generally are the same or similar to those used in MLRA 78. Other important crops adaptable to conservation tillage are peanut and soybean in MLRA 80A and oat in MLRA 80B. Research results involving these crops are not available from MLRAs 80A and 80B.

Soybean often is grown after wheat at various locations using conservation tillage practices, either double-cropped or in rotation. Similar conservation tillage practices should be suitable for soybean in MLRAs 80A and 80B. Peanut is highly susceptible to various diseases. At Yoakum, TX, in MLRA 87 (Texas Claypan area), moldboard plowing resulted in better disease control and greater yields than discing or no-tillage. Weed control was difficult with no-tillage, especially for grassy species.[44] At several locations in MLRA 78, weed control with herbicides was satisfactory and no increase in disease incidence occurred when peanut was strip-till planted into wheat residues, as compared to planting after moldboard plowing. Strip-till planting reduced soil erosion, but it reduced yields 10 to 15%.[45]

14.4. CRITICAL FACTORS RELATED TO ADOPTION OF CONSERVATION TILLAGE

Most soils of the SGP are suitable for crop production by conservation tillage methods and, unless high residue conditions prevail, do not have a cold temperature limitation (slow warming in the spring) due to surface residues.[46] However, lower soil temperatures with no-tillage were beneficial for wheat production where grazing by livestock was involved.[41]

The extensive area of sandy soils with low water-holding capacities may not benefit as much as finer-textured soils from conservation tillage with respect to water conservation efforts. Sandy soils also are subject to compaction by normal cultural operation and by animal trampling when crops are grazed by them.[47] However, provided adequate residues are available initially, these soils can be loosened if needed, for example, by chiseling, and still retain enough residues on the surface for the system to qualify as conservation tillage.

Overall, climatic conditions in the SGP permit the use of conservation tillage systems. In fact, climatic factors of limited and erratic precipitation, high potential evaporation, high summer temperatures, and high wind speeds are reasons why conservation tillage should be used in the SGP.

14.4.1. Crop Residue Level Effect

Although suitable systems are available for most SGP crops, problems remain that will delay the adoption of conservation tillage in some cases. These include inadequate residue production by dryland crops in drier portions of the region, high residue production at more humid locations, and low residue production by cotton and peanut. The potential for greater disease incidence in peanut also is of concern in conservation tillage. To make conservation tillage more widely adaptable, practices, equipment, or materials needed include those that (1) retard residue losses where residue production is limited, (2) hasten residue decay or provide for the removal of some residues where large amounts are produced, (3) provide for satisfactory crop establishment under high-residue conditions, and (4) allow low-residue-producing crops such as cotton and peanut to be grown economically in conjunction with crops that produce adequate residues for soil and water conservation purposes.

On dryland, residue production by crops may not be adequate to provide enough residue for management by conservation tillage methods. Residue production by cotton and peanut crops is especially limited. Even if tillage practices that retain all residues of these crops on the soil surface are used, sufficient amounts may not be available for the practices to qualify as conservation tillage. These crops, however, can be grown in rotation or in strip-cropping systems involving other crops that provide more residues, thus affording soil and water conservation benefits.

A major limitation to enhanced water conservation with no-tillage and conservation tillage in general in the SGP is low-residue production by dryland crops, especially in the drier western portion of the region. Because of limited residues, stormwater runoff is greater from no-tillage than from conventional (SM) tillage watersheds. Runoff is greatest during fallow after sorghum when surface cover by residues is least. However, although runoff is greater with no-tillage, soil water contents due to tillage methods at planting of either crop usually are similar, apparently because soil disturbance by tillage increases evaporative losses of soil water. This negates the advantage that SM tillage provides with respect to runoff. Because soil water contents at planting often were similar, differences in dryland winter wheat and grain sorghum yields due to tillage also were slight.[48]

Surface residues resulting from the use of conservation tillage led to increased water storage,[9,10,28] reduced evaporation,[49] moderation of summer maximum soil temperatures,[46] and reduced wind speeds at the soil surface.[50] Undoubtedly the greatest climatic limitation to the adoption of conservation tillage in most of the SGP is limited precipitation. Precipitation is often too low to provide enough water for good growth of dryland crops, thus not producing adequate residues to be managed for conservation purposes.

14.4.2. Crop Rotation Effect

Winter wheat and grain sorghum are two major crops of the SGP and can be successfully grown by conservation tillage methods when grown in rotation. Weed control with herbicides or a combination of herbicides and tillage often is satisfactory in this rotation. However, this rotation results in only two crops in 3 years, which may not be satisfactory for producers that require full utilization of land resources for their farming enterprise to be economical.

Growing either winter wheat or grain sorghum annually by conservation tillage is possible, but weed and volunteer crop plant control problems occur in some cases, especially when a no-tillage system is used. The problem generally is most severe when the crops are irrigated or grown in the more humid eastern portion of the SGP. Under such conditions, planting and seedling vigor problems may be encountered. Occasional clean tillage or use of reduced tillage systems minimizes these problems.[30,31,40,51,52] In other cases, annual wheat has been grown successfully by no-tillage methods under these conditions.[41]

In efforts to overcome the problem of severe wind erosion on sandy soils in which cotton is grown, Keeling et al.[53] and Lyle and Bordovsky[54] grew cotton in rotation with sorghum and wheat, which provided adequate cover to control erosion. Lint yields of irrigated cotton generally were not affected by tillage method (conventional, minimum, or no-tillage). However, dryland cotton usually yielded more with minimum and no-tillage than with conventional tillage when the cotton was rotated after sorghum or wheat and when planted after wheat used as a cover crop, then killed. Economic returns were greater with no-tillage, which should make the system acceptable to producers.[53-55]

Harman et al.[56] reported greater economic returns for no-tillage than for conventional tillage dryland cotton grown in rotation with irrigated barley (*Hordeum vulgare* L.). The soil was Sherm clay loam (fine, mixed, mesic torrertic paleustoll) at Etter, TX. Herbicide costs were $155/ha greater with no-tillage than with conventional tillage, but long-term annual profits were $82/ha greater with no-tillage because of an average lint yield increase of 110 kg/ha and lower machinery depreciation costs.

14.4.3. Economic Considerations

Most crop producers operate under some constraints with respect to capital, land, and equipment resources. For the production of crops, these resources are managed primarily to meet their immediate needs and, secondarily, to meet their perceived long-term needs. To achieve these ends, producers generally select crop production options that involve the least risk. Because conservation tillage is a relatively new practice, at least under many circumstances, producers may avoid using this practice and continue to use practices that have proven adequate through past experiences. Adoption of conservation tillage may also require the purchase of new or different equipment, an investment that producers may not care to make unless there is little or no risk involved with respect to meeting their needs.

A major consideration is the producer's managerial ability. Conservation tillage requires a relatively high level of management because it is a system that does not begin and end with a given crop. For satisfactory crop production by conservation tillage methods, practices applied to the current, or even the previous, crop may affect future crops, especially when weeds are controlled with herbicides. Also, producers must be capable of dealing with unexpected problems.

The use of economically feasible practices is essential for long-term economically sound crop production, regardless of the tillage system employed. Studies have shown that various conservation tillage systems adaptable to the SGP are economically feasible.[55-57] A major advantage often was shown for conservation tillage, especially when long-term equipment costs and depreciation were considered in the analyses. When only short-term costs and returns were considered, conservation tillage sometimes was less economical because herbicide costs are high for some production systems. High short-term costs could cause producers that have limited capital or credit available to them to opt for systems that involve lower short-term costs.

Commodity price support programs administered by governmental agencies usually are based on the production of certain crops grown on pre-established areas of land. Under such constraints, producers may lose flexibility of land use for other crops, which may also thwart implementation of conservation tillage systems. This is especially true if the conservation tillage system would involve two or more crops in a rotation.

Some crops are well-adapted to certain soils and climatic conditions, for example, cotton on portions of MLRA 77. Other crops are less economical when grown in place of cotton. Hence, cotton remains the choice of producers, even though severe wind erosion often occurs in the cotton-growing region. Unless suitable alternative crops become available or strict erosion control regulations are imposed, cotton will remain the producers' choice crop for the region.

Adoption of conservation tillage in the SGP has suffered in part from a lack of strong and effective leadership on the part of agency personnel responsible for implementing sound soil and water conservation practices. Conflicting reports concerning conservation tillage may also cause producers to doubt the effectiveness of these systems for their production situation. Unless producers are provided with sound information and strongly encouraged to use it, many producers may not adopt this soil and water conservation practice that is so needed in the SGP.

14.5. ACCEPTANCE OF CONSERVATION TILLAGE

Acceptance of conservation tillage by farmers varies with crops being grown and areas within the region. Some form of conservation tillage (usually SM tillage) often is used for winter wheat in the drier western areas, but it is much less common in the more humid eastern areas, especially where wheat is grown continually (annual cropping). Major reasons for low acceptance of conservation tillage in the more humid wheat-growing areas are the continuing problem with cheatgrass control, difficulty in crop establishment where large amounts of residue remain on the surface, poor vigor of plants growing in residues, and a general perception that it is not economical.

Similar problems are encountered for irrigated wheat in MLRA 77, except that cheatgrass problems are slight. Discing and chiseling of annually cropped, irrigated wheat, can result in retaining adequate residues for conservation purposes, provided that these implements are used wisely. Subsequent ridge-tillage for furrow-irrigated wheat may result in surfaces virtually devoid of residues.

Some producers view surface residues as a hindrance to successful and economical wheat production and thus burn the residues. Fortunately, residue burning does not necessarily increase the potential for erosion on the fine-textured soils irrigation is practiced because tillage can provide a rough surface on which to control wind erosion and water erosion is not a serious problem. In addition, irrigation can be used to assure crop establishment when timely precipitation does not occur. Unfortunately, residue burning increases the soil organic matter decline rate,[58] which may have long-term implications with respect to sustaining crop production in the region.

In a broad sense, acceptance of conservation tillage for sorghum production has followed patterns similar to those for wheat. SM tillage often is used for dryland sorghum in drier areas, especially when it is grown in rotation with wheat. In more humid areas, where it is grown annually, conservation tillage is rarely used. For irrigated sorghum, few residues remain at planting because of one or more discings, possibly a chiseling, and tillage that forms ridges on which the sorghum is planted and furrows for irrigation water flow. Some producers have accepted conservation tillage for irrigated sorghum, especially when it is grown in rotation with wheat.

Few producers practice conservation tillage for annually cropped cotton, mainly because cotton does not produce enough residue to qualify as conservation tillage, even if all residues were retained on the surface. In any case, to reduce problems with cultural operations for the next crop, cotton stalks are usually shredded, then incorporated by tillage that forms ridges to help control erosion and on which the cotton is planted. Some producers, however, use conservation tillage when cotton is grown in rotation with grain (wheat or sorghum) crops or when cotton follows a winter cover crop. In both cases, the goal is improved control of erosion, mainly that caused by wind.

For the SGP region as a whole, the Conservation Technology Information Center (CTIC) national survey[59] indicated that some form of conservation tillage was used on 30% of the cropland in 1985. Of the total area devoted to conservation tillage, no-, ridge, minimum, and reduced tillage were used on 4, 1, 55, and 40% of the area, respectively.[60] The 30% use value for conservation tillage, however, may be too high because major portions of Colorado and Kansas were included in the SGP for the CTIC survey. Areas of those states are not considered a part of the SGP for the purposes of this chapter. Wheat and sorghum are major crops in Colorado and Kansas, and some form of conservation tillage often is used for their production. In contrast, conservation tillage rarely is used for cotton, which is not grown in Colorado and Kansas. Hence, the adoption value undoubtedly is biased in favor of conservation tillage when those states are included in the SGP.

Much information regarding conservation tillage systems has become available in recent years, and satisfactory systems are now available for many crops in the SGP region. These systems, when properly implemented, have the potential to greatly reduce soil erosion and improve water conservation. This potential, along with the growing emphasis on protection of the environment, should lead to greater adoption of conservation tillage by producers in the region. Major recent advances in equipment suitable for crop production by conservation tillage methods will provide an impetus for adoption of this resource-conserving tillage method. In addition, strong education and/or demonstration programs should be implemented to apprise producers of conservation tillage practices and the value of those practices for conserving soil and water resources and for protecting the environment.

14.6. ENVIRONMENTAL ISSUES RELATED TO CONSERVATION TILLAGE

14.6.1. Use of Agrochemicals and Fertilizers

Adoption of conservation tillage systems, especially those involving no-tillage or herbicide-tillage combinations, usually requires applications of more herbicides than used with conventional tillage systems, mainly to control weeds before crops are planted. When applied according to manufacturers' directions, within safe limits (wind speeds, sprayer calibrations, etc.), and according to other acceptable practices, most herbicides commonly used in the SGP pose no serious threat to the environment. However, some herbicides degrade slowly and may be transported by water or by soil that erodes from the point of application. This could result in water contamination or adverse effects on nontarget plants under some conditions, especially where runoff and soil losses are large as they are in the more humid areas of the SGP.

In general, transport of herbicides is less under conservation than under conventional tillage conditions because water and soil movement across the land is less with conservation tillage. An exception would be where residues amounts are too low to adequately protect the surface, which can result in soil surface sealing and, consequently, greater runoff[48] and potentially greater herbicide losses. Preliminary results, however, indicate that less than 0.1% of applied atrazine is lost in runoff water from no-tillage watersheds and that losses of other herbicides are greater with conventional than with conservation (no-) tillage from a clay loam soil.[61] Deep percolation of water is negligible on fine-textured soils of the SGP.[62]

Chemical use for controlling insects usually is similar under conservation and conventional tillage conditions. In some cases, however, insect problems have been less severe with conservation than with conventional tillage. For example, greenbug (*Schizaphis graminum* Rondani) infestations and damage to grain sorghum[63] and wheat[64] were less under conservation than under conventional tillage conditions in the SGP. Lower infestations suggest that fewer chemicals would be required to control greenbugs in locations in which conservation tillage is practiced.

Fertilizer applications for crops under conservation and conventional tillage conditions usually are similar in the SGP. However, a response to additional fertilizer may be obtained where large amounts of residue are retained on the soil surface, as with no-tillage. For such conditions, an increase of 20 to 25% in applications of nitrogen fertilizer was recommended in some cases at more humid locations.[65] Responses to additional fertilizer, therefore, may occur in more humid areas of the SGP.

As for pesticides, losses of nutrients could occur in runoff or with eroded soil, especially in the more humid areas. Soluble nutrients could also move to groundwater on highly permeable soils. On fine-textured soils of the SGP, nutrient losses in runoff have been slight[66,67] and deep movement is negligible because percolation of water is slight.[62]

14.6.2. Groundwater Quality

Deep percolation of water on the fine-textured soils of the SGP, mainly those of MLRA 77, is negligible.[62] The groundwater (specifically, the Ogallala Aquifer) that underlies most of MLRA 77 is at a depth of about 60 m below the surface. Because of these conditions, the potential for groundwater contamination by pesticides and nutrients is slight on fine-textured soils. On more permeable soils and where a water table occurs at shallower depths, percolating water containing nutrients or pesticides could potentially contaminate the groundwater. Nitrate-nitrogen levels ranged from 2.1 to 34.0 mg/L in the groundwater under a no-tillage watershed in MLRA 80A at El Reno, OK. The range was from 0.05 to 8.8 mg/L with conventional tillage and 0.02 to 2.5 mg/L under native grass. In MLRA 78 at Woodward, OK, the ranges were 0.2 to 12.2 and 1.4 to 9.5 mg/L under no-tillage and native grass watersheds, respectively.[67] Nitrate-nitrogen also increased in groundwater under minimum tillage watersheds, apparently due to increased nutrient transport to the groundwater because of reduced runoff (greater infiltration) with no-tillage.[68] Greater nitrogen mineralization with conservation tillage also may have been involved.[42] Improved practices such as precise timing and placement of nitrogen fertilizer with respect to crop needs should result in reduced nitrate-nitrogen levels in the groundwater.[69]

As for nutrients, soluble pesticides could be transported to the groundwater on permeable soils where the water table is at a shallow depth. However, tests have detected no pesticides in groundwater at several SGP locations.[69] Soil conditions undoubtedly play a major role in pesticide transport, but interception and retention by crop residues that reduce pesticide (herbicide) transport in runoff across the surface and to groundwater also are involved. In addition, microbially active soil beneath the residue mulch retains the herbicides and enhances their transformation rates.[70]

14.6.3. Soil Erosion

The potential for erosion and, hence, for environmental pollution exists on all soils of the SGP. In MLRA 77, where the surfaces are relatively flat, water erosion generally is slight. In addition, most runoff is into shallow, flat-bottomed lakes (playas) that dot the region. Runoff into these lakes does not enter major streams; therefore, sediments in the runoff have a limited effect on the environment.

In MLRAs 78, 80A, and 80B, runoff water from croplands flows into streams. Thus, if erosion occurs, sediments are carried into the streams, which would affect water quality. However, sediment losses in these MLRAs were less from no-tillage and reduced-tillage than from conventional tillage watersheds that were cropped to winter wheat and grain sorghum.[66,67] For example, annual sediment losses at El Reno, OK, were 6.4 and 0.4 Mg/ha with conventional and no-tillage treatments, respectively. At Woodward, OK, losses with the respective treatments were 15.9 and 0.9 Mg/ha annually.

Wind erosion affects the environment well beyond the site at which the erosion occurs. For example, dust carried aloft by storms in the Great Plains was found in the eastern United States during the 1930s,[2] and dust from storms in recent years often has darkened the sky many kilometers from where the dust originated. Wind erosion is possible on most SGP soils if wind speeds are adequate; surfaces are dry, smooth, and unprotected; and soil materials are finely divided or structureless and lack cohesion. On most fine-textured soils, emergency tillage that roughens the surface usually controls wind erosion.[71] However, on sandy soils that lack cohesion, wind erosion control is more difficult. Such soils are common at various locations in the SGP, and environmental concerns are causing an intensification of efforts to devise improved measures for controlling wind erosion.

14.6.4. Resilience of Tillage Systems Without Pesticide Inputs

Within the last few decades, crop production in the SGP has become highly dependent on pesticide use, not only with conservation tillage, but also with conventional tillage. Herbicides are the main pesticides, but insecticides are used frequently for some crops. Occasionally, chemicals are used to control plant diseases and other crop-damaging organisms. Except for herbicides, pesticide requirements are virtually identical for crop production under conventional and conservation tillage conditions. Hence, crop production in either case often would suffer if pesticide inputs were eliminated. Undoubtedly, pesticide use could be reduced by the adoption of integrated pest management strategies,[72] but such strategies are not available for all crops under all conditions in the SGP.

Herbicides often are used for crop production by conventional tillage methods, primarily for growing season weed control. Many weeds can be controlled by cultivation, but hand weeding may be required as well. In such cases, the cost of cultivation and hand weeding may exceed the cost of herbicides, which would reduce profits for producers. In addition, labor for timely hand weeding may not be available, which could result in lower yields and further reduce profits. Frequent cultivations may adversely affect soil conditions and increase the potential for erosion on some soils.

Use of some herbicides is relied upon for weed control for most conservation tillage systems. For the no-tillage system, all weed control is with herbicides. Use of herbicides could be reduced for all except the no-tillage system if more intensive tillage would still provide for retention of adequate residues on the surface for effective soil and water conservation. Another requirement would be that cultivation equipment be available for effective growing-season weed control. Cultivation is difficult when relatively large amounts of crop residues are present on the soil surface. No-tillage, which is the conservation tillage system that has shown the greatest promise for conserving soil and water resources when adequate residues are retained on the soil surface, would not be an option if the use of herbicides were eliminated.

14.7. GENERAL SUMMARY AND CONCLUSIONS

Contrary to the opinions of the early explorers, the SGP has become an important agricultural region of the United States. In general, low and erratic precipitation and soils that are subject to both wind and water erosion, ensure that improved soil and water conservation are major goals of conservation tillage in the SGP.

Most soils of the SGP are suitable for crop production by conservation tillage methods. However, precipitation in the SGP is highly variable within and among years. As a result, residue production by dryland crops often is low, which is a major limitation of conservation tillage in the drier western portion of the SGP, even for grain crops such as winter wheat and grain sorghum that produce more residues than cotton. Cropping systems involving fallow are sometimes used to increase SWS for subsequent crops at the drier locations. In the more humid eastern portion, high residue production by crops such as annually grown winter wheat may result in crop establishment and plant vigor problems. Under the higher precipitation conditions, weed control problems also may be more severe, and clean tillage may be used to minimize the weed control problems.

Conservation tillage systems are available for most crops of the SGP. One of these is SM tillage, which was developed to help control wind erosion and is now widely used in the SGP. However, adoption of herbicide-based conservation tillage in the SGP has been limited. In addition to those already mentioned, constraints to adoption of conservation tillage include producer needs, preferences, and managerial ability; economic considerations; lack of alternative crops; limitations imposed by governmental price support programs; and lack of strong and effective leadership on the part of agency personnel responsible for implementing sound soil and water conservation practices.

14.8. REFERENCES

1. Webb, W. P. *The Great Plains* (Boston: Ginn and Company, 1931).
2. Bennett, H. H. *Soil Conservation* (New York: McGraw-Hill, 1939).
3. Conservation Technology Information Center. "Tillage Definitions," *Conserv. Impact* 8(10):7 (1990).
4. Soil Conservation Service, U.S. Department of Agriculture: "Land Resource Regions and Major Land Resource Areas of the United States," Agric. Handb. 296 (Washington, DC: U.S. Government Printing Office, 1981).
5. Unger, P. W., and F. B. Pringle. "Pullman Soils: Distribution, Importance, Variability & Management," *Tex. Agric. Exp. Stn. Bull.* B-1372 (1981).
6. Johnson, W. C., C. E. Van Doren and E. Burnett. "Summer Fallow in the Southern Great Plains," in *Summer Fallow in the Western United States,* Conserv. Res. Rep. No. 17, (Washington, DC: U.S. Government Printing Office, 1974), pp. 86–109.
7. Musick, J. T. Unpublished results (1991).
8. Stewart, B. A. "Resources and Problems in the Southern Great Plains Area," in *Proc. Workshop on Planning and Management of Water Conservation Systems in the Great Plains States* (Lincoln, NE: U.S. Department of Agriculture, Soil Conservation Service, Midwest National Technical Center, 1985), pp. 56–71.
9. Unger, P. W. "Tillage and Residue Effects on Wheat, Sorghum, and Sunflower Grown in Rotation," *Soil Sci. Soc. Am. J.* 48:885–891 (1984).
10. Unger, P. W., and A. F. Wiese. "Managing Irrigated Winter Wheat Residues for Water Storage and Subsequent Dryland Grain Sorghum Production," *Soil Sci. Soc. Am. J.* 43:582–588 (1979).
11. Jones, O. R., and W. C. Johnson. "Cropping Practices: Southern Great Plains," in *Dryland Agriculture,* Agron. Monogr. 23, H. E. Dregne, and W. O. Willis (Eds.) (Madison, WI: American Society of Agronomy, 1983), pp. 365–385.
12. Greb, B. W., D. E. Smika, and J. R. Welsh. "Technology and Wheat Yields in the Central Great Plains — Experiment Station Advances," *J. Soil Water Conserv.* 34:264–268 (1979).
13. Allen, R. R., and C. R. Fenster. "Stubble-Mulch Equipment for Soil and Water Conservation," *J. Soil Water Conserv.* 41:11–16 (1986).
14. Fryrear, D. W. "Tillage Influences Erodibility of Dryland Sandy Soils," in *Proc. Conf. Crop Production with Conservation in the 80's* (St. Joseph, MI: American Society of Agricultural Engineers, 1980), pp. 153–163.

15. Bilbro, J. D., and D. W. Fryrear. "Pearl Millet Versus Gin Trash Mulches for Increasing Soil Water and Cotton Yields in a Semiarid Region," *J. Soil Water Conserv.* 46:66–69 (1991).
16. Bilbro, J. D., and D. W. Fryrear. "Wind Erosion Control with Annual Plants," in *Proc. Int. Conf. Dryland Farming, Challenges in Dryland Agriculture,* P. W. Unger, T. V. Sneed, W. R. Jordan, and R. W. Jensen (Eds.) (College Station, TX: Texas Agricultural Experiment Station, 1988), pp. 89–90.
17. Bilbro, J. D., D. W. Fryrear and T. M. Zobeck. "Flexible Conservation Systems for Wind Erosion Control in Semiarid Regions," in *Proc. Natl. Symp. Conservation Systems, Optimum Erosion Control at Least Cost* (St. Joseph, MI: American Society of Agricultural Engineers, 1987), pp. 114–120.
18. Koshi, P. T., and D. W. Fryrear. "Effect of Seedbed Configuration and Cotton Bur Mulch on Lint Cotton Yield, Soil Water, and Water Use," *Agron. J.* 63:817–822 (1971).
19. Russel, J. C. "The Effect of Surface Cover on Soil Moisture Losses by Evaporation," *Soil Sci. Soc. Am. Proc.* 4:65–70 (1939).
20. Wiese, A. F., and T. J. Army. "Effect of Tillage and Chemical Weed Control Practices on Soil Moisture Storage and Losses," *Agron. J.* 50:465–468 (1958).
21. Wiese, A. F., and T. J. Army. "Effect of Chemical Fallow on Soil Moisture Storage," *Agron. J.* 52:612–613 (1960).
22. Wiese, A. F., J. J. Bond and T. J. Army. "Chemical Fallow in the Southern Great Plains," *Weeds* 8:284–290 (1960).
23. Unger, P. W., R. R. Allen and A. F. Wiese. "Tillage and Herbicides for Surface Residue Maintenance, Weed Control, and Water Conservation," *J. Soil Water Conserv.* 26:147–150 (1971).
24. Unger, P. W. "Tillage Effects on Wheat and Sunflower Grown in Rotation," *Soil Sci. Soc. Am. J.* 45:941–945 (1981).
25. Unger, P. W. "Wheat Residue Management Effects on Soil Water Storage and Corn Production," *Soil Sci. Soc. Am. J.* 50:764–770 (1986).
26. Musick, J. T., A. F. Wiese and R. R. Allen. "Management of Bed-Furrow Irrigated Soil with Limited- and No-Tillage Systems," *Trans. Am. Soc. Agric. Eng.* 10:666–672 (1977).
27. Baumhardt, R. L., R. E. Zartman and P. W. Unger. "Grain Sorghum Response to Tillage Method Used during Fallow and to Limited Irrigation," *Agron. J.* 77:643–646 (1985).
28. Unger, P. W. "Straw-Mulch Rate Effect on Soil Water Storage and Sorghum Yield," *Soil Sci. Soc. Am. J.* 42:486–491 (1978).
29. Jones, O. R. Unpublished results (1991).
30. Allen, R. R., J. T. Musick and A. F. Wiese. "Limited Tillage of Furrow Irrigated Winter Wheat," *Trans. Am. Soc. Agric. Eng.* 19:234–236, 241 (1976).
31. Unger, P. W. "Tillage Effects on Winter Wheat Production Where the Irrigated and Dryland Crops are Alternated," *Agron. J.* 69:944–950 (1977).
32. Gerard, C. J., and D. G. Bordovsky. "Conservation Tillage Studies in the Rolling Plains," in *Proc. Great Plains Conservation Tillage Symp., Conservation Tillage, North Platte, NE,* Publ. No. 110 (North Platte, NE: Great Plains Agriculture Council, 1984), pp. 201–216.

33. Unger, P. W., C. J. Gerard, J. E. Matocha, F. M. Hons, D. G. Bordovsky and C. W. Wendt. "Water Management with Conservation Tillage," in *Conservation Tillage in Texas,* F. M. Hons (Ed.) Res. Monogr. No. 15 (College Station, TX: Texas Agricultural Experiment Station, 1988), pp. 10–15.
34. Clark, L. E. "Response of Cotton to Cultural Practices," *Tex. Agric. Exp. Stn. Progr. Rep.* PR-4175 (1981).
35. Unger, P. W. "Ridge Height and Furrow Blocking Effects on Wheat and Grain Sorghum Production and Water Use," *Soil Sci. Soc. Am. J.* 56:1609–1614 (1992).
36. Clark, L. E., H. T. Wiedemann, C. J. Gerard and J. R. Martin. "A Reduced Tillage System with Furrow Diking for Cotton Production," *Trans. Am. Soc. Agric. Eng.* 34:1597–1603 (1991).
37. Stiegler, J., W. Downs and F. Hawk. "Lo-Till System," in *Proceedings Great Plains Conservation Tillage Symp., Conservation Tillage, North Platte, NE,* Publ. No. 110 (North Platte, NE: Great Plains Agricultural Council, 1984), pp. 197–200.
38. Heer, W. F., and E. G. Krenzer, Jr. "Soil Water Availability for Spring Growth of Winter Wheat (*Triticum aestivum* L.) as Influenced by Early Growth and Tillage," *Soil Tillage Res.* 14:185–196 (1989).
39. Dao, T. H., and W. Lonkerd. "Soil Water Conservation Premise for Alternate Cereal-Crop Sequences in Continuous Wheat Areas," *Agron. Abstr.,* p. 312 (1990).
40. Stiegler, J. H. Personal communication (1991).
41. Dao, T. H., and H. T. Nguyen. "Growth Response of Cultivars to Conservation Tillage in a Continuous Wheat Cropping System," *Agron. J.* 81:923–929 (1989).
42. Dao, T. H. Personal communication (1991).
43. Epplin, F. M., T. F. Tice, S. J. Handke, T. F. Peeper and E. G. Krenzer, Jr. "Economics of Conservation Tillage Systems for Winter Wheat Production in Oklahoma," *J. Soil Water Conserv.* 38:294–297 (1983).
44. Boswell, T. E., and W. J. Grichar. "Comparison of Land Preparation Methods in Peanut Production," *Tex. Agric. Exp. Stn. Bull.* PR-3860 (1981).
45. Colburn, A. E. Personal communication (1991).
46. Unger, P. W. "Straw-Mulch Effects on Soil Temperatures and Sorghum Germination and Growth," *Agron. J.* 70:858–864 (1978).
47. Krenzer, E. G., Jr., Chaw Foh Chee, and J. F. Stone. "Effects of Animal Traffic on Soil Compaction in Wheat Pastures," *J. Prod. Agric.* 2:246–249 (1989).
48. Jones, O. R., W. L. Harman and S. J. Smith. "Agronomic and Economic Performance of Conservation-Tillage Systems on Dryland," in *Proc. Natl. Symp. Conservation Systems* (St. Joseph, MI: American Society of Agricultural Engineers, 1987), pp. 332–341.
49. Steiner, J. L. "Tillage and Surface Residue Effects on Evaporation from Soils," *Soil Sci. Soc. Am. J.* 53:911–916 (1989).
50. Smika, D. E. "Soil Water Change as Related to Position of Wheat Straw Mulch on the Soil Surface," *Soil Sci. Soc. Am. J.* 47:988–991 (1983).
51. Allen, R. R. "Reduced Tillage-Energy Systems for Furrow Irrigated Sorghum on Wide Beds," *Trans. Am. Soc. Agric. Eng.* 6:1736–1740 (1985).
52. Allen, R. R., J. T. Musick and D. A. Dusek. "Limited Tillage and Energy Use with Furrow-Irrigated Grain Sorghum," *Trans. Am. Soc. Agric. Eng.* 23: 346–350 (1980).

53. Keeling, J. W., C. W. Wendt, J. R. Gannaway, A. B. Onken, W. M. Lyle, R. J. Lascano, and J. R. Abernathy. "Conservation Tillage Cropping Systems for the Texas Southern High Plains," in *Proc. 1988 Southern Conservation Tillage Conference,* Spec. Bull. 88-1, K. H. Remy (Ed.) (Mississippi State, MS: Mississippi Agriculture and Forestry Experiment Station, 1988), pp. 19–21.
54. Lyle, W. M., and J. P. Bordovsky. "Integrating Irrigation and Conservation Tillage Technology," in *Proc. Southern Region No-Tillage Conference, Conservation Tillage: Today and Tomorrow,* Misc. Publ. MP-1636, T. J. Gerik and B. L. Harris (Eds.) (College Station, TX: Texas Agricultural Experiment Station, 1987), pp. 67–71.
55. Keeling, W., E. Segarra and J. R. Abernathy. "Evaluation of Conservation Tillage Cropping Systems for Cotton on the Texas Southern High Plains," *J. Prod. Agric.* 2:269–273 (1989).
56. Harman, W. L., G. J. Michels and A. F. Wiese. "A Conservation Tillage System for Profitable Cotton Production in the Central Texas High Plains," *Agron. J.* 81:615–618 (1989).
57. Harman, W. L., and J. R. Martin. "Economics of Conservation Tillage Research in Texas," in *Proc. Southern Region No-Tillage Conference, Conservation Tillage: Today and Tomorrow,* Misc. Publ. MP-1636, T. J. Gerik, and B. L. Harris (Eds.) (College Station, TX: Texas Agricultural Experiment Station, 1987), pp. 24–37.
58. Unger, P. W., R. R. Allen and J. J. Parker. "Cultural Practices for Irrigated Winter Wheat Production," *Soil Sci Soc. Am. Proc.* 37:437–442 (1973).
59. "1985 National Survey: Conservation Tillage Practices" (West Lafayette, IN: Conservation Technology Information Center, 1986).
60. Allmaras, R. R., G. W. Langdale, P. W. Unger, R. H. Dowdy and D. M. Van Doren. "Adoption of Conservation Tillage and Associated Planting Systems," in *Soil Management for Sustainability,* R. Lal, and F. J. Pierce (Eds.) (Ankeny, IA: Soil and Water Conservation Society, 1991), pp. 53–83.
61. Jones, O. R. Personal communication (1991).
62. Aronovici, V. S., and A. D. Schneider. "Deep Percolation Through Pullman Soil in the Southern High Plains," *J. Soil Water Conserv.* 27:70–73 (1972).
63. Burton, R. L., O. R. Jones, J. D. Burd, G. A. Wicks and E. G. Krenzer, Jr. "Damage by Greenbug (*Homoptera: Aphididae*) to Grain Sorghum as Affected by Tillage, Surface Residues, and Canopy," *J. Econ. Entomol.* 80:792–798 (1987).
64. Burton, R. L., and E. G. Krenzer, Jr. "Reduction of Greenbug (*Homoptera: Aphididae*) Populations by Surface Residues in Wheat Tillage Studies," *J. Econ. Entomol.* 78:390–394 (1985).
65. Phillips, S. H., and H. M. Young, Jr. *No-Tillage Farming* (Milwaukee, WI: Reiman Associates, Inc., 1973).
66. Sharpley, A. N., S. J. Smith, J. R. Williams, O. R. Jones and G. A. Coleman. "Water Quality Impacts Associated with Sorghum Culture in the Southern Plains," *J. Environ. Qual.* 20:239–244 (1991).
67. Smith, S. J., A. N. Sharpley, J. W. Naney, W. A. Berg and O. R. Jones. "Water Quality Impacts Associated with Wheat Culture in the Southern Plains," *J. Environ. Qual.* 20:244–249 (1991).

68. Naney, J. W., D. C. Kent, S. J. Smith and W. A. Berg. "Characterizing Ground-Water Quality Impacted by Tillage and Cropping Practices," in *Proc. 1990 Cluster of Conferences,* Kansas City, MO, (February 1990), pp. 97–111.
69. Smith, S. J. Personal communication (1991).
70. Dao, T. H. "Field Decay of Wheat Straw and Its Effects on Metribuzin and S-Ethyl Metribuzin Sorption and Elution from Crop Residues," *J. Environ. Qual.* 20:203–208 (1991).
71. Lyles, L., and J. Tatarko. "Emergency Tillage to Control Wind Erosion: Influences on Winter Wheat Yields," *J. Soil Water Conserv.* 37:344–347 (1982).
72. Adkisson, P. L. "The Value of Integrated Pest Management to Crop Production," in *Proc. Int. Conf. Dryland Farming, Challenges in Dryland Agriculture,* P. W. Unger, T. V. Sneed, W. R. Jordan and R. W. Jensen (Eds.) (College Station, TX: Texas Agricultural Experiment Station, 1988), pp. 905–906.

CHAPTER 15

Conservation Tillage in Eastern Europe

Andelko Butorac
University of Zagreb; Zagreb, Croatia

TABLE OF CONTENTS

0-87371-571-3/94/$0.00+$.50

15.1. INTRODUCTION

It is well established that there is no universal tillage system, due to regional differences in habitats, primarily in soils and climate. Conservation tillage in different regions will have specific characteristics in line with particular ecological conditions and plant production characteristics. Thus, soil tillage has to be approached differentially. Agriculture as a whole and field crop production in particular changes the mechanism and functioning of natural ecosystems. As the population of only a few species is mainly grown in an agroecosystem, application of differential agricultural technology is required. A soil tillage system becomes an efficient tool for the growth and development of crops only when it corresponds to their variable requirements and climatic conditions.

Despite its primary natural and scientific dimension, conservation tillage cannot be regarded apart from the current social and economic trends in eastern Europe, which have a direct influence on the adoption and application of technological innovations. Not contesting the basics of conservation tillage primarily set, by American and Canadian science and practice, it should be pointed out that the development of the system in Europe, and especially in eastern Europe, proceeded differently. This is a result of ecological differences as well as of the prevailing conceptions in soil science in Europe, mostly from the end of the 19th century and the first few decades of the 20th century. All this can probably explain the often contradictory attitudes toward the role of conservation tillage with respect to its introduction into general agricultural production. Here, practice may sometimes precede science. Despite scientifically proven advantages of conservation tillage, it is very difficult for this technology to be transferred into practice. Whether it is possible to form tillage management regions for the wider area of Europe with all of its ecological diversities, as has been done in the United States,[1] remains to be established in terms of the results of scientific research and practical experience. If such classification proves to be justified, conservation tillage will certainly find its place. If this should happen, attention should be focused on the habitat, i.e., its essential elements — soil and climate and, naturally, the prevailing systems of crop production.

15.2. DESCRIPTION OF THE EASTERN EUROPEAN AGROECOSYSTEM

15.2.1. Soil Types

The area of eastern Europe is very heterogeneous with regard to prevailing soil type. Soils belong to almost all of the land capability classes, which underlines the importance of applying different forms of conservation tillage, based on climate and crop requirements. Regarding the region as a whole, the following prevailing soil types can be separated: gray forest, chestnut-arid brown, chernozem (northern, deep, normal, and southern), meadows and bog, podzol, and carbonate saline soils.[2]

According to Andreae,[3] by far the largest parts of Poland and Russia consist of podzolized leached soils, which prevail in both the Subarctic and continental cool-summer climatic zones. Adjacent to this podzolized area, there is in the south an extremely wide and fertile belt of chernozem (black earth) and chernozemic soils. This band surrounds the Black Sea to the north and runs in an east-northeast direction. Adjoining this fertile area in the south, chestnut-brown, brown, and gray soils are found, distributed in the Danube area. Brown and yellow-brown soils prevail in the western Balkans, where red soil is also found in places. Within individual countries, there is also a mosaic of soil types (Figure 15.1). This imposes special requirements on soil tillage because of the constraints of some soils, primarily expressed in adverse physical properties, which naturally leads to certain constraints regarding the application of some forms of conservation tillage, notably reduction of the depth and number of tillage events. Different forms of conservation tillage are applicable to soil types belonging to higher land capability classes. This particularly applies to chernozem and all chernozem-type soils, but also to a number of other soil types. In other words, higher natural fertility of a soil, but also newly created technological fertility along with favorable soil stratigraphy, provide conditions for simplification of tillage and even plowless tillage.

15.2.2. Climatic Factors

Eastern Europe shows considerable climatic heterogeneity (Figure 15.2). There is a clash of various climatic types, partly influenced by the humid Atlantic climate of western Europe, partly by the cold Arctic climate of northern Europe, and especially by the eastern European continental climate. Effects of the Mediterranean climate are felt in smaller parts of some countries.

According to Andreae,[3] the following climate zones prevail in eastern Europe:

1. The Arctic climate belt begins north of Archangel and runs up to the northern part of Russia.
2. The Subarctic climate belt starting from the White Sea with its southern boundary approximately from Karelia to Sverdlowsk.
3. A continental cool-summer climate zone which differs from the Subarctic climate belt by starting out with a wide western part and gradually narrowing to the east. The delimitation of this climate wedge is given approximately by the three points of St. Petersburg, Bucharest and Novosibirsk.
4. There is a continental warm-summer climate region (thus a pronounced grain-corn climate) in eastern Europe which stretches over the largest part of the Balkans, i.e., the Danube Basin south of Budapest and the bordering mountain regions.

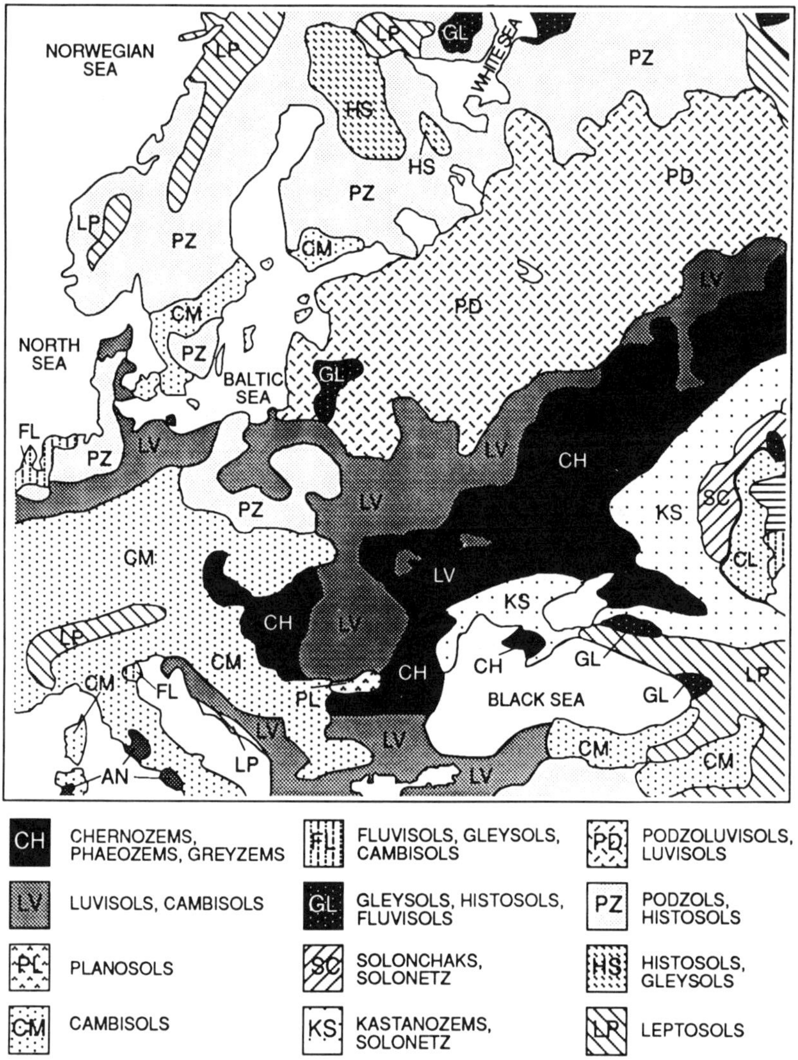

Figure 15.1. Soil map of eastern Europe. (Adapted from *FAO-UNESCO Soil Map of the World-Revised Legend,* World Soil Resources Rep. 60 (Rome: Food and Agriculture Organization, 1988). With permission.)

5. Still further south are the bordering zones with subtropical warm-summer climate. They include the area on the Adriatic as well as parts of the Black Sea coast.

As regards the average annual precipitation, most of eastern Europe is either in the 300 to 500 mm or 500 to 1000 mm bracket.

Climate zones, ecologically regarded as zonobiomes of eastern Europe, cover in part the iv. zonobiome of the winter rain region, vi. zonobiome of the

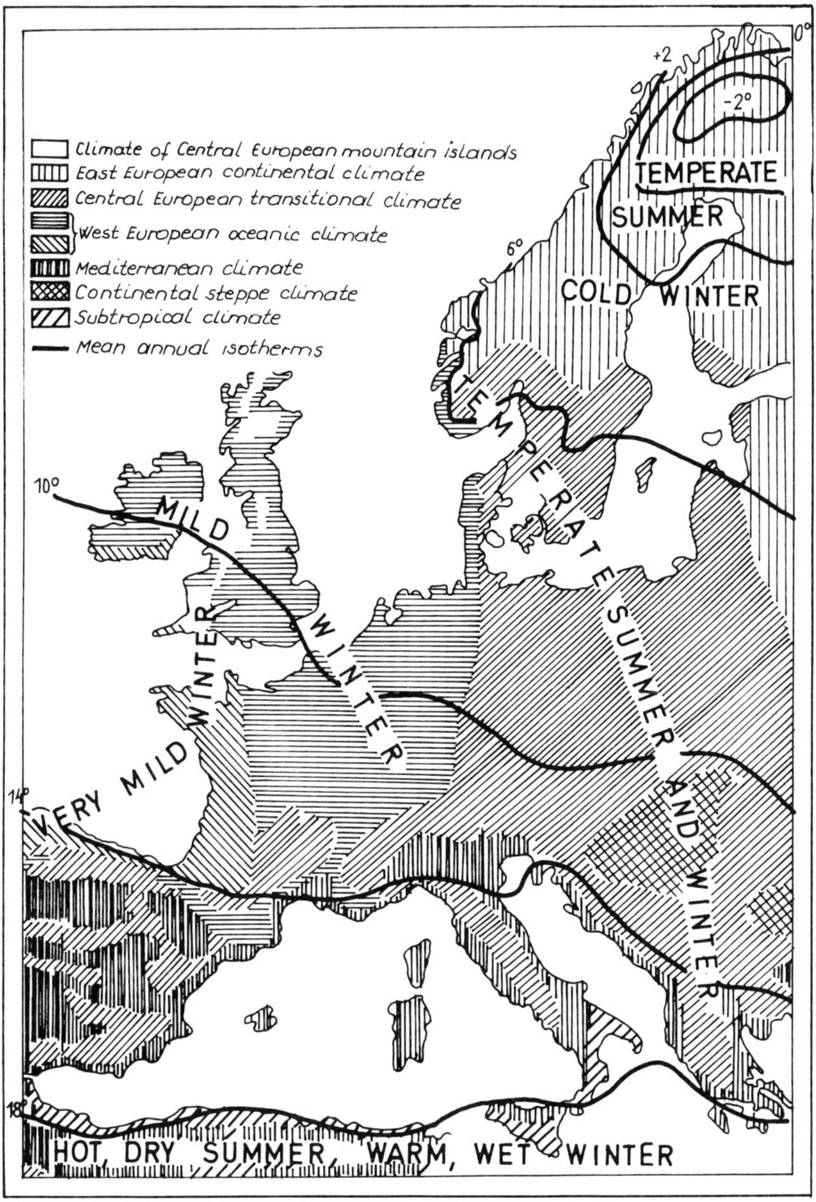

Figure 15.2. Climatic areas of Europe. (After Sydow-Wagner; from Andreae, B., *Farming Development and Space* (Berlin: Walter de Gruyter, 1981). With permission.)

temperate nemoral climate, and vii. zonobiome of the arid temperate climate with possible variations of the main climatic characteristics toward smaller regions.[2]

15.2.3. Cropping Systems

Large physical geographical differences in eastern Europe are not reflected in the farming types and agricultural zones as would be expected.[3] The reason for this is that the socialization of agriculture has led to a definite leveling of the farm size structure; thus farm size cannot exert a decisive influence on agricultural differentiation as it does in large parts of the western world. In the centrally planned economies of eastern Europe, insufficient attention was paid to habitat factors, such as climate, soil, and relief. This does not apply equally to all eastern European countries. Within the scope of the social and economic processes taking place in eastern Europe, its agriculture is subjected to considerable changes. This primarily refers to ownership transformation. However, considering the existing agricultural zones in eastern Europe, and the cropping systems within them, five important farm enterprises or farm enterprise groups can be distinguished: grain cropping, fodder cropping, hoe cropping, specialty crops (grapes, fruit, and other bush and tree crops) and dairying.[3] Secondary enterprises were assigned to these main enterprises, as far as possible, according to the closeness of their farm enterprise relationship. Agricultural zones, for instance, extending from the Baltic to the Adriatic Sea are made up of the following nine agricultural regions (Figure 15.3):[3]

- hoe crops - dairying regions
- dairying - hoe crops - grain regions
- dairying - hoe crops - fodder cropping regions
- special crops - grain - hoe crops regions
- grain - special crops - dairying regions
- special crops - dairying - grain region
- special crops - dairying - fodder cropping region
- dairying - fodder cropping - hoe crops region
- fodder cropping - dairying - hoe crops region

Limited space does not allow closer elaboration of the space distribution and major characteristics of the mentioned agricultural regions. The diversity of cropping systems naturally calls for a specific approach to soil tillage. For the former Soviet Union, notably for Russia and the Ukraine, special types of crop rotation were utilized on the basis of production and economic data and ecological characteristics of discrete regions.[4] This approach was more or less present in other eastern European countries, together with its positive and negative implications for production, economic, and ecological issues.

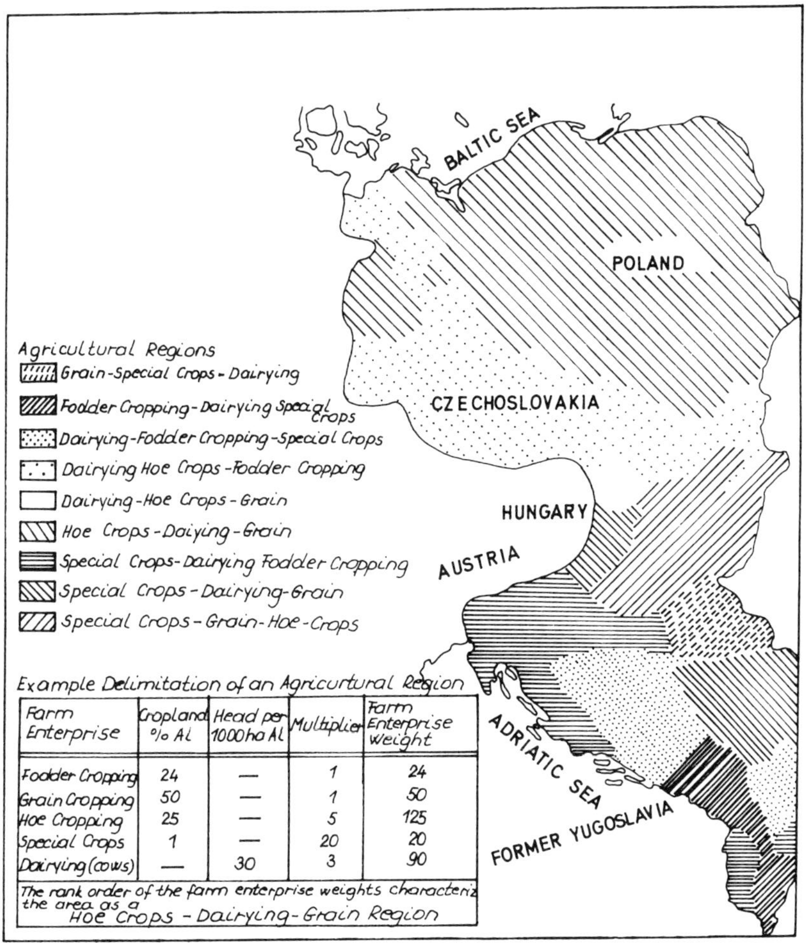

Figure 15.3. Agricultural zones in the Baltic Sea-Adriatic Sea area. (From Andreae, B., *Farming Development and Space* (Berlin: Walter de Gruyter, 1981). With permission.)

15.3. SOIL AND CLIMATE CONSTRAINTS FOR CONSERVATION TILLAGE

15.3.1. Soil Characteristics

When assessing the suitability of soils for conservation tillage, it is necessary to consider the potential constraints resulting from soil properties. In addition to practical experience, which has to be respected to a certain extent, emphasis should be placed on scientific methods. However, due to lack of coordination in the investigations done so far, as well as ecological diversity in individual regions, there are no uniform methods for this assessment. Still, some general criteria can be established, taking into account the ecological specifics and plant requirements. The greatest advances in this respect have been made in Great Britain. Some initial efforts have been also made in eastern Europe.[5] All of these methods mainly take account of soil properties in terms of fertility and effects of possible reductions in tillage. In principle, these methods should be primarily based on experimental results. In this respect, there is the Romanian experience with soil classification for tillage requirements.[5] Climate, slope, soil texture, mechanical properties, soil structure, compaction, permeability, drainage, bearing capacity, presence of coarse fragments, and occurrence of rocks are some of the factors that must be considered when choosing the most efficient methods of soil tillage. The author maintains that the three principal soil factors decisive for tillage are slope, saturation with water, and the soil profile. By combining these three criteria and by adding special cases (such as sandy, saline and alkaline soils, damaged soils, organic soils, and coarse fragments and rocks), 21 classes are obtained. On the basis of such classification, it is possible to separate not only soils that are suitable for minimum tillage, but also those requiring deeper tillage. There have also been other attempts to devise adequate classification aimed at assessing the possibilities for the application of conservation tillage, as well as attempts to adapt some British classifications to specific regional conditions.[6] The search for new criteria for such classifications is ongoing, but it is inconceivable that any of them should claim to be universal. Nevertheless, further investigations, primarily in the field of soil mapping and soil physics, along with experimental verification through different systems of tillage and cropping systems, will result in a comprehensive classification that would enable regionalization for soil tillage in Europe. The basic criteria for such a classification remain to be determined, however, it would include soil and climate types, land use, cropping systems, etc. A further division would be called for into lower production and ecological units, taking into account all the potential constraints to the application of conservation tillage. In a sense, this would be an atlas of soil tillage regions.

15.3.2. Climatic Constraints

Climatic constraints to the application of conservation tillage in the wider area of eastern Europe are primarily manifested through unfavorable hydrothermic characteristics. This is reflected in either an excess or a shortage of precipitation, especially in the vegetation period. Due to the wide range of extremely light to extremely heavy soils, climatic aberrations, and even meteorological aversions, climate may become the decisive factor in some eastern European regions for the application of conservation tillage. The main global climatic feature of most of eastern Europe is its continentality, which has suffered marked aberrations in the last decade through increased aridity, a consequence of global climatic changes. Such a complex climatic system and unfavorable hydrological situation may have direct consequences on present soil tillage systems. In those parts of eastern Europe where soil tillage could function in climate correction, in terms of water conservation (i.e., deeper tillage), there either are no conditions for a lasting application of conservation tillage or such tillage may be out of the question. Some countries have a high proportion of such regions, especially those in which the effects of Mediterranean climate prevail during the vegetation season.

Regarding both positive and negative interactions between soil and climate, it should be stressed that eastern Europe has extensive areas where both the soils and the climate are favorable and provide a sound basis for the application of conservation tillage. In contrast, there are also considerable areas with unfavorable soils and climate, in which conservation tillage is *a priori* ruled out. Long-term experience and practice in such regions undoubtedly favor conventional tillage. Conservation and conventional tillage could be alternated in some regions of eastern Europe. This assumption certainly requires scientific verification through field experiments. In some countries, this process is underway, while in almost all countries its dimensions can be worked out on the basis of soil and climate characteristics, as well as crop requirements.

15.4. ADOPTION OF CONSERVATION TILLAGE SYSTEMS

In accordance with the prevailing types of soil and climate and physiognomy of field crop production, conventional tillage systems differ widely from one another. Here, traditional and practical experience play an important part. More than in some other agricultural spheres, the safety factor is the main reason that conventional tillage is not easily abandoned. Different forms of conservation tillage are looked upon with distrust. In contrast to conservation tillage, conventional tillage systems have been precisely elaborated for discrete groups of crops, separately for winter and for summer crops and separately for sequential cropping.

15.4.1. Critical Factors Related to Acceptance of Conservation Tillage

In most eastern European countries, the development of conservation tillage systems proceeds more in a scientific than in a practical direction. In different regions, various forms of this system are investigated more or less successfully. It can even be said that the social and economical conditions of a country are more significant in this respect than the natural ecological conditions. Conservation systems have become part of the agricultural practice in some European countries, depending partly on the objective potentials for adoption of new scientific achievements and technological innovations, partly for ecological and economic reasons, and partly on different concepts in the field of soil tillage. Conservation tillage is certainly penetrating into European agriculture, partly as a system of minimum tillage, partly as zero tillage, and partly as reduced tillage. However, conventional soil tillage, often tending toward deep tillage, characterizes the agriculture of southern and southeastern Europe, sometimes with good reason and sometimes as tradition. Thus, these two conceptions are bound to clash, but the final outcome is easily predictable. Though not yet widespread, conservation tillage has a certain advantage owing to the possibilities that it offers. Consideration of this problem per discrete regions, or countries, often reveals essential differences not only in its particular aspects but also conceptual aspects.

In Hungary, the history of conservation tillage goes back to the beginning of the last century when Pethe, in 1818, constructed the "Hungarian plow-plant," which was used for seedbed preparation, seeding and harrowing, all in one operation.[7] The advantages of the Hungarian plow-plant were elimination of trampling, higher yields and reduced consumption of time and energy. Jori's and Soos'[8] investigations were aimed at uncovering new soil tillage methods, including omission of plowing. Birkas et al.[9] gave an extensive survey of conventional and reduced tillage in Hungary, emphasizing that new methods of soil tillage were applied where they could result in lower production costs without any hazards to yields. Conventional tillage dominated up to the end of the 1970s. Introduction of new soil tillage systems was stimulated by reduced energy costs, soil protection, and a more economic soil moisture management. However, the no-till system is still rare in Hungary.

The results of the investigation on reduced tillage in Bohemia and Slovakia point to their rational approach to this problem, and numerous scientific institutions took part in the research. Special technology was worked out for the application of reduced tillage which was recommended by the Ministry of Agriculture. Testing the technology in different climate-soil regions of the country showed that the highest yields of winter crops were obtained with direct drilling, which can be used for cereals after all annual crops. Based on experimental results as well as on soil classification investigations, farming systems were worked out for particular ecological units. Parameters limiting the application of minimum tillage were also determined.[10] New methods of soil tillage were investigated from different aspects-physics, chemistry, and biology of the soil, plant development, yield, weed infestation in different climate, and soil conditions.[11] More recent investigations also used this approach.[12]

Positive experience with reduced tillage was also obtained in Romania.[13-15] Particular crops responded differently to reduce tillage in different climate-soil conditions, but occasional deeper tillage in alternation with reduced tillage was necessary. Herbicide application enabled reduction of tillage for corn (*Zea mays* L.), sunflower (*Helianthus* sp.), and soybean (*Glycine max* L. Merrill). Emphasis was often laid on a combination of some practices-seedbed preparation, application of herbicides, insecticides and fungicides, fertilization, and seeding, which allowed tillage reduction.

The first investigations of minimum tillage in Bulgaria determined the preconditions needed for adoption of conservation or reduced tillage. However, the result of these investigations did not correspond to the expectations based on habitat conditions, especially for regions with more fertile soil types. In accord with the ecological characteristics of some Bulgarian regions, reduced tillage systems have been characterized on the basis of scientific findings and experience.[16] They involve occasional plowing at different depths, loosening of soil, shallow tillage, direct drilling, and specific tilling practices on excessively moist and inclined terrains.

In Poland, investigations of conservation or reduced tillage started in the mid-1960s. They were later intensified and often proceeded in the direction of testing the efficiency of direct drilling. A relatively large number of Polish researchers are concerned with this issue. Radecki[17] reported on the results of direct drilling on black soils, which illustrates the general approach to direct drilling in Poland. It seems that the main problem relating to direct drilling that remains to be solved is construction of adequate tools and machinery for seeding and herbicide application.

The issue of conservation tillage in the former Soviet Union shows some specific features. It is known as "pochvozashchitnaia obrabotka," which is based on the notion that tillage does not affect only chemical processes and weed and pest elimination, but also the entire soil ecosystem, including biological, physical, and chemical properties. This led to the postulate: soil tillage should not be abandoned but rather should be improved and made more efficient and cheaper.[18] It was found that frequent tillage caused soil compaction and a deterioration in soil structure, which in turn developed the need for more tillage.[4] These two postulates are flexible enough to allow a differential approach to soil tillage. Thus, a special form of conservation tillage, particularly on Ukrainian chernozem, was developed primarily for cereals, the basic tool of which is the "ploskorez," a rigid cultivator. There is no inversion of soil in this practice, its essence being soil loosening at various depths. In addition to its main effect on the soil, including erosion control, such tillage has indirect influence on climate correction by conserving soil water, because lack of precipitation and its irregular distribution throughout the year is one of the major characteristics of the climate in that region. Numerous papers are dedicated to this specific form of conservation tillage.[19] Some earlier dilemmas regarding the application of this tillage system have been overcome, though some open questions still require approaches and improvement. This tillage system originates from Ovsinski's system, which was theoretically and practically elaborated at the end of the 19th and at the beginning of the 20th century.[4] In the theoretical exposition of his system, Ovsinski starts from the assumption that each soil, with rare exceptions, in its natural state, is permeated with plant roots and earthworm burrows, which allows air movement down to a considerable depth, along with sufficient water permeability. In contrast, soil tillage destroys the pore network formed by roots and earthworms and turns the soil into a powdery mass. Shallow tillage leads to "underground moistening" due to penetration of atmospheric air into the soil. The lower layers of the soil are colder in summer, and this causes condensation of water vapor from the soil. This improves the nutrient regime and creates conditions for the activity of useful soil bacteria, destroys weeds, and forms a loose surface cover layer.

Certain prerequisites must be met for the application of conservation tillage without the plow, as it is understood by researchers in the former Soviet Union. Among other things, they refer to stricter weed, disease, and pest control; a more exactly defined farming system in terms of crop rotations for greater regions of the country; utilization of stubble and plant residues to keep snow on the fields; increased circulation of organic matter in the soil; and procurement of energy sources for microbial activity. A detailed description of this tillage system, which includes tillage without soil inversion and special cultivators that leave stubble and plant residues on the surface for soil protection from wind erosion, is given in the monograph *Pochvozashchitnoe bespluznoe zemledelie*.[4]

Investigations of reduced tillage systems in the former Soviet Union, which in a certain sense can be regarded as part of conservation tillage, were carried out on particular soil types from light to heavy, for particular crops or crop rotations, but mostly for cereals.[20-23] Attempts to minimize tillage go back to the end of the 19th century, as reported by Mendeliev and Kostichev.[21] Results of long-term trials that served as the basis for the theory of minimum tillage are of special interest. They are indispensable for studies of physical, chemical, and biological processes that slowly take place in the soil ecosystem. Some of these processes include the quantitative and qualitative changes of organic matter, compaction, and structure formations of the soil, and composition of pollutants, including their cumulative effects on soil conditions. Intensive soil tillage increases humus mineralization processes, which is its essential negative element, while surface tillage fosters better humus balance. On more fertile soils, particularly those of "light" texture, plowing is necessary only at certain time intervals. Differential approaches to the application of reduced tillage methods and technology also take into account the specific requirements of advanced farming to reduce excessive compaction and pulverization of the soil through successive use of heavy machinery. Briefly, different forms of reduced or minimum tillage, which also means conservation tillage, have become deeply rooted in the agricultural practice of the former Soviet Union.

Investigations of conservation or reduced tillage in the former Yugoslavia are conceptually close to European trends, although they are not always based on adequate analytical data due to a lack of appropriate machinery and equipment. Diversity of ecological conditions and a relatively large number of field crops grown often result in a large divergence of investigation results. Though some issues have been solved successfully, notably the method and depth of plowing for some crops and certain ecological units, it is not yet known exactly which physical, chemical, and biological changes occur in the soil under the influence of conservation or reduced tillage let alone efficient weed control. The dynamics of plant diseases and pests under conservation or reduced tillage also have not been investigated.[6,19]

In eastern Europe, the development of minimum tillage can be traced to the second half of the 19th century, and into the first half of this century. In the author's opinion, minimum tillage in all its numerous forms cannot be understood as part of conservation tillage, although some of its forms are similar. Minimum tillage in Europe in general, and in eastern Europe in particular, has been influenced not only by natural factors and field crop production, but also to a considerable extent by social factors, productive forces, and tradition in soil tillage. Scientific investigations in this field have so far proceeded quite independently in eastern and western Europe. The first mention of this tillage system in eastern Europe is found in the works of a number of authors in the early 1970s.[6,19] The conceptions are more or less identical, the differences being only in nuances. According to our own conceptions, in cropping systems on areas that primarily require amelioration, minimum tillage could be justified only after the soils have been transformed into a higher state of so-called physical fertility.[24-28] This is needed because minimum tillage cannot positively influence hydraulic conductivity and thus help correct climatic deficiencies such as enhancing water storage in cases of adverse rainfall distribution. However, compensation for fewer tillage practices and shallower plowing depth was found in more efficient fertilization.

The very conception of minimum tillage in western, and particularly in southern Europe, showed different trends from the very beginning. Minimum tillage has not been well accepted in countries with a tradition of deep tillage, such as Italy. Overall, minimum tillage tends to combine tillage operations to one pass of the tractor, reduces some tillage practices from the tillage system, reduces plowing depth, and replaces some tillage practices by herbicide application. Minimizing soil tillage is possible only when loosening involves no inversion of soil by plowing, in which case a protective mulch layer of plant residues is formed on the surface.

15.4.2. Future Directions

In the 1990s, the use of the plow may be put to a severe test in eastern European agriculture. Adoption of the most rigorous type of conservation tillage, i.e., the no-tillage system, will minimize use of the plow in agricultural practice. It is further foreseen that conservation tillage will at least partialy play the same role in the future that the plow had in the past. It is still necessary to scientifically verify the advantages and disadvantages of conservation tillage in its broader sense, the effect of climate on its application, suitability of soils for conservation tillage, including factors such as drainage, water, and wind erosion, etc. Of no less importance is soil moisture, fertilization, and liming from the aspect of the required rates and application methods. There is also an energy side to conservation tillage, related to fertilizer application on the one hand, and to the use of machinery on the other. Weed control and disease and pest management are seen in a new light in conservation tillage. These problems will become increasingly evident with the expansion of conservation tillage in agricultural practice, and have been a recent subject for investigation in almost all eastern European countries.

15.5. ENVIRONMENTAL ISSUES IN CONSERVATION TILLAGE ADOPTION

15.5.1. Weed Control

Though investigations of conservation tillage in eastern Europe, similarly to western European countries, have dealt with the consequences of the application of agrochemicals and fertilizers on the environment, their scope and intensity are still limited. Eastern European researchers accept that weed control is a vital component of the success of conservation tillage and that the application of agrochemicals plays an important role in crop production.

It is a common belief in the former Soviet Union that weed control in different forms of conservation, or reduced tillage, is much easier than in tillage systems that use the plow. This is explained by the fact that plowed-under weed seeds are brought up to the surface layer by subsequent plowing.[29-31] In contrast, in different forms of conservation tillage, two thirds of the potential germinating seeds are placed in deeper layers, where their germination does not result in optimum weed establishment. The aim of soil tillage is to eliminate potential weediness from the surface 10 cm, which is easier to control than the entire plow layer.[4] Overall, the best combination of crop rotation, tillage, and herbicides must be chosen in order to achieve high yields, taking into consideration weed ecology and biology as well as potential interactions between crops, weeds, herbicides, and tillage.[19]

15.5.2. Fertilizer Use

Interrelationships between conservation tillage and fertilization are very important. There is, however, no evidence that phosphorus and potassium requirements of crops grown in no-tillage systems differ from those for which soil tillage is performed. However, in some conditions, nitrogen behaves differently. This seems to be a consequence of the restricted release of nitrogen from organic matter through mineralization in the untilled plow layer. In continuous direct drilling, a considerable amount of nutrients remain in the surface layer. With changes in the tillage system, notably in a no-till system, changes also occur in the proportions of nitrogen, phosphorus, and potassium applied as well as the methods of application.

The experience of nitrogen application in conservation tillage may be less favorable than in conventional tillage. This was confirmed by investigations of several nitrogen rates in no-till conditions and various methods and depths of conventional tillage.[25-28] Intensified acidification may also result from higher nitrogen fertilization, which is necessary because of the increased accumulation of organic matter in some forms of conservation tillage.[19]

15.5.3. Groundwater Quality

According to the available information, groundwater quality does not seem to have been investigated within conservation tillage studies in eastern Europe from the aspect of its effect on the environment. However, considerable research has been done in terms of the chemical as well as physical and biological changes in the soil under the influence of conservation tillage.[19]

15.5.4. Soil Erosion

Soil tillage has a certain influence on soil erosion. Soil degradation, manifested through erosion, is present in eastern Europe, though with regional differences and differences between water and wind erosion. In the last few years, investigations have been carried out with the aim of checking the efficiency of different forms of conservation tillage upon soil erosion rates, soil condition, and crop yields.[30,32,33] The results achieved indicate that the success of the system largely depends on the soil type, efficacy of weed control, as well as disease and pest management. In general, it has been determined that well-drained soils, light to medium in texture with a low humus content, respond best to conservation tillage. In contrast, soils with unfavorable physical properties (especially poor drainage), with a high humus content, as well as heavy soils, require plowing. Use of mulching for erosion control may have adverse effects in cold climatic regions, notably in northern Europe because of lesser warming of the soil under the mulch, especially in poorly drained soils.

15.6. GENERAL SUMMARY AND CONCLUSIONS

The eastern European approach to soil tillage in general, and to conservation tillage in particular, differs from the approach to tillage in other parts of Europe. This is not so much due to specific ecological features as to general social, economic, and production-technological conditions. The differences are greater in practical than in scientific approaches to the problem. Taking into consideration the natural, scientific, and social conditions, some specific features of the agroecosystem of eastern Europe are pointed out, as well as the approach, achievements, and results attained in conservation tillage, including possible positive and negative implications for crop yields and environment.

In spite of its application in agricultural practice, conservation tillage has not reached the expected scope in all eastern European countries, though the results achieved to date are promising. It has been generally comprehended that this tillage system is not a transitory trend but is an everyday practice, spreading increasingly because of its evident advantages and also because of ecological and economic pressures. In some countries, a wider application of conservation tillage must be preceded by hydro- and agro-ameliorative measures of land development. Thus, ameliorative practices and conservation tillage are not incompatible opposites but a unity that must be optimally coordinated.

Conservation tillage systems have been adopted to different extents in eastern European agricultural practice. Naturally, some important aspects of conservation tillage still require scientific verification and justification in specific ecological conditions of particular regions in eastern Europe. In contrast, regions of good soil types and favorable climate are also widespread in eastern Europe, which provide almost unlimited potential for the application of conservation tillage.

Production-ecological conditions, reduction of production costs, and the adverse effects of conventional tillage and high technology upon soil fertility (i.e., primarily anthropogenic soil compaction caused by heavy machinery and tools, and increasing tractor power), as well as constraints that appear along with tillage simplification, call for new solutions in all eastern European countries. This whole issue cannot be separated from the level of their technical development and economic power. Therefore, all the above-mentioned factors largely govern the extent to which conservation tillage is applied in agricultural practice and its role in the future could be the same as that of the plow in the past.

15.7. REFERENCES

1. Allmaras, R. R. and R. H. Dowdy. "Conservation Tillage Systems and Their Adoption in the United States," *Soil Tillage Res.* 5(2):197–221 (1985).
2. Walter, H. *Vegetation of the Earth and Ecological Systems of the Geo-Biosphere* (New York: Springer-Verlag, 1979), p. 274.
3. Andreae, B. *Farming Development and Space* (Berlin: Walter de Gruyter, 1981), pp. 197–267.
4. Morgun, F. T., and N. K. Shikula. *Pochvozashchitnoe bespluznoe zemledelie* (Moscow: Kolos, 1984), p. 275.
5. Canarache, A. "Romanian Experience with Land Classification Related to Soil Tillage," *Soil Tillage Res.* 10:39–54 (1987).
6. Butorac, A., I. Zugec and F. Basic. "Stanje i perspektive reducirane obrade tla u svijetu i u nas," *Poljopr. Aktual.* 1-2:159–262 (1986).

7. Egerszegi, S. "Development of Minimum Cultivation and the Role of Tillage in Chemotechnique," *6th Intr. Conf. Soil Tillage, Summaries* (Wageningen, 1973), pp. 19(1)–19(2).
8. Jori, I., and S. Soos. "Technologien und Gerate der energiesparenden, pfluglosen Bodenbearbeitung in Ungarn," *Vortr. Anlaslich Wiss. Tag. Forschungzent. Bodenfruchtbarketi* pp. 23–31 (1989).
9. Birkas, M., J. Antal and I. Dorogi. "Conventional and Reduced Tillage in Hungary," *Soil Tillage Res.* 13:233–252 (1989).
10. Simon, J. "The Limitations of Growing Cereals under Zero Tillage in Czechoslovakia," *Proc. 9th Conf. ISTRO* (Osijek, Croatia, 1982), pp. 142–146.
11. Cerny, V. "Development of Tillage in Czechoslovakia," *Proc. 8th Conf. ISTRO* Vol. 1 (Hohenheim, Germany, 1979), pp. 36–38.
12. Suskevic, M. "Bodeneigenschaften und Ertrage bei langjahrig reduzierter Bodenbearbeitung auf unterschiedlichen Standorten der CSSR," *Vortr. Anlaslich Wiss. Tag. Forschungzentr. Bodenfruchtbarkeit (Muncheberg)* pp. 76–84 (1989).
13. Pintilie, C., Sin, Gh. and Damian, L. "Influenta lucrarilov superficiale si adince ale solului porumb," *An. ICCCPT* 37:221–225 (1979).
14. Pintilie, C., Gh. Sin, A. Arfire, H. Nicolae, I. Bondarev, Fl. Ionescu, E. Timirgazin and M. Les. Lucrarile minime ale solului se perspectiva lor in Romania. *Prob. agrofitotehnie teoret. apl.* 1:97–116 (1979).
15. Sin, Gh., C. Pintilie, H. Nicolae, C. Nicolae and Gh. Eliade. "Some Aspects Concerning Soil Tillage in Romania," *Proc. 8th Conf. ISTRO* Vol. 1 (Hohenheim, Germany, 1979), pp. 39–44.
16. Stojnev, K. "Bodenokologish und agrotechnisch begrundete Methoden und Systeme zur Reduzierung der Bodenberbeitung in der V.R. Bulgarien," *Vortr. Anlasslich Wiss. Tag. Forschungszentr. Bodenfruchtbarkeit* (Muncheberg) pp. 123–132 (1989).
17. Radecki, A. *Studia nad mozliwoscia zastosowania siewu bezposredniego na czarnych ziemiach wlasciwych* (Warsaw, 1986), p. 86.
18. Narcisov, V. P. *Nauchnie osnovi sistem zemledelia* (Moscow: Kolos, 1982), p. 326.
19. Butorac, A. "Conservation Tillage Systems in Europe," (in press).
20. Kiver, V. F. and A. P. Pogrebniak. "Minimalizacia obrabotki poimennih pochv Moldavii," *Zemledelie* 4:33–35 (1978).
21. Dospehov, B. A. "Minimalizacia obrabotki pochvi: napravlenia issledovanii i perspektivi vnedrenia v proizvodstvo," *Zemledelie* 9:26–31 (1978).
22. Puponin, A. I., I. A. Cudanov and V. P. Vasiljev. "Sovershenstvuiutsia Sistemi Obrabotki Pochvi," *Zemledelie* 2:39–45 (1988).
23. Arlauskas, M. P., I. S. Kochetov, R. J. Ramazanov, I. I. Popov, J. Logachev and N. I. Kartanushev. "Razrabotka i primenenie minimalnoi technologii obrabotki pochvi," *Zemledelie* 10:58–66 (1989).
24. Butorac, A., L. Lackovic and T. Bestak. "Comparative Studies of Different Ways of Seedbed Preparation for Maize (*Zea mays* L.) in Combination with Mineral Fertilizers," *Proc. 7th Conf. ISTRO,* Vol. 5 (Uppsala, Sweden, 1976), pp. 1–7.
25. Butorac, A., L. Lackovic, T. Bestak, D. Vasilj and V. Seiwerth. "Interrelationship of Soil Tillage and Fertilizing in Growing Main Field Crops on Hypogley," *Proc. 8th Conf. ISTRO,* Vol. 2 (Hohenheim, Germany, 1979), pp, 359–364.

26. Butorac, A., L. Lackovic, T. Bestak, D. Vasilj and V. Seiwerth. "Efficiency of Reduced and Conventional Soil Tillage in Interaction with Mineral Fertilizing in Crop Rotation Winter Wheat-Sugar Beet-Maize on Lessive Pseudogley," *Poljopr. Znan. Smotra* 54:5–30 (1981) (in Croatian with English summary).
27. Butorac, A., L. Lackovic, T. Bestak, D. Vasilj and V. Seiwerth. "Istrazivanje sistema reducirane i konvencionalne obrade tla u kombinaciji s mineralnom gnojidbom za glavne oranicne kulture na hipogleju srednje Podravine, Zbornik Radova sa Savjetovanja, " *Aktual. Probl. Poljopr Mehanizacije* pp. 129–145 (1981).
28. Butorac, A., L. Lackovic, T. Bestak, D. Vasilj and V. Seiwerth. "Investigations of the Interaction of Minimalized Tillage and Mineral Fertilizing on Lessive-Brown Soil," *Poljopr. znan. smotra,* 55:137–156 (1981) (in Croatian with English summary).
29. Kiver, V. F., P. A. Melua, A. D. Pilipenko, M. I. Bondarenko, V. V. Labunskij and A. P. Pogrebniak. "Zasorenost posevov pri minimalnoi obrabotke pochvi na oroshaemih zemlah Moldavii," *Zemledelie* 3:38–41 (1979).
30. Beljauskas, P. M. "Minimalnaia obrabotka pochvi na erodirovanih sklonah," *Zemledelie* 7:17–18 (1981).
31. Krutt, V. M., V. P. Krotinov and A. I. Gorbatenko. "Minimalnaia obrabotka pochvi i zassorrenost posevov," *Zemledelie* 6:16–18 (1984).
32. Christov, A., N. Onchev and E. Tzvetkova. "Antierosion and Agrotechnical Efficiency of Zero and Subsurface Basic Tillage of Different Soil Types," *Proc. 9th Conf. ISTRO* (Osijek, Croatia, 1982), pp. 91–96.
33. Stoichev, T., N. Onchev, A. Hristov and E. Tzvetkova. "Protivoerozionna i ottokonamaliavashcha efektivnost na tri nachina na osnova obrabotka i mulchiraneto na karbonaten chernozem," *Pochvozn. Agrokhim.* 18(5):128–136 (1983).
34. *FAO-UNESCO Soil Map of the World—Revised Legend*, World Soil Resources Rep. 60 (Rome: Food and Agriculture Organization, 1988).

Index

INDEX

A

B

C

D

E

F

G

H

I

N

O

P

Q

R

S

Y

Z